Sportswriter

SPORTSWRITER

The Life and Times
of Grantland Rice

CHARLES FOUNTAIN

New York   Oxford
OXFORD UNIVERSITY PRESS
1993

Oxford University Press

Oxford   New York   Toronto
Delhi   Bombay   Calcutta   Madras   Karachi
Kuala Lumpur   Singapore   Hong Kong   Tokyo
Nairobi   Dar es Salaam   Cape Town
Melbourne   Auckland   Madrid

and associated companies in
Berlin   Ibadan

Library of Congress Cataloging-in-Publication Data
Fountain, Charles.
Sportswriter : the life and times of Grantland Rice /
Charles Fountain.
p.   cm.
Includes bibliographical references and index.
ISBN 0-19-506176-4
1. Rice, Grantland, 1880-1954.
2. Sportswriter--United States--Biography.
I. Title.
GV742.42.R53F68   1993
070.4'49796'092--dc20   [B]   92-46146

The author wishes to express his gratitude to the following for permission to quote from copyrighted material:

Macmillan, Inc., for permission to quote excerpts from "Champion" by Ring Lardner (copyright © 1916 by Ellis A. Lardner) and "I Can't Breathe" by Ring Lardner (copyright © 1926 by International Magazines Co., Inc.; renewed 1954 by Ellis A. Lardner.

New York Post for permission to quote from a Jimmy Cannon column of July 16, 1954; copyright © 1954 by the New York Post.

Random House, Inc., for permission to quote from "A Very Pious Story" by Red Smith, included in The Red Smith Reader by Red Smith. Copyright © 1982 by Random House, Inc. Reprinted by permission of Random House, Inc.

2 4 6 8 9 7 5 3 1

Printed in the United States of America
on acid-free paper

For Cathy

Contents

Sportswriter

1

How He Played the Game

His greatest line has become sport's most battered cliché: "For when the One Great Scorer comes to write against your name, He marks—not that you won or lost—but how you played the Game."

He wrote, and through his writing inspired, some of journalism's most purple prose: "Outlined against the blue-gray October sky, the Four Horsemen rode again. In dramatic lore they are known as Famine, Pestilence, Destruction and Death. These are only aliases. Their real names are Stuhldreher, Miller, Crowley and Layden. They formed the crest of the South Bend cyclone before which another fighting Army football team was swept over the precipice at the Polo Grounds yesterday as 55,000 spectators peered down on the bewildering panorama spread on the green plain below."

For more than fifty years he filled his column with this sort of saccharin rhyme and hero worship. Were he writing today his florid style and unfailingly upbeat assessment of all that he witnessed would doom him to deserved obscurity at some weekly newspaper buried deep in the bowels of the Heartland, writing high school sports in a way that would make the parents proud. Yet in his time—and one cannot separate Grantland Rice from his time—he was the best.

Less celebrated than Ring Lardner or Damon Runyon, Rice was nevertheless the first important American sportswriter. For, while Lardner and Runyon used sportswriting to shape their own talents and careers, leaving their legacies in other areas, Rice used his talent and career to shape American sportswriting. He was its pre-eminent voice in the decades when sport was coming to the fore of American society, a time when—to the newspaper reader—the sportswriter was as central to the game as the athletes themselves. Rice found nobility and gentility in sport and chronicled it in noble and gentle language, and in doing so fashioned our perceptions of what sportswriting should be—and our perceptions of what sport should be as well.

When the historians and the revisionists find an error in Grantland Rice's work they call attention to it and correct it with a particular vigor. Why? It is not, surely, because Rice's mistakes were numerous—to the contrary, the evidence suggests that he was a very careful reporter. Nor is it because Rice's mistakes are particularly egregious. A kind man, he damned no reputations, and his culpability came in giving a man a break he might not have deserved. No. It is rather because what Rice wrote is embedded in the legend of the times. He is the Matthew, Mark, Luke and John of American sport. What he said has endured as surely as if it had been chiseled in stone.

He was an astoundingly prolific writer; he published more than sixty-seven million words during a fifty-three-year career. Much like the the national debt, these are numbers too vast to comprehend; they must be put in human terms. Doing the arithmetic, sixty-seven million words spread out over a fifty-three years averages out to about 3500 words—or ten to fifteen typewritten pages—each and every day. Rice published verse and a handful of modest books, but most of his work came in his syndicated daily column and his regular articles in the slicks, the glossy mass-circulation magazines that set the agenda of American popular culture in the decades before television.

He was the most respected writer of his day. Grantland Rice's All America football team—published in *Collier's* for more than twenty years—was *the* All America football team, successor to Walter Camp's original. Speculated upon throughout the season, eagerly anticipated as the mid-December publication date approached, it was accepted as the standard thereafter. When Grantland Rice began writing about golf, the game enjoyed its first rush of popular appeal, the heretofore hallowed lawns of the elitist clubs suddenly the venue of spectators interested in this chap from Georgia—Jones—of whom the great Rice spoke so highly. Athletes as well as fans accorded him this respect, and they always had a moment for Granny, even if otherwise occupied with matters of import. At the 1932

Olympics in Los Angeles, Rice sent word from the press box to the stadium floor to Babe Didrikson that he'd like to speak to her; he wanted to arrange a golf game. Didrikson, who was in the process of standing the Games on their ear—she won two gold medals and one silver—didn't think of questioning the summons. "She trotted up to the press box immediately," said Braven Dyer of the *Los Angeles Times*. "I think Babe would have done anything for Granny Rice in those days, it was a case of absolutely mutual admiration."

Respect, while generally conveyed grudgingly, is not uncommon for a writer of Rice's stature. Genuine affection, however, is another matter entirely; it is most rare; yet Grantland Rice was as loved as he was respected. In the half-century he spent at the summit of a generally catty and sometimes ruthless profession, nobody ever heard an unkind word spoken about Grantland Rice. "Every time Grant Rice meets a man, Grant Rice's circle of friends and well-wishers has been increased by one," said Irvin S. Cobb. Someone once came up to sportswriter Richards Vidmer and bubbled: "I've just met the greatest guy you ever saw! Grantland Rice."

"That," replied Vidmer, "is the most unoriginal remark I ever heard."

"He had pure courtesy," said Bruce Barton, writer, politician, advertising executive and a long-time friend and confidant of Rice. "Courtesy is no easy virtue. It means, first of all, being instinctively and sincerely aware of the other person, with spontaneous respect and consideration for his feelings and the instinct to react always appropriately."

Rice enjoyed a celebrity in his time the likes of which no other sports journalist enjoyed until Howard Cosell thundered onto the stage in the 1970s. Yet he was a remarkably self-effacing man. Red Smith told the story of the year Rice's press credential for the Army-Notre Dame game got lost in the mail. "He went down to Broadway and bought a ticket from a scalper and watched the game from the stands," wrote Smith, "with his typewriter between his knees. A friend who heard of this was aghast. 'Why didn't you throw some weight around?' he demanded. 'Tell you the truth,'" Granny said, 'I don't weigh much.'"

For all his accomplishments, Rice is dismissed today as a sentimental anachronism. Or worse. Robert Lipsyte, speaking of Rice and his influence in his seminal 1975 book *Sportsworld: An American Dreamland,* wrote: "Painting the lily is not only presumptuous, but ultimately destructive. The flower dies. By layering sports with pseudo-myth and fakelore, by assigning brutish or supernatural identities to athletes, the Rice-ites dehumanized the contest and made objects of the athletes." Ask a student of sports journalism today: Who is the most gifted and important sportswriter of all time? and you will likely get an unhesitating "Red Smith" in response. But had you put the same question to the late Red Smith, the erudite and Pulitzer Prize-winning stylist of the *New York Herald Tribune* and the *New York Times,* Smith would have told you—as he told the countless people who asked during his lifetime—that the greatest of them all was Grantland Rice.

"Grantland Rice was the greatest man I have known," said Smith, "the greatest talent, the greatest gentleman. The most treasured privilege I have had in this world was knowing him and going about with him as his friend. I shall be grateful all my life."

Trying to compare Rice with Smith, or anyone else, and labeling someone the greatest is akin to speculating as to whether or not Dempsey could have licked Ali. It is entertaining diversion, but ultimately pointless. It is also unnecessary, for however Grantland Rice may fare against his colleagues and his successors in the judgment of history, his own history cannot be denied; he is a man who influenced mightily the ebb and flow of sportswriting, and colored and shaped our perceptions of an entire era of American history as well.

Let's begin with it's-how-you-played-the-game. Like all clichés, it came into the language because it is a forceful and compelling articulation of a shared value or belief. This was how Rice saw it. "I learned a lot more from defeat than I ever did from winning," he wrote of his days as a college athlete. "I remember very little about the games I won or lost, but I always had a deep feeling of enthusiasm for the contest itself." And because Rice saw it this way, American society came to see sport this way. Our collective memory is hazy on who won those Army-Notre Dame games in the 1920s. But our conviction is certain that these games represented the best of sport—two noble institutions, proudly represented by men of talent, courage and character, fighting determinedly for victory, yet knowing all along that the honor is in the struggle and not the triumph. We see the games this way because this is the way Rice presented them; and his presentation was so vivid, his assertions so convincing, his arguments so eloquent.

The twenties are the "Golden Age" of sport because Rice saw them as golden. He was not a fool. He knew athletes and knew their failings. He knew that Babe Ruth was a sot and that Ty Cobb was a boorish churl. But he chose instead to celebrate their enormous athletic gifts, and in Granny's eyes Ruth's dissipation was merely exuberance, Cobb's misanthropy dismissed as tenacity. "If he couldn't say something nice about an athlete, he was likely to write about another athlete," said a colleague after Rice's death.

The hindsight of history might allow us to dismiss this as convenient journalism—few sins in journalism today are as heinous as the shameless boosterism and the *nil-nisi-bonum*-of-the-living essays that Rice so regularly trafficked in. But understanding Rice's attitude and philosophy is essential to understanding why it was left to the revisionists to reveal that our heroes of the age had feet of clay. Journalism has long been held to be the first draft of history, and in the matter of American sport in its formative decades, Grantland Rice is its Gibbon. Had Ring Lardner and Damon Runyon—contemporaries and close friends of Rice—not turned their attentions from sportswriting to other pursuits, our sense of sport would almost surely be different, filtered, as it would be, through the prism of Lardner's cynicism or Runyon's irreverence, instead of Rice's compassionate optimism.

Grantland Rice was courtly in both personal style and professional product—this at a time when both the dugout and the press box were peopled by hooligans. He brought respectability to the previously seedy craft of sportswriting. His soft gentlemanly bearing and unfailing courtesy in dealing with coaches and athletes brought a dignity to the craft that it didn't always deserve. "Even jealous or disdainful colleagues had to be grateful for the positive image of the sportswriter he was projecting to the public," said Lipsyte.

His column was as gentle as his demeanor. It would invariably begin with a verse; verse that was sometimes ironic, often whimsical, sometimes sober, but always tenderly bespeaking an exhilaration that the reader could not help but share:

> Another new May morning—I've known plenty—
> Since I first saw the dogwood's snowy flame;
> But I am younger than I was at twenty,
> Because I've seen so much more of the game;
> There is no heart that can grow old at playtime,
> Nor whither on the trail of Jones or Ruth,
> So when I mark the golden flush of Maytime
> I throw off years in the company of youth.
>
> So here's to sport—let those who will decry it—
> I'll sing "the fountain of eternal youth";
> And if some day its critics would but try it,
> They, too, at last might find the golden truth;
> We have a way of frowning at all pleasure,
> If it is fun—why then it must be wrong;
> But just as long as youth remains a treasure,
> I'll yield the humble tribute of a song.

Later, this would become popularly known as the "Gee Whiz!" school of sportswriting, to distinguish it from the "Aw Nuts!" school, embodied perhaps most vividly—surely most enduringly—by Rice's pal Ring Lardner.

Rice saw the sports page as a grand canvas on which to sketch for his readers the abundant glory he found. But there was something that distinguished Rice's work from that of his fellow gee-whiz practitioners. His words weren't mere gimmicks; they conveyed image and feeling, owing their genesis to his fine classical education and bespeaking a sensitivity as well as a fondness for the games he was chronicling.

Rice whet his readers' appetites not only for the games but also for his own distinctive approach to reporting them. For a nation thirsting for entertainment in its every endeavor, Rice entertained; for a nation consumed with heroes, Rice packaged heroes. In the process he became as famous as the men and events he was writing about. For Americans of his generation he *was* the game—his story became the event, much like the telecast has so often become the event in the eighties and nineties.

And along the way he elevated his craft—yes elevated, there can be no argument on that. Whether his prose was overwrought or his verse was good or bad (and it's had its supporters as well as its critics) is not the issue. For the question is not whether he was the equal of William Faulkner or John Masefield, but whether his verse and his literary approach to journalism made for a better newspaper article—Rice was after all a journalist, not an artist—a more compelling, more engaging morning read than the industry standard: "The Giants got six-hit pitching from Christy Mathewson and defeated the Pirates 3–0 this afternoon, before a crowd of 20,000 at the Polo Grounds." The answer is: Of course it did.

For in addition to breaking up the monotony of the sports page it also liberated the sports-page from the limits and rigors of inverted-pyramid journalism. Because Rice was so enormously popular, his work became the industry standard. Writers imitated Rice; editors encouraged it. The resulting product was often unfortunate—pockets of this wretched excess persist to this day. But so too did Rice's influence establish a sports-page environment that allowed followers like John Kieran, Jimmy Cannon, Red Smith and Frank Deford to flourish.

It was also during this era that the sports page grew to be an intergral and indispensable part of the American newspaper. Two days before the first Dempsey-Tunney fight in 1926, the sports section of the Sunday *New York Herald Tribune* ran ten full pages for the first time. *Tribune* publisher Ogden Reid—athlete, fan, patron of sport—was on hand as the paper was going to bed that Saturday night and was aghast at the thoroughness of his sports section. "Grant, you're making the *Tribune* more of a sports paper than anything else!" he said to Rice. "At this rate, we're becoming all sports and damn the rest of the world!"

"You could do worse," said Rice in return.

Rice loved sport—he loved the contests, loved the atmosphere, loved the camaraderie of the press box, of a passenger-train drawing room. And he loved sharing it all with the millions who read his work. "He was the evangelist of fun," said Bruce Barton in an eloquent eulogy delivered at Rice's funeral, "the bringer of good news about games. He was forever seeking out young men of athletic talent, lending them a hand and building them up, and sharing them with the rest of us as our heroes. He made the playing fields respectable. Never by preaching or propaganda, but by the sheer contagion of his joy in living, he made us want to play. And in so doing he made us a people of better health and happiness in peace: of greater strength in adversity. This was his gift to his country; few men have made a greater."

2

The Blue-gray October Sky

*I*t *is October 18, 1924,* a Saturday. The heart of the day, mid-afternoon, will be mild and pleasant in Manhattan, despite an intermittently cloudy sky. The edges of the day, however, early morning and especially the hours at dusk, carry the unmistakable and invigorating snap of the season.

On Broadway, *Abie's Irish Rose* is in its third year at the Republic Theater on 43rd Street, and George White's Scandals is one block south at the Apollo. The Marx Brothers' "laugh a minute review" is at the Casino on Broadway and 59th; Will Rogers is featuring a new monologue nightly in the *Ziegfeld Follies* at the New Amsterdam; while at the Plymouth on West 45th Street, the sobering Maxwell Anderson-Laurence Stallings drama *What Price Glory* is reminding theater-goers that there was nothing really romantic about the Great War.

The city is still all atwitter over the German ZR-3 Zepplin, which set down in Lakehurst, New Jersey, this past Wednesday after a three-day crossing of the Atlantic. These are gay, spirited times in New York; the city is flushed with wealth and full of itself. Seven rooms on Park Avenue run a pricey $425 a month, but there seems no shortage of interested renters in this time of egalitarian prosperity. Just as there is no shortage of customers at the Cutting Larson Co. on Broadway at Columbus Circle,

10

where a new Oldsmobile coach is selling for $1065 (tax and spare tire extra). Lord and Taylor on Fifth Avenue is selling raccoon coats for $250–$550 and is deluged with buyers, because a raccoon coat gives one the bearing and air of a character in a Fitzgerald story, and to be a character in a Fitzgerald story is as good as it gets in trendy Manhattan in 1924.

While these accoutrements of the good life may still be beyond some, a radio is becoming as common a fixture in a Manhattan parlor as a sofa and an easy chair. At Ludwig-Baumann on Eighth Avenue at Thirty-fifth Street, a Freed-Eisemann four-tube Neutrodyne radio is selling for $100, or just $1.50 per week. A Ware, three-tube Neutrodyne—"equal to most four-tube sets"—is $65, a mere one dollar per week.

This afternoon, the sounds of college football will crackle from those Freed-Eisemanns and Wares. It has been eight days since Walter Johnson came in with just one day's rest to pitch four innings of relief and give the Senators an extra-inning, game-seven victory over John McGraw's Giants in the world series★; and from now until the Tournament of Roses Bowl on New Year's Day, the nation's sports pages will be given over to college football.

In Champaign-Urbana, Illinois, Bob Zuppke's Fightin' Illini—undefeated since 1922 and boasting of a halfback named Red Grange, already being hailed as the finest runner to ever pull a jersey over a pair of pads—is playing host to Fielding "Hurry Up" Yost's Michigan Wolverines, themselves undefeated since 1921. Seventy thousand fans will watch Grange score four touchdowns in the first twelve minutes of an Illinois romp. It is the largest crowd yet to see a college football game in the Midwest. But the center of college football remains the East. The Michigan-Illinois crowd will be equaled in New Haven, where Dartmouth visits Yale. And Pennsylvania Station is jammed on this morning as Columbia fans make their way to Philadelphia, part of the 50,000 who will watch the Lions take on Penn this afternoon at Franklin Field.

New York's fancy has been caught by a game that is being played right here in Manhattan, at a venerable, queer-shaped stadium shoe-horned into a sliver of flat earth at the edge of the Harlem River, beneath a craggy face of rock known as Coogan's Bluff. The stadium is the Polo Grounds and the game is the twelfth meeting between Army and Notre Dame.

Until two years ago this game had been played on the Plains of West Point, where admittance was free and most of the fans—at least in the early years—had come specifically to watch the parade of cadets on Saturday morning and happened upon the football game rather by accident. But interest had grown—particularly in 1919 and 20, when Knute Rockne's team from Indiana featured a halfback named George Gipp—and the modest facilities of the Academy could not contain the crowds. Also, as one

★They don't yet capitalize newspaper references to the World Series in 1924.

Notre Dame man put it, it had grown "much too exciting to give away any longer." In 1923, the out-of-town reviews having pronounced the show a resounding hit, it moved to Broadway.

Actually it moved first to Brooklyn, to Ebbets Field, because the Yankees and Giants had the Polo Grounds booked for the World Series. The cozy 35,000-seat home of the baseball Dodgers had its capacity swelled for the football game with the construction of wooden bleachers along the sidelines, but even that wasn't enough to handle the throngs that descended on Flatbush—a number of newsmen and football people, including Grantland Rice and Walter Camp, spent the game standing on the sidelines. This year, officials at the two schools pushed the game back in the schedule, so that they might avail themselves of the larger capacity of the Polo Grounds.

There are still tickets available, though as game time nears they are not selling as many as they might have, because, despite the protestations of the New York Giants—in control of the Polo Grounds and thus the people hawking the tickets for this game—the rumors since mid-week have insisted there is not a ticket to be had. There is foundation for the rumor. The advance sale for this game has been the largest in New York Giant history—larger even than the advance sale for Yankee-Giant World Series games in 1921–22–23. By 2:30—game time—it is sold out, and those not lucky enough to be among the 55,000 inside will have to content themselves with their radios. This lovely work of the sorcerer for three years now has been bringing important sporting events to those who, for whatever reason, have been shut out of seeing it for themselves; and for this game, listeners will have a choice between two different New York broadcasts. Ennis Brown and J. Andrew White will be on WJZ; and the redoubtable Graham McNamee will be at the microphone for WEAF.

The Army-Notre Dame game is also commanding the attention of the country's elite sportswriters. On this Saturday the press box in the Polo Grounds is to sportswriting what Sylvia Beach's Shakespeare Book Store is to expatriate literature in Paris, or the Algonquin Hotel is to the literary lunch crowd. It generally takes something along the lines of a world series game or championship fight to attract a crowd with this sort of sportswriting pedigree.

From the New York *World* comes Heywood Broun. Broun spends most of his time these days on his daily literary and general interest column. He has not been assigned regularly to sports since he went off to cover the Great War. (His reporting of the story of the chaos and crippling ineptitude of the supply system for the AEF brought about the forfeiture of the $10,000 bond his newspaper posted with the War Department to insure that he would abide by the rigid censorship regulations.) Noted for his frumpiness—he dresses like a vagrant—the thirty-six-year-old Broun is still a decade away from founding the Newspaper Guild—the reporters' union—and his saber-rattling on behalf of unionism in general, and he still takes the opportunity to indulge himself with some work for the sports page.

Broun's sartorial opposite is forty-year-old Damon Runyon of the *New York American*. Dressed, as has become his habit, in a two-hundred-dollar tailored suit, spats, and a fifteen-cent silk cravat, Runyon's black hair is slicked down against his scalp, and wire rimmed glasses sit upon his long, slender, prominent nose. Like Broun, Runyon spends much of his time these days on matters other than sports. He covers all the sensational trials for the Hearst papers, and he has of late been making a cottage industry of Broadway, bringing to life the midnight denizens of the city's main street—the showgirls, the tycoons, the gamblers and panhandlers. People are stunned to discover that Runyon was born in Kansas and raised in Colorado; nobody knows New York or can bring its people alive in a newspaper paragraph better than Runyon. Friends and publishing house types have been urging him to write some fiction—a novel or some short stories based on his Broadway adventures and characters. He's resisted up until now, but won't for much longer. His first collection of stories will finally be published in 1931; it'll be called *Guys and Dolls,* and it will make him one of the richest and most famous writers in America. But he'll never give up the newspaper game. Though he modestly denies it is true, he revels in William Randolph Hearst's assertion that he "is the best reporter in the world."

At twenty-seven, Paul Gallico of the *Daily News* is already among the city's most powerful sportswriters. As sports editor of the *Daily News* he has turned the tabloid's sports page into one of the most solid and comprehensive in the city; as a writer he has the Runyon-like capacity to weave an unmistakable literacy into the short, punchy, hard-hitting tabloid sports form. The son of a concert pianist, fluent in German and French, Gallico is a burly man—six-feet, three-inches and 190 pounds—with dark, handsome features. Captain of the varsity crew at Columbia during his senior year, he is proud of his own athletic wherewithal. He has pioneered the sort of participatory journalism that George Plimpton will refine a half-century from now; his first big story in the *News* was a first-hand account of sparring a minute and a half with heavyweight champ Jack Dempsey. "It was as if a building had fallen on me," he said. Gallico will remain at the *News* for another twelve years, when he will abruptly walk away from one of the richest salaries in American journalism to go to Europe and write fiction. He will write fiction successfully for more than thirty years; his canon of work will exceed thirty books, including the tender and critically acclaimed *The Snow Goose* and the commercial blockbuster *The Poseidon Adventure*.

Other notables in the press box include Gene Fowler, future novelist and scenarist, backing up Runyon for Hearst; Frank Wallace, future novelist and scenarist, reporting for the Associated Press; and Joe Vila, sports editor and columnist for *The Sun*. Missing today is Ring Lardner, who's on his way back from Europe. The brilliant satirist and story writer hardly bothers with sports at all these days, but he'd been a fixture at the Army-Notre Dame since the days of George Gipp. Lardner carries a soft spot in his heart for Notre Dame—he too is a Midwesterner, and had gotten his journalistic start covering Notre Dame for the *South Bend Tribune*.

Finally, covering the game for the *New York Herald Tribune* and the Tribune syndicate is Grantland Rice. His presence alone is enough to grant an event significance, for he is unquestionably the best-read and most-admired sportswriter working today. He travels more than twenty thousand miles a year, an itinerant witness to world series games, big stakes races, important fights, the National Open golf championships, appealing college football games like this, and myriad other events of a far-flung readership's fantasy. His vision of the games America plays in the twenties appears six days a week in more than one hundred newspapers across the country.

Just two weeks short of his forty-fourth birthday, Rice has lost most of his once-blond hair, and his athlete's physique has begun to soften a little. But he still carries his two hundred-plus pounds gracefully on his six-foot frame. It is not hard to imagine the college athlete he was, or to understand why he hits a golf ball 260 yards off the tee. He cuts an imposing figure, particularly when you couple his fame with his physical presence, his booming Southern voice and crunching handshake. But there is a gentleness and warmth in that big voice, a depth of feeling transmitted in the handshake. He is a man who listens as well as he talks, and a man that others naturally gravitate towards, fame and stature notwithstanding. "Wherever Grantland Rice sits," his friend Red Smith★ will say some years later, "that's the head of the table."

Twenty minutes before the start of the game, as the Army band plays a march and 1200 West Point cadets parade once around the field to the enthusiastic applause of the crowd, the talk in the press box concerns the Notre Dame backfield. Rice and the other big-name writers are pressing George Strickler, the student publicity assistant at Notre Dame, and Frank Wallace, his predecessor in that job, for details on Notre Dame. While this game has been big news in the New York papers all week, Rice, Broun, Runyon, and Gallico haven't been writing the pre-game stuff. That task has fallen to younger staffers while the varsity writers have concerned themselves with their columns and such. So, having arrived on the story just this morning in a sense they are anxious for all the "dope." But they

---

★On this day, Red Smith is a Notre Dame sophomore, no doubt gathered with the other students in the gymnasium on campus, following the progress of the game on a device called the Grid-Graph, where a little white light marks the progress of the ball.

don't need much filling in. They have all seen and written about this back-field before.

A senior unit, the members of the Notre Dame backfield have been together since their sophomore year; and after the events of this day they will be linked for all eternity. In their careers they have lost just two games—both to Nebraska in Lincoln, 14–6 in 1922 and 14–7 in 1923. There is also one tie marring the record of these four remarkable athletes—a scoreless dual with Army at the last game played on the Plains in 1922, a tie made possible when Notre Dame safety Harry Stuhldreher soared over a wall of Army blockers and made a head-high tackle on an Army halfback named Timberlake at the Notre Dame goal line in the waning moments of the game.

If there is a *primo inter pares,* a first among his equals, in this backfield, it is Stuhldreher, the five-foot, eight inch, 154-pound quarterback. Not because he is the most gifted, or productive, but simply because in the Rockne scheme of things the quarterback is Rockne's "vicar on the field," an extension of Rockne himself. Stuhldreher was the first of his class to attract the coach's notice—the only one in fact, to attract his notice while playing on the freshman team. "He was a good and fearless blocker," said Rockne, "sounded like a leader on the field, and as he gained in football knowledge, showed signs of smartness in emergencies." Courage, leader-ship, intelligence. Combine them with quickness, which Stuhldreher pos-sessed in abundance, and you have the consummate Rockne football player. To ensure that he had not guessed wrong with his quarterback, Rockne had not started him immediately, but kept him at his side during the first few games of his sophomore year, giving him the Rockne tutorial.

By the Georgia Tech game, the sixth game of his sophomore year, Stuhldreher had supplanted senior Frank Thomas as the starter, and was giving the wags in the press box as much trouble as he was the George Tech defense. "Give Miller credit for all his plays," one exasperated writer told the telegraph operator after tiring of trying to spell Stuhldreher's name. It was not long before sportswriters and telegraph operators across the land had learned how to spell it.

Stuhldreher and his coach went back almost a decade; they had first met when Stuhldreher was a teenager in Massillon, Ohio, and Rockne was augmenting his Notre Dame salary by playing some professional ball with the Massillon Tigers; Stuhldreher used to earn admission to the games by carrying Rockne's uniform into the stadium. His decision to attend Notre Dame was something of a foregone conclusion, though not as automatic, perhaps, as that of his teammate, Don Miller.

Miller, the right halfback, an All-America selection in 1923, hails from what is surely the most remarkable family in Notre Dame football his-tory. His oldest brother, Harry or "Red," as he was known to the fans, captained the team in 1908 and made third team on Walter Camp's All-America list a year later. Another brother, Ray, was the back-up end to Knute Rockne in 1913, and would later win election as mayor of Cleveland. A third brother, Walter, was a teammate of George Gipp's in the 1919–21 era, and lost just once in his varsity career. Finally a fourth brother, Gerry, entered Notre Dame in 1921 with Don as the more heralded of the two, but would be beset with injuries and on this day is listed as a back-up halfback on the Notre Dame roster.

Though Don Miller would ultimately become the most famous of his family, he was most insecure of his status at the start of his Notre Dame career. He was at the end of the queue his freshman year when they passed out uniforms; and in those days before recruiting and coddling of prospects, freshman players were issued uniforms on a first-come, first-served basis. They had run out by the time Miller got to the front of the line. "It took me three weeks to get a uniform as a freshman," he would explain much later. "I offered up Daily Communion during that time to get a uniform." It was not until the spring practice of his freshman year—after a winter of indoor track had improved his speed—that he caught Rockne's eye. And by the fall of 1922, he was the first of this famous backfield to earn a starting position.

He might not have had the chance to sparkle in spring practice had not another freshman halfback been bounced out of school, giving Rockne the chance to look a little longer at those buried a little deeper on the depth chart. Jimmy Crowley, the five-foot, eleven-inch, 165-pound left half-back, had been caught in a nickel crap game his freshman year and suspended for the spring semester. He left the campus and took a job as a soda jerk in Indianapolis. He wrote letters home to his mother in Green Bay, but instead of sending them from Indianapolis he sent them along to his ex-roommate at Notre Dame, so that they might carry a South Bend postmark. In the summer, he returned to campus and made up the work he had missed in the spring, regained his eligibility, earned the starting halfback spot, and succeeded in keeping his gaming transgression from his family. Still the team's blithe spirit—he has declared himself a candidate for president in this election year—he had been steered to Notre Dame by his high school coach, Curly Lambeau, who had played one year at Notre Dame, during the George Gipp era, before returning to Green Bay to put together a professional team he called the Packers.

Elmer Layden is the fullback at 162 pounds. He had actually enrolled at the University of Iowa, but the Iowa football coaches lost interest when they learned that he had hurt his knee playing high school basketball. Discouraged, Layden sought the counsel of his high school coach from Davenport, Walter Halas, the older brother of the Chicago Bears' George Halas. Halas was just about to leave Davenport for Notre Dame and a job as

Knute Rockne's assistant. He brought Layden along. The knee gave him no trouble in South Bend.

Layden began his Notre Dame career in Crowley's shadow. They were both left halfbacks—Gipp's old spot, the featured spot in the Notre Dame offense, the triple-threat man who was expected to run, kick and pass. Layden was a step faster than Crowley, and decidedly the better punter and drop kicker. But Crowley was an unsurpassed open-field runner, a better passer than Layden and the best blocker of the four. During their sophomore years, Crowley and Layden split the position but Crowley was clearly playing with greater distinction. Against Purdue in the third game of the season, "Crowley astonished Purdue a great deal and me a great deal more with the liveliest exhibition of cutting, jumping, side-stepping, change of pace and determined ball-toting that I had seen in many a day," said Rockne. By the Indiana game three weeks later, Layden saw just a few minutes of playing time, and the next week went into Rockne's office and told him he was quitting. Rockne told him not to do that—he was starting against Army on Saturday. This was a nice problem for a coach to have—two stellar athletes backing each other up at a key position. But it was a dicey problem as well—two stellar athletes wanting, and deserving, more playing time than they were getting. Two weeks later, Rockne found a way to solve it when All-America candidate Paul Castner was lost for the season with a broken pelvis.

Notre Dame mavens will argue that Castner's injury cost the team an undefeated season in '22; it was two weeks later that the Irish lost to Nebraska in a lackluster game. But those same mavens will continue the story without pausing for breath, explaining that the Castner injury was the most serendipitous calamity in Notre Dame history, for it prompted Rockne to make a move, and the move he made shored up his reputation as a football genius.

The Rockne offense was built on speed and quickness rather than brute force, a novel approach to the game in the early 1920s. He had devised a shift that was revolutionary in concept and in effect, and was opening the game up. The backfield set itself, and then, before the ball was snapped, moved, in unison, a step or two to the left or right, set themselves for just a fraction of a second before the ball was snapped and they were off. Opposing coaches were crying foul, Rockne's tactics were circumventing the intent of the game, they cried; and they clamored for the National Collegiate Athletic Association to put a stop to it, asking that if the shift was not declared outright illegal, that at least the players should be forced to come to a set position for a full second before the ball was snapped. In 1922, however, that legislation was a half-dozen years in the future, and in the meantime nobody was able to devise a defense able to check the gazelles that always seemed to dominate a Notre Dame backfield.

Castner had possessed adequate speed and quickness, but he was also decidedly slower than Stuhldreher, Crowley and Miller. And Castner's three back-ups, gifted athletes all, were archetypal fullbacks: sturdy, lumbering, straight-ahead, guts-and-muscle runners—good men to hand the ball to when you needed a short yard deep in the opposition's territory, but less suited to the wide-open game of speed and deception that was the Notre Dame offense. Not only had Rockne lost his most reliable performer when Castner went down, he was in jeopardy of losing the personality of his offense.

But what if, thought Rockne, he put Elmer Layden at fullback. That would give him a fullback even quicker than Castner, while at the same time eliminate the problem of having to force Layden and Crowley to share a position when both deserved to play full time. Layden was amenable, but proffered the caveat to Rockne that perhaps he was too small—everybody on everyone's opposing line was going to outweigh him by at least twenty pounds. Rockne told him he was going to put into the game a quick opening attack for the fullback—the play that in later decades would come to be known as the "quick hitter"—and that as a track man who ran the hundred-yard dash in less than ten seconds, Layden was perfect for the position. Rockne proved prophetic. "Compared with the orthodox fullback, he was a saber instead of a club," said the coach.

Against Carnegie Tech on November 25, 1922, the ninth game of their sophomore season, Stuhldreher, Crowley, Miller and Layden appeared as a unit for the first time. That first day Miller scored from nine yards out and Stuhldreher threw a twelve-yard touchdown pass to Layden as the Irish won 13–0. And over the next thirteen games, leading up to this Saturday in New York, they had grown in each other's presence. The whole was so much greater than the sum of its parts. Each was instinctively aware of the others, complemented their strengths and compensated for their weaknesses. From high in the stadium it was pure ballet—first the set, then the shift, the four moving as one, gliding to their side, a natural, instinctive motion, seemingly without thought, surely without effort. And then the snap of the ball and the explosion. One play it would be Crowley around right end, with Stuhldreher, Miller and Layden roaring in front. Next perhaps it was Miller's turn to sweep the left end, with Crowley this time joining in the wave of blockers. Then it was Layden or Stuhldreher quickly through the line, suddenly alone in the defensive backfield, which was where they functioned best—juking, faking, spinning their way to extra yardage.

Each man was capable of passing and catching the ball, and each did. Each was likely to punt. It was impossible for defenses to key on any single man, or prepare for any set pattern of plays. Stuhldreher was more than Rockne's "vicar on the field"; he was his alter ego, calling signals without a huddle, orchestrating and embodying the Rockne philosophy that to beat the other guy you've got to out-think the other guy, and the way to do it is with the unusual and the unpredictable.

The fans loved it. This was so different, and so much more fun, than ramming the ball into the scrum every time and laboring it up the field three and four yards at a time. But the magic was in more than just the achievement—the mere presence of these four young men brought the electricity of anticipation to the stadium. Whenever the Irish had the football the crowd dared not turn its head for fear of missing a sliver of history—it was the same electricity that was there when Ruth stepped into the box in the late innings of a close game; or when Jones addressed the ball on the eighteenth tee of an even match; or when Man O' War passed the sixteenth pole a length behind and closing.

Notre Dame roared through its 1923 schedule, not allowing a touchdown until its fifth game, allowing only five touchdowns all season, while the offense averaged nearly thirty points a game. Only the unaccountably flat performance against Nebraska—where a muddy field slowed their fleetness afoot and changed the character of their attack—cost them a perfect season, as the now-junior backfield grew in stature and notoriety as well as competence and poise. Sportswriters who witnessed their heroics were moved to paeans and hyperbole. Even Rockne, ordinarily a fount of eloquence and insight, sounded a little bit like Casey Stengel with the grammar cleaned up when talking about his prize backfield:

> How it came to pass that four young men so eminently qualified by temperament, physique, and instinctive pacing, complement one another perfectly and thus produce the best coordinated and most picturesque backfield in the recent history of football—how that came about is one of the inscrutable achievements of coincidence of which I know nothing save that it's a rather satisfying mouthful of words.

Stuhldreher, Crowley, Layden and Miller have Notre Dame off to a 2–0 start in 1924; but they've beaten nobody of consequence, routing Lombard 40–0 and Wabash 34–0. Neither team provided any kind of test of how strong the Irish really are. Army will be different. The Cadets are big, quick and experienced. Grantland Rice, in his Saturday morning column of predictions, rated the game even.

Rockne, in public at least, is picking Army. On Thursday, in his nationally syndicated column, he said, "Our team has not shown anything this fall except a backfield and no backfield can go without a line to lift the enemy out of the way. No forward pass attack is good unless accomplished [sic] by a strong running attack. The Notre Dame line will have to show vast improvement this week or the Army will triumph."

Knute Rockne had never intended to make football a career. He was going to be a doctor. He was a brilliant student, particularly in the sci-

ences, which comprised the bulk of his curriculum—of his thirty undergraduate courses, twenty-five were in chemistry, physics, botany, physi-
ology or pharmacy, and his four-year average was 90.52. He was accepted
at St. Louis University medical school, and Jesse Harper, his Notre Dame
coach, had found him a high-school teaching and coaching job in the area
to help offset his tuition and living expenses. But the school said no—no
first year medical student could balance the exigencies of studies and a full-
time job. Without the job there was no money for tuition, so Rockne's
dream of becoming a doctor was deferred, and the autumn of 1914 found
him right back at Notre Dame, as assistant football coach, head track coach,
and chemistry instructor.

As a coach, Rockne proved not only a teacher and shaper of men, but
also an evolutionary who tailored a sport so that it was the perfect mirror
of his times—footloose, brassy, glamorous, and above all, fun to be around.
And Rockne quickly came to see football as a career that provided not
only fulfillment but challenge—physical and emotional challenge, and even—
and this is what no doubt sold him on the idea—yes, it even provided
intellectual challenge.

As his team makes ready to play Army on this day, Rockne boasts of
a record of forty-nine wins, four losses and three ties since taking over as
head coach at the start of the 1918 season. He has innovated and he has
inspired—some of his innovation has come from the inspiration he has
been able to instill in his charges; likewise, some of the inspiration was
born of his innovations—his players had the utmost confidence in his abilities; Rock would think of something, they believed; all we need do is just
what he tells us.

As the Notre Dame success mounted and the stories have grown of
Rockne's abilities as a motivator and orator; he has became one of the
most sought-after men in the nation—for football clinics (naturally), after-
dinner speeches, business conventions, newspaper and magazine articles,
radio broadcasts. He has been able to increase his $12,000 Notre Dame
salary manyfold from these outside engagements, all to his rather pleasant
astonishment. ("Can you imagine them paying me $300 just to make a
speech?" he'd ask incredulously.) He is a colossal national figure—the very
embodiment of the game on the national stage. So when the Rock comes
to New York, particularly to play Army, it is big news, and all the press
box stars—the Brouns, Runyons, Gallicos, and Rices—check in and further the legend.

As a by-product, Rockne has spread the popularity of Notre Dame.
Notre Dame has become a true national team with a national following.
Subway alumni, they call them, the multitudes of clerks and shopkeepers,
civil servants and day-laborers, immigrants and parish priests, even a sizable number of alumni from other colleges that have found a special attraction in Notre Dame football, and have packed stadiums from the Mid-

west to the deep South to the Polo Grounds in northern Manhattan. Each football win has been like a stone tossed into a pond, expanding the school's legions of followers relentlessly outward. Catholic families who cannot tell you whether South Bend is in Iowa, Indiana or Illinois dream of sending their sons to Notre Dame.

So, as many eyes will be on Rockne, at least at the start, as will be fixed on the talented Irish backfield this afternoon. Many fans will no doubt be disappointed when they first spy the famous coach. He is shorter than they expected—heroes are nearly always physically smaller than the figures who bestride our imaginations—barely five feet, eight inches tall, barrel-chested and spindly legged, dwarfed even by his undersized backfield. If they are close enough to the sideline to get a close look at him, the fans will notice that he seems much older than his thirty-six years. His generous nose has been distorted by a childhood run-in with a baseball bat, his jowly face creased and leathered by the sun and winds of fifteen Indiana autumns. Beneath his brown fedora, his head is round and bald. As the players go through the last of their warm-ups on the field he makes his way to the black canvas chair where he will sit, outwardly calm, and watch the game.

When Notre Dame kicks off to Army to start the game, the Irish starters are on the bench. This is a favorite ploy of Rockne's, starting the second string, or what he calls the "shock troops." His hope is to get a full period out of the second team, and then send in a fresh varsity. "We know by experience that two football teams generally hit the hardest in the first quarter," says Rockne. "With a light backfield . . . , why not save them from this bruising and unnecessary thumping?" The strategy gives Harry Stuhldreher a chance to study the Army defense and plan his attack; and, ever the psychologist, Rockne has found that this respite from action also gets the first string a little restless and sends them out just a wee bit more eager for the fray than they might have been at the start of the contest. But these benefits are incidental, he insists. "The main reason is to save them as much as possible from Saturday to Saturday, for boys will be boys—and often they must be guarded against their own zeal for play."

Rockne has cause to be a little nervous about his strategy as Army takes the kick-off and rips off three quick first downs, bringing the ball into Notre Dame territory. But the shock troops collect themselves in time and stall the Army drive, and when Stuhldreher, Crowley, Layden and Miller enter the game in the closing moments of the first period, the game is still scoreless. The quartet begins its romp towards immortality early in the second period. Jimmy Crowley opens the drive from the Notre Dame fifteen with a fifteen-yard gain around left end. Layden rips up through the middle for six, and then Don Miller circles right end for ten more. Stuhldreher's blocking has led both end sweeps, the intrepid quarterback hurling his frail frame headlong—as is his wont—into the knees of a prospective Army tackler.

Next, Army stops a Layden dive for no gain, then Stuhldreher gets himself into the offense, opening it up a bit with a twelve-yard pass to Jimmy Crowley—first down on the Army forty-four. Don Miller gets the call from here, and with Stuhldreher again leading the interference with reckless abandon, Miller breaks clear around the right side and seems headed for a touchdown before Army halfback Harry Wilson catches up to him and drags him out of bounds at the ten. Crowley and Layden bang out five yards, and then, on third and goal, Layden explodes up the middle and into the end zone standing up—6–0, Notre Dame.

Layden misses the extra point but it seems of little consequence. The drive had been absolute artistry—perfection even. It had been not so much football as it was choreography—eighty-five yards in eight plays, a pleasing mix of sweeps and dives and a pass play thrown in for variety, with all four backs sharing equally in the show. Each was perfectly in sync with the others and as a unit they were seemingly a step or two faster than anyone else on the field. They were playing their own game, levels above what was being played by their eighteen supporting actors. In the press box the wags take note, and at halftime the talk centers on the Notre Dame backfield. Sportswriters are a clannish lot, given to much in-game kibbitzing on what the angle of the story should be. It is agreed, that barring some dramatic and unexpected second-half reversal, the angle on this day's story is going to be this incomparable Notre Dame backfield.

For though they lead only 6–0, Notre Dame had clearly outclassed the Cadets, who failed to gain even a single first down in the second quarter. Harry Wilson, the Penn State alumnus and the bellwether of the Army backfield, had been particularly frustrated. Time and again, as the play was set, Adam Walsh, the Notre Dame center and captain, would scream out: "Watch Wilson! Watch Wilson!" And sure enough, Wilson would get the call and would be met at the line by a gaggle of Notre Dame tacklers. Some time later, Knute Rockne would reveal that Walsh's uncanny intuition had been assisted by an observation the coach had made in an earlier game that Wilson tended to blush just before he was set to carry the ball. He was generally a devastating open-field threat, but as Grantland Rice points out in the press box: "You can't run through a broken field until you get there."

After the half, during which the crew of the ZR-3 zeppelin changed sides of the field to show their support for both clubs, Notre Dame resumed control of the game. On the first Army drive, Elmer Layden intercepted a pass on the Army forty-eight. Crowley started the drive with a fifteen-yard gain around right end. Then it was Miller for seven and Layden for eight, bringing the ball to the Army twenty. A Don Miller dive into the line was met at the line and stopped for no gain. But on the next play Jimmy Crowley swept the left end, then cut back across the middle and clear to the other side of the field, diving into the corner of the end zone for Notre Dame's second touchdown. Elmer Layden place-kicked the extra point and it was 13–0, Notre Dame.

The Irish then threatened to break it wide open. Twice more before the close of the third period, they brought the ball down inside the Army ten-yard line. The drives again showcased the speed, agility, versatility, precision, cohesion and overall dominance of Messrs Stuhldreher, Crowley, Layden and Miller. The one element missing from the Notre Dame attack was the pass. After the twelve-yard completion to Jimmy Crowley in the first scoring drive, Stuhldreher called for a pass just three more times. One was completed for a short gain in a drive that stalled at midfield; another fell incomplete, and the last was intercepted by Army safety Mike Yeomans on the Army five-yard line. That ended the third Notre Dame scoring threat. A last threat, late in the third period, came up dry when Rockne eschewed a field goal attempt on a fourth-down-and-five and saw his boys stopped at the nine.

Despite the missed opportunities to put it away, the Notre Dame players are cocky by now, and unable to resist needling the Army players when they get the chance. Miller and Crowley particularly delight in giving it to Ed Garbisch, Army's All America center—a defender who has been making their afternoons miserable for three years now, but who, on this day, is being frustrated by center Notre Dame Adam Walsh and the rest of the Irish line. "Is that the great Mr. Garbisch," one asks the other with affected incredulity as they walk past a prostrate Garbisch.

"Yes, that's the great Mr. Garbisch," comes the solemn reply.

And later, when another play has gone for a big gain through Garbisch's hole and he is once again picking himself up off the Polo Grounds grass, Crowley and Miller stroll past and Crowley asks with utter amazement:

"You don't mean to say that's the great Mr. Garbisch?"

"If the number's correct," replies Miller, "it's none other than the great Mr. Garbisch."

But the Notre Dame cockiness is premature. They get a little careless in the fourth period and it puts Army right back into the game. It's all set up with a Harry Wilson dash off-tackle for thirty-four yards, bringing the Cadets from a deep hole to midfield. The offense sputters there, but the ensuing punt leaves Notre Dame in a deep hole on their five, and Layden is forced to punt from behind his own goal line. Layden, who has hit a couple of eighty yarders this season, mis-hits this one and gets it out only to the Notre Dame thirty-five, where the Notre Dame defense interferes with Mike Yeomans's fair catch, giving the Army a first down on the Notre Dame twenty. Four plays later, on a fourth-and-two from the twelve, quarterback Neil Harding freezes the Notre Dame defense with a fake to the line and then skirts the right side untouched for a touchdown. Ed Garbisch drop-kicks the extra point, and Army is suddenly within a touchdown of tying the game—a touchdown and an extra point of winning.

But the balance of the game is played out uneventfully near midfield. The Army defense, inspired by the touchdown, finally puts a lock on the Irish backfield, blocking off the ends and plugging up the line. But the Notre Dame defense is similarly steeled, and the game ends 13–7.

In the vanishing daylight and gathering chill of the press box the reporters get to work on their morning stories. They work quickly, for most want to get down to the locker room and visit with Rockne. They'll not be scavenging for quotes for their stories; it'll be another quarter-century before anyone gives a thought to working this post-game intelligence into his account of the game. Today, they're just looking to be sociable and perhaps collect something they can use in a future piece. Their task in the press box now is to tell the readers what happened here at the Polo Grounds this afternoon—to tell them about this ethereal Notre Dame backfield and the gritty Army refusal to give up despite having been outclassed and outplayed. And to tell them about it quickly, for the bulldog editions of most of the papers will be on the streets in just a couple of hours. Down on Park Row the editors will have noted the Western Union ticker report that the game is over and that Notre Dame had won; they expect the copy from their reporters to follow forthwith.

Some of the reporters are likely to check to see how Grantland Rice is coming along, for most of them are invited to the customary Saturday evening open house at his apartment on Fifth Avenue. There'll be good food and a supply of the choicer prohibition liquor and some ample and splendid football conversation, for all of the local coaches will be there, and Rockne is sure to put in a brief appearance before Notre Dame's train leaves for South Bend. And the party begins as soon as Rice files his story, hence the interest in his progress. They note with satisfaction that, as usual, his story seems to be offering scant resistance. His hat is pushed back slightly off his brow; his meaty fingers have the keys of his Royal clattering.

True to the conversation during the game, and true to the events on the field, the Notre Dame backfield is providing the heart of the story as one by one the reporters bring their copy to the Western Union operator to be filed back to their newspapers.

Young Frank Wallace is especially anxious to file his story and get down to the locker room; he has a lot of friends still on the team. His story has to service a broad range of Associated Press clients, most of them indifferent to his Notre Dame allegiance. But on this day he is able to wear his heart on his sleeve and still deliver a direct-and-to-the-point story in the best traditions of the Associated Press:

> The brilliant Notre Dame backfield dazzled the Army line today and romped away with a 13 to 7 victory in one of the hardest-fought of the intersectional series between the two teams. More than 50,000 people saw the game at the Polo Grounds.

Alison Danzig of the good, gray *New York Times* opened his story in good, gray *Times*-style, choosing a sedate, understated metaphor of Notre Dame football as modern mechanical wonder:

> Moving with speed, power and precision, Knute Rockne's Notre Dame football machine, 1924 model, defeated the Army 13–7, before 60,000★ at the Polo Grounds yesterday.

After saluting the "pluck" of the Army team, suggesting that "An epic might be written about the Army's brave stand and gallant counter-attack," Danzig turned to the Notre Dame backfield in paragraph three:

> If an epic could be written about the Army, there was also material for several poems in the swift, dashing play of the men from Indiana. Notre Dame's backfield attack had some of the poetry of motion about it.

While it was the practice of the conservative *Times* to eschew attempts at writing epics and or lesser prose poems, most of the other writers in the press box tried—believing, as sportswriters believe to this day, that an extraordinary athletic endeavor merits an extraordinary rendering in the next morning's newspaper. So the writer exerts a little extra effort, reaching, as the athlete does, down within himself, hoping that the effort will produce a result that transcends previous limits. Usually, then as now, the sweat shows.

Nowhere is this more evident than in the piece that Damon Runyon wrote for the *New York American* and for the Hearst syndicate. Knowing that everyone was going to be playing some variation of the Notre Dame-backfield theme, Runyon opted for something entirely different. During the waning moments of the game, his eye fixed upon the the West Point mascot, the Army mule. He rather anticipated that the Notre Dame fans would try to snatch the mule's blanket at the close of the game, a favorite Navy ploy whenever they defeated the Army. And Runyon rather imag-

---

★Attendance figures were largely a matter of the writer's speculation in 1924. It was thus not uncommon to have "official" attendance figures that varied from Frank Wallace's 50,000 to Grantland Rice's 55,000 to Alison Danzig's 60,000.

ined that the mule was anticipating being shorn of its blanket, too, and was "looking mighty peevish toward the finish." Notre Dame and its fans made no attempt to snatch the blanket, but Runyon elected to stay with this angle anyway. He opened his piece with six longish and rather peculiar paragraphs on the Army mule. He then gives a couple of perfunctory paragraphs to the game and the Notre Dame backfield—"Give a team a Don Miller, a Layden and a Crowley and it really doesn't need anybody else"—before turning to what it was he really did and loved best, giving the readers a feel for the New York people who were in the stands:

> It wasn't a college crowd, but it was a well dressed, high toned crowd who paid plenty of money at the gate to see this game. Naturally the Army had the heavier following, numerically and vocally, but after the New Yorkers got a load of the Notre Dame's football playing, the visitors had plenty of supporters. One thing about New York, it is always with a winner.

If Damon Runyon reached a bit far and stumbled, Paul Gallico in the *Daily News* reached so far that he lost his balance completely and fell flat on his face:

> The great error that the Army made was in not taking advantage of a time-honored institution of the service, shooting at sunrise. Early yesterday morning the Army should have sent emissaries to arrest Messrs. Don Miller and Crowley.
> Then, as the sun rose over the horizon, the Army should have shot these two young men full of holes.
> As it was they didn't, and these two young men yesterday afternoon did just that thing to the Army team—shot it full of holes.

Heywood Broun hit upon a cavalry metaphor in his sparse but solid accounting of the game on the front page of the New York *World:*

> The Army thought the attack would come from the sky, but Notre Dame switched tactics and defeated the West Pointers with sweeping cavalry charges around the ends. . . .
> They were light horsemen, these running backs of Notre Dame. but they swung against the Army ends with speed and numbers.

But Broun, Runyon, Gallico and the others will be merely voices in the chorus when the readers turn to their Sunday morning papers, for gentle Grantland Rice, in the thirty or so minutes allotted him and in the clamor of the press box has on this day produced a piece of journalism history. The opening lines are destined to find a place in *Bartlett's Familiar Quotations,* change forever the lives of the four men they honor, and to endure until this day as the most famous "lead" in sports journalism history:

> Outlined against a blue-gray October sky, the Four Horsemen rode again. In dramatic lore they are known as Famine, Pestilence, Destruction and Death. These are only aliases. There real names are Stuhldreher, Miller, Crowley and Layden. They formed the crest of the South Bend cyclone before which another fighting Army football team was swept over the precipice at the Polo Grounds yesterday afternoon as 55,000 spectators peered down on the bewildering panorama spread on the green plain below.
>
> A cyclone can't be snared. It may be surrounded, but somewhere it breaks through to keep on going. When the cyclone starts from South Bend, where the candle lights still gleam through the Indiana sycamores, those in the way must take to storm cellars at top speed. Yesterday the cyclone stuck again as Notre Dame beat the Army, 13 to 7, with a set of backfield stars that ripped and crashed through a strong Army defense with more speed and power than the warring cadets could meet.
>
> Notre Dame won its ninth game in twelve Army starts through the driving power of one of the greatest backfield that ever churned up the turf of any gridiron in any football age. Brilliant backfields may come and go, but in Stuhldreher, Miller, Crowley and Layden, covered by a fast and charging line, Notre Dame can take its place in front of the field.

The story appeared on page one, column one (the top left-hand corner) of the *New York Herald Tribune,* and in the more than one hundred other newspapers across the country that subscribed to Rice's work through the New York Tribune syndicate. Upwards of ten million people had access to Rice's story that Sunday, and those who read it, savored it—for it was Grantland Rice as they loved him. The piece was replete with the imagery and poetry they'd come to expect from Rice's work—at other points in the story the backfield had "the mixed blood of the tiger and the antelope," and "The Army line was giving it all it had but when a tank tears in with the speed of a motorcycle, what chance has flesh and blood to hold?" But the beauty of Rice's work was also that he never let the artistry overwhelm his duty as a journalist, as Runyon, Gallico, and to a lesser extent, Broun, did on this day. Rice was a very good reporter, and after the semantic crescendo of the opening, he settled into the then-common practice of telling the story chronologically, and his story told his readers exactly what they needed to know about what happened at the Polo Grounds. He described the game's ebb and flow in the first quarter; described in relevant detail the three drives for scores; paid homage to the Army defense; provided some atmosphere—" . . . the first big thrill of the afternoon set the great crowd into a cheering whirl and brought about

the wild flutter of flags that are thrown to the wind in exciting moments."
He kept it all in perspective with detail and observation—"In that second
period Notre Dame made eight first downs to the Army's none . . . "
and "Up to this point the Army had been outplayed by a crushing mar-
gin." At the close of his story he summed up—"The Army brought a fine
football team into action but it was beaten by a faster and smoother team.
. . . It was the Western speed and interference that brought about Army
doom. The Army line couldn't get through to break up the attacking plays,
and once started the bewildering speed and power of the Western backs
lashed along for eight, ten and fifteen yards on play after play."

And it was all tied up in a tidy, smooth-flowing package. The Four
Horseman reference is really more of an epigraph than a lead. The cyclone
metaphor is the one Rice sustains thoughout. The cyclone "opened like a
zephyr." Later, it "struck with too much speed and power to be stopped."

It is impossible to read Rice's account of this game and not come
away knowing what happened, and how and why it happened. This is
true whether you read the story amidst all the others on the morning after
the game, or whether you read it in a vacuum after the passage of more
than sixty-five years. It was not even necessary to have read, or even heard
of, Vicente Blasco-Ibanez's novel, *The Four Horsemen of the Apocalypse,* or
to have seen the Rudolph Valentino movie of the same name, or to be
versed in the Book of Revelation to comprehend and enjoy Rice's simple
and clearly stated reference to the Four Horsemen. Thus, Rice the sportswriter was, in this story and in so much of his work, all things to all
people. There were readers who cared not a whit for sports who enjoyed
his writing; there were others who wouldn't know a poem from a point-
after-touchdown who were equally avid.

Why is the Four Horsemen story the most famous in the history of
sportswriting? Why indeed, argue the cynics who dispatch with Rice by
asking: From what angle had he been watching the game that he saw the
runners outlined against a blue-gray sky? Why, he would have had to have
been lying flat on his back in the middle of the field to gather such a
perspective. When you add to that the fact that the story has more metaphors than verbs—"Rockne's light and tottering line was just about as
tottering as the Rock of Gibraltar"—and abounds in breathless hyperbole—" . . . hammered relentlessly and remorselessly . . . "—it is easy
to dismiss the entire package as the overwrought, overblown, meretri-
cious utterings of an out-of-touch sentimentalist.

So why *has* it endured? Well, to begin with, this was an event that
was entitled to some hyperbole—it was a hell of a football game; and the
Four Horsemen were every bit as good as Rice and his cronies in the press
box made them out to be. Second, and more important, the story has
endured because Grantland Rice fashioned a phrase that was clever, mem-

orable, and fit precisely with the event and the times. It was not that Rice's story stopped people in their tracks—readers across America were not grasping Rice's story in wonder and telling each other that this was the greatest story in sportswriting history, in the same way they were telling each other that the Four Horsemen were the greatest backfield in college football history. It was merely that the event had captured their attention, and the phrase lingered in their consciousness. And gradually, as the details of the event faded from the consciousness, as they inevitably do, the lingering phrase transcended and supplanted the event itself.

A contemporary analogy might be made with Al Michaels, the ABC sportscaster, and his utterance in the closing seconds of the improbable triumph of the United States hockey team over the Russians at the 1980 Winter Olympics in Lake Placid. Who among the millions who were transfixed in front of their televisions is likely to forget Michaels's exuberant shout: "Do you believe in miracles? Yes!" as the seconds ticked away in Lake Placid? Michaels was articulating the exhilaration and incredulity we all felt at what was—by the standards of sports and games—a searing moment. Like the Rice lines on the Four Horsemen, Michaels's comment was clever, memorable, and perfectly suited to the times. As with the Notre Dame-Army game, it was the *event* that commanded our attention, but it is Michaels's phrase that lingers—the phrase that is surely destined to outlive its speaker, and seems likely to transcend the details of the event and become the memory we carry of that February evening in Lake Placid.

Make no mistake about it: the immortality of Al Michaels's words has been helped immeasurably—guaranteed even—by the millions of times the words have been replayed on videotape in the years since Michaels spoke them; and, likewise, the immortality of Grantland Rice's story, and of the Four Horsemen themselves, got an immeasurable boost from another. His name was George Strickler, Notre Dame's student publicity assistant, later destined to enjoy a long career in sports, first as a reporter for the *Chicago Tribune,* then as the publicity director for the National Football League, and finally as sports editor of the *Chicago Tribune.*

Strickler claims to have given Rice the idea for the Four Horsemen lead. According to Strickler, he had just seen the Ruldolph Valentino film that week on campus, and when he heard Rice and Damon Runyon, Davis Walsh, and Jack Kofeod talking during halftime about the dominance of the Notre Dame backfield, he interjected: "Just like the Four Horsemen."

Rice remembered it differently in his memoirs. He said the idea first came to him during the 1923 Army-Notre Dame game at Ebbets Field, which he watched from the sideline with Brink Thorne, one-time Yale football captain. "In one wild end run," wrote Rice, "the Irish backfield of Harry Stuhldreher, Jim Crowley, Don Miller and Elmer Layden, swept off the field over the sideline. At least two of them jumped over me, down on my knees.

" 'It's worse than a cavalry charge,' I said to Brink. 'They're like a wild horse stampede.' "

Whether or not Rice owed Strickler for the genesis of his famous lead, he certainly did owe him a debt of gratitude for what he did next. When he saw the Grantland Rice story on Sunday morning, Strickler was hit with the idea of putting these guys up on some horses and snapping a picture. He immediately wired his father, who worked for Notre Dame, and told him to arrange to get ahold of four horses. He then wired a South Bend photographer named Christman and told him to be at practice at Cartier Field on Monday afternoon at three o'clock.

At practice that Monday, Strickler showed up with one photographer and four horses in tow and apprised Rockne of his plans. "He just stared at me like he couldn't believe it," said Strickler, "and who could blame him? I talked fast, telling him what a great publicity stunt it would be. Well, when it came to public relations and what was good for Notre Dame, Rock had real sharp instincts."

Still, Rockne gave his permission grudgingly. "Okay," he said, eyeing the horses narrowly, according to Strickler. "You can put the boys aboard, but God help you if they fall off and get hurt." Stuhldreher, Crowley, Layden and Miller, in full football uniform, climbed aboard Strickler's four nags happily if not terribly deftly. Strickler handed them each a football and Christman took the picture.

"If you look closely at the picture, you will note that Crowley is sitting a little cock-eyed in the saddle," said Strickler. "That's because he had the makings of a fine boil on one side of his bloomers."

It's unrecorded how many people looked closely enough at the picture to notice Crowley's discomfort, but an awful lot of people saw that picture. Strickler sent it out over the wires and newspapers all over the country picked it up. He sold copies of it to students on campus and all told earned himself more than $10,000 from that picture.

The picture served to complement and reinforce Rice's eloquent lead, and by the time of the the Notre Dame-Princeton game the following Saturday, every sportswriter in America was referring to the Notre Dame backfield as the Four Horsemen. They remain the Four Horsemen to this day. Whenever one is mentioned apart from the others, he was quickly identified as "one of the Four Horsemen."

Immortality requires three elements: Great deeds. A great stage. Great good fortune. The Four Horsemen performed the great deeds; they swept through the balance of the 1924 season undefeated, avenging their sophomore and junior year losses to Nebraska—the only blemishes on their varsity record—with a 34–6 thrashing of their nemesis in front of their home fans in South Bend. On New Year's Day, 1925, they defeated Pop Warner's Stanford team, 27–10 in Pasadena, in what was then called the Tournament of Roses Bowl.

They performed these deeds on the necessary great stage—the Rose Bowl; the dedication game in Chicago's Soldier Field, against Northwestern before 100,000 spectators; and especially at the Polo Grounds in front of the New York writers, who in those days had the same power to confer instant notoriety and credibility that network television has today.

Last—particularly—they were blessed with the great good fortune of Grantland Rice's story and George Strickler's picture. The four men formed a remarkable bond that continued until their deaths in the 1970s and 1980s. They all tried coaching for a while, but only Layden stayed with football, coaching Notre Dame from 1934 until 1940 and then serving as commissioner of the National Football League. The rest drifted into other professions—Don Miller became a lawyer and then later U.S. Attorney in Cleveland; Crowley became general manager of a Pennsylvania television station; and Stuhldreher became a high-ranking executive with U.S. Steel. But they remained the Four Horsemen, even long after time and nature had stripped them of their stallion speed and physiques. They were inducted in the college football Hall of Fame as a unit—they remain the only complete backfield to earn that honor. And they remained as inseparable as they were on that October afternoon in the Polo Grounds; their friendship grew to be as legendary as their football. They spoke to one another weekly, and appeared together as the Four Horsemen some thirty times a year. And they never lost sight of the fact that their lives would have been dramatically different, and markedly less fulfilling, had it not been for Grantland Rice's story in 1924. Shortly before Rice's death, he had lunch with Crowley, Layden and Miller at Toots Shor's in New York. After an evening of reminiscing, Don Miller, right halfback and Cleveland attorney, made certain Rice knew what it all meant, and how much they appreciated it.

"Granny," he said. "Rock put us together in the same backfield but the day you wrote us up as the Four Horsemen, you conferred an immortality on us that gold could never buy. Let's face it. We were good, sure. But we'd have been just as dead two years after graduation as any other backfield if you hadn't painted that tag line on us. It's twenty-nine years since we played. Each year we run faster, block better, score more TD's than ever! The older we are, the younger we become—in legend. Another thing. In business, that tag line has opened more doors . . . has meant more to each of us in associations, warmth, friendship and revenue, than you'll ever know."

The modest Rice was touched by Miller's remarks. "That's as nice a compliment as a fellow can receive," he said.

But Rice had been well served by the Four Horsemen, too. By 1924, he already had the fame and the "associations, warmth, friendship and revenue" that the Four Horsemen would come to know. But the Four Horsemen article became his signature piece. With each appearance, each accomplishment, each parcel of distinction that the Four Horsemen brought upon themselves, Rice's position as the man who christened them was reinforced, his position as the pre-eminent American sportswriter made just a little more secure.

And, of the sixty-seven million words he published in his lifetime, the few dozen words he wrote on the Four Horseman—together with the immortal couplet on "how you played the game"—are the words for which he's best remembered. "As long as there is football, or a Notre Dame, the Four Horsemen will ride," said Frank Wallace, "with Rice in the carriage behind."

3

Tennessee: Where the Softer Dreams Remained

In Murfreesboro, Tennessee, on the southeast corner of Spring and College streets, one block off the Town Square, there stands a plaque marking the site where Henry Grantland Rice was born on November 1, 1880.

The aluminum and steel marker was put there by the Tennessee Historical Commission on a radiant May afternoon in 1956, two years after Rice's death. Bill Corum of the *New York Journal American,* Arthur Daley of the *New York Times,* Fred Russell of the *Nashville Banner* and other erstwhile colleagues came down to join local officials in the speechmaking and celebration as the Murfreesboro Central High School band played and young boys in blue jeans and crew cuts haunched down on the sidewalk to gaze at the oil portrait of Rice that rested against a vase of flowers at the base of the marker.

"Though he traveled far afield, he was a Tennessean," said Tennessee governor Frank Clement. It was a statement of simple eloquence and powerful insight, cutting right to the soul of Grantland Rice's life and career. For though he came to embody the worldliness of New York and the sophistication of the World Series, the Rose Bowl or a championship fight, he did so without ever losing sight of the values of the bluegrass country of Middle Tennessee—the simple values of small-town America—

33

he had acquired during his youth, and had come to hold as fervently as a true believer holds his belief in God. As the plaque noted, this son of Murfreesboro had become "internationally famous for his influence for sportsmanship and fair play."

Rice also had a tender and abiding affection for the verdant hills of Middle Tennessee. He had left them eagerly as a young man to seek his destiny in distant and larger cities, only to return after five years. When he left a second time, at the age of thirty, he did so reluctantly, and only after it had become apparent that his abundant talent needed a grander theater. And until the end of his life he talked of returning, even long after he realized he never would.

> "I'm going home some day"—
>   So moves the dream of all the roving world—
> The seekers of far lands who've lost their way—
>   God's countless aliens by the current whirled
> From out the harbor, and by the tempest tossed
>   To unknown lands where they must ever roam—
> And this is all that makes life worth the cost—
> This endless dream—"Some day I'm going home."

The house in which Grantland Rice was born had long-since been razed by 1956. It was a two-story frame colonial, and Rice had been born in an upstairs rear bedroom on the north side of the house, the first child of Bolling and Beulah Rice.

The Rice presence in the American South dates back at least as far as the Revolutionary War. Lieutenant John Rice—patriarch of a far-flung Rice family—fought with a North Carolina regiment. John Rice's son Hopkins Rice—Grantland's great-grandfather—was born in 1785 in Nash County, North Carolina, and eventually came to raise a family of eleven children on a 2000-acre antebellum plantation just south of Eutaw, Alabama. One of the eleven was Grantland Rice's grandfather, John.

How and why John Rice's son Bolling came to find himself in Murfreesboro in the 1880s is not clear; it is likely that the young Grantland never knew or cared, for the dominant family—and thus the dominant family history—in his life was his mother's family—the Grantlands.

Henry Grantland Rice was named for his maternal grandfather, Henry Grantland, the man he called the central figure in his family and a "100-proof individualist." Henry Grantland had come to America from England in 1835, at the age of fifteen. He settled first in Fairfax County, Virginia, then worked his way south to his own cotton plantation on the banks of the Tennessee River in northern Alabama. The outbreak of the Civil War found Grantland on this plantation in Alabama, trying to ward off a Federal gunboat with rakes, spades and a single shotgun. "The gunboat won that one," reported his grandson.

Shortly thereafter, Grantland left his home to fight with the Confederate forces under General Braxton Bragg, first with the Army of Alabama and West Florida, and then with the Army of Tennessee, seeing action in the battles of Shiloh, Murfreesboro, and Chattanooga.

Grantland, a major by the war's end, later told his grandson that he'd had little respect for General Braxton Bragg. His hero, he explained, was General Nathan Bedford Forrest, the swashbuckling cavalryman that Robert E. Lee had called "my finest general." Forrest thus became the first boyhood hero of the young Grantland Rice, as he was to so many other Tennessee boys. Forrest held a place in Tennessee legend and folklore akin to that of Davy Crockett and Andrew Jackson, earned with his daring raids, his individual heroism and colorfully stated and oft-quoted secret to victory: "To git thur the fustest with the mostest men."

Major Grantland loved to regale his grandchildren with stories of the war and his past; and the story his namesake liked the best was the story about the cotton. Before he left his Alabama farm in '61, Grantland presciently stashed six 500–pound bales of cotton in a cave near his home as a hedge against the vicissitudes of war. A quarter-century later, the young Grantland Rice used to love to prompt his grandfather into telling him what happened when the war was over; it was a story Rice loved repeating in his later years.

"Well," begins Henry Grantland in his grandson's telling of the story, "when the war was over, I threw down my musket, grabbed my horse and galloped back to the farm. I didn't forget those bales. There they'd been—for almost four years—in a damp, dripping cave . . . known only to God and me. The selling price in 1864 *[sic]* was one dollar a pound, but all the water had swollen those rotting old bales till they weighed about a half ton—one thousand pounds apiece! When the Yankee carpetbagger, Butler the Buyer, came through, he paid me a thousand dollars for each waterlogged bale."

Henry Grantland took his 6000 Yankee greenbacks to Nashville and established himself in the cotton business, and by the mid-1880s he was one of the city's leading citizens. He was the president of the Grantland Cotton Company, his farm and brokerage business in East Nashville; he was a director of the Nashville Cotton Mills Company, a manufacturing concern that turned the raw cotton into cloth and yarns; and he was the cashier at the First National Bank of Nashville.

The presence of Henry Grantland, and the opportunity he provided, lured Bolling and Beulah Rice from Murfreesboro to Nashville when young Grantland was four. There was little between the two towns in the mid-1880s except for forty miles of a loose gravel and crushed-stone road called the Murfreesboro Pike, and the adjacent tracks of the Nashville, Chattanooga and St. Louis Railroad—the N. & C., as it was known to the citizens of Middle Tennessee. The Rices, a family of some means, would have no doubt eschewed the dusty, bouncy day-long ride on the Pike in favor of the infinitely swifter and decidedly more pleasant—if somewhat sooty—trip on an N. & C. day coach. At the passenger depot on Walnut Street, two blocks from the Tennessee state capitol, the Rices would have no doubt been met by a carriage and driver sent by Henry Grantland, to bring them the last few blocks across the Cumberland River to the Grant-land home at 614 Woodland Avenue in East Nashville.

The Rices remained under the Grantland roof for four years. For a time, Bolling Rice ran a produce business with his brother, called Rice and Brother, Produce and Commission ("Dried Fruit a Speciality") but it was short-lived; and for the rest his life he was connected with the Grant-land Cotton Company in a variety of front-office and bookkeeping positions, including secretary-treasurer.

It can be said with reasonable confidence that the large, high-ceilinged Grantland home on Woodland Avenue was a most pleasant place to spend a young boyhood. There were servants aplenty; there were the colorful stories of a doting Grandfather Grantland and his romantic and only partially explained wanderings—he would be gone sometimes for five and six weeks, off to Pittsburgh by a combination of boat, train, and horse and wagon, "to buy tools for the farm." There was also the bounty of Grand-mother Grantland's table. "There were hog brains—the most magnificent of dishes," recalled Rice late in his life, "hominy grits, ham and ham gravy, waffles, fried sliced apples, corn pone, fried and scrambled eggs—food I've dreamed about but have seldom seen for half a century, at least in such profusion."

When Rice was eight the family moved out into a home of their own, on Vaughan Pike, further out in East Nashville, and Rice continued to have a childhood that could well have served as the backdrop for a juvenile fiction series in a popular magazine of the time. It began with a hearty dose of the old-fashioned work ethic. When the family moved out to Vaughan Pike, Bolling Rice gave his three sons—Grantland was followed in quick succession by his brothers Bolling and John—a parcel of land on the property to farm. The idea—and presumably the lesson—that Bolling Rice had in mind for his sons was "to make the land yield a profit."

With the assistance of a hired hand named Horace the boys grew "tomatoes, potatoes, beans, asparagus, onions, beets, peas cabbage, lettuce and practically everything else that grows including all kissin' kin of the worm and grub family."

During the harvest, Rice worked a sixteen-hour day, beginning at 3:00 a.m., when he would drive a wagon loaded with produce into market, returning to his land by 5:30 or 6:00 and spend the day loading the wagon for another pre-dawn trip into Nashville. "Yes, I knew hard work, at twelve," he said. "After this type of training, no amount of toil seemed hard or long." Certainly no one who knew Grantland Rice the writer and had witnessed him dispatch a 3000–word magazine article or a week's worth of columns at one sitting ever doubted his extraordinary capacity for work.

But life was hardly drudgery. The first Christmas on Vaughan Pike, Rice found a football, a baseball and a bat under the tree. "Those three presents were the sounding instruments that directed my life," he wrote much later. "They were the Pied Piper in my march through life."

As Rice remembered it, the area around the Vaughan Pike house also "had trees to climb and room to roam," and the possibilities for fun were constrained only by the limits of a boy's imagination. In the summer, after the muddy spring flood waters had subsided, the Cumberland River—in those days before the river was dammed—ran clear and cool and shallow, and the bottom was a clean and firm gravel, perfect for swimming. And in the winter the northern Tennessee climate was just cold enough to freeze Shelby Pond for skating and drop an occasional layer of soft snow for sledding. Kites were big in Nashville in the 1890s. The warming breezes of March would send the boys scurrying to the neighborhood corner store to buy balls of cord and sheets of colored tissue paper, and the homemade kites were as much a contest of creativity as they were of engineering—the boys would fashion their kites from contrasting or complementary colors of tissue paper. By the time the March breezes began to subside and the boys' interests began to turn in other directions "nearly every tree along the sidewalk would have a kite with its tangled rag tail caught up in the upper reaches," remembered a childhood companion of Rice's. "However, the leaves soon came out and covered up the mess, and by fall, rain and wind had about disposed of the remnants of the kites."

A game that consumed many hours for young Nashvillians—year round—before the turn of the century was a long-jump, triple-jump, leap-frog combination called "Foot and a Half." There were assorted local rules by neighborhood to this game but essentially it began with one boy who was "it" or "down," who would bend over on the starting line—sometimes simply a line drawn in the dirt but more often a piece of tin laid on the ground. A leader would then leapfrog over the boy who was it and draw a line in the dirt at the point where his heels landed. Everybody else would then have to leapfrog over the boy who was down and clear the line of the leader. Anyone who couldn't make it past the leader's line became "it." Once this preliminary round was complete, the boy who was down moved out to the second line and bent over. Starting from the original line the others then made another round of jumps, the leader again marking his landing place to provide the target. The line—and the boy who was "down"—thus kept moving further away from the original tin-strip starting point. When the boy who was down had moved too far away to reasonably clear in a single leap, the leader would call for a "one and over," and then a "two and over," and so on, which meant that each boy could take one standing broad jump, and then leapfrog the down boy. An elaborate system of dares and challenges to the leader would keep the leader's position shifting and keep the game going long past suppertime into the Tennessee twilight.

Grantland and his brothers were avid players in all of it. Grantland's zest for play led to a daring and fearlessness that some of the boys marveled at and all of them respected. Collecting bird eggs (another consuming passion of Nashville boys in the 1880s and 1890s; the corner stores all sold twelve-by-twenty-four-inch wooden trays especially made for collecting and displaying bird eggs) one Saturday in the vicinity of what is now Shelby Park, Grantland climbed a tree and shinnied out onto a slender branch to inspect and rifle a promising-looking birds' nest. A cracking sound caused the other boys to turn suddenly and look, in time to see Grantland plunge to the ground, still clinging to the shorn branch. He suffered a broken arm. Edward Webb, one of the boys with him, remembered they had a difficult time getting him home.

It would not be his last broken bone. But neither this nor other mishaps instilled any sense of caution in the young Grantland, or kept him from putting himself in harm's way. He was particularly fond of football. A mere wisp of a boy—he stood nearly six feet tall by the age of thirteen, yet he weighed but a hundred pounds—he was particularly ill-suited for the game. But only in body, not in mind. For though it would exact its price, it was a game that he doggedly stuck with—he played not only on the vacant lots of Nashville with his neighborhood chums, but throughout high school and college as well—his tenacity and pluck allowing him to at least hold his own against the stouter lads.

Baseball was a different matter altogether. His gangly frame belied a quickness, a coordination, and a strength that put him among the most graceful practitioners of the game in the city of Nashville. And if you could excel at just one game in America in the 1890s, baseball would surely be it. Particularly in a city like Nashville, where the spring arrived early and the temperate autumn lingered.

The game that had been adapted and standardized from the English games of cricket and rounders and the various other crude facsimiles that were played in the early decades of the nineteenth century, was fully a half-century old as the 1890s began. It had lost most of its rough edges—nowhere did they pitch underhand anymore, or use posts driven into the ground for bases, or put a man out by hitting him with a thrown ball. Even in the most remote outposts and in the smallest towns, fathers no longer taught their sons to field their positions by keeping one foot on the base until the ball was hit, as was the common practice in the 1860s and 1870s.

Baseball in the 1890s did not mean professional baseball. It meant town fields in North Brookfield, Massachusetts; Gilmore, Ohio; Factoryville, Pennsylvania; Narrows, Georgia; Humboldt, Kansas; Nashville, Tennessee, and thousands of other cities and towns and villages across the land, where a dozen different adult teams would play a forty- or fifty-game schedule from May through September—three evenings a week, with doubleheaders on Decoration Day, Independence Day and Labor Day. Sometimes it would be town against town, and the covered wooden grandstand would be filled, and never mind what Cap Anson would be doing with the White Stockings or what John McGraw was up to with the Baltimore club, for we have our own heroes here to talk about: "Hasn't young Burke—the one who works at the feed store—been tearing the cover off the ball lately; and it's a good thing too because Shea hasn't hit a lick since he married the O'Sullivan girl; and how about Samuels, the new dentist? Got a heck of an arm out there in right, don't he?"

Mostly, baseball was vacant lots and young boys:

Just at this time every season, when the March sun warms the town;
When the little green leaves peep shyly from the stark, bare limbs of brown;
When the voice of the rooter rises in the roll of a rippling cheer,
The winds of another springtime blow back from another year.
The cry of the barefoot legions, the shouts of the tattered host
As twinkling feet raced madly in a dash for the telephone post,
To a wagon wheel "for second base," with never a touch of fatigue,
When I was one of "The Ragged Stars," the champs of the Alley League.

Nashville had its professional team—the Americans, born in 1885. They played on a field known as Athletic Park in a section of town known as Sulphur Springs Bottom. * As a boy, however, the baseball that concerned Grantland Rice was not the baseball played at Sulphur Springs Bottom, but the baseball that was played in the fields along Vaughan Pike and later, after the family moved to West Nashville when he was a teenager, in the vacant lots along Broad Street and West End Avenue.

Pick-up games and games of "Five Hundred" or "One-Eyed Cat" or "Hit the Bat" are as rare as vacant lots these days, but in baseball's first century they were ubiquitous. "One-Eyed Cat" ** was the game of choice

---

*Later, when he was a sportswriter on the Nashville *Tennessean,* Grantland Rice would begin calling the stadium Sulphur Dell, believing it had a more lyrical ring to it. The name stuck, and soon all of Nashville knew it as Sulphur Dell.

**The game is also referred to as "Scrub."

in Grantland Rice's Nashville when there were not enough boys for two sides. When the boys had gathered, somebody—often the boy who owned the ball or bat—would begin by shouting out "One-two-three. One!" The rest of the boys would count off: two, three, four, and so on. Number one was the batter, number two the catcher, number three the pitcher, number four the first baseman and so on around to right field. The hitter would hit until he made an out. Then the catcher would take over as the batter and everyone would move up one position.

Often the action would have to stop to repair the ball. The most popular ball was a model called the "Home Run." It was a good quality ball in its day, well-made with a stitched horsehide cover. Still, it would accept only limited punishment, and since there was seldom money enough for a new ball the boys always brought along some linen thread, a curved needle, and some beeswax for the necessary surgery to the precious base-ball. Grantland never had to be asked twice to partake of any of this. Even after he was a member of the Vanderbilt University varsity, he could be found at the vacant lot at the confluence of Broad Street and West End Avenue, lofting long fly balls from the base of the triangular lot, out to his brothers Bolling and John and the other younger neighborhood kids guarding the distant reaches of the lot along Addison Avenue.

During the Vaughan Pike days, Grantland split his school time between two military schools—the Tennessee Military Institute and the Nashville Military Institute. When the family moved to West Nashville, Rice was enrolled at the Wallace University School. As its name implies, the function of the Wallace University School was to prepare its students for a classical university education, accomplished by "the mental, moral, and physical education and training of youth." Headmaster Clarence B. Wallace, a spindly, bespectacled man in his mid-thirties, rooted the curriculum in Latin and Greek. So intense was his belief in the importance of the classical languages—particularly Latin—that he did not even bother to teach English grammar. His feeling was that the students would learn the fundamentals of English from their study of Latin. His system apparently worked. Generation upon generation of students swore that Wallace was the most gifted teacher they had ever encountered. Rice, when he was at the height of his career, wrote to Wallace and said: "If I have gained any measure success in writing, I owe it all to you." The puckish Wallace, ever the schoolmaster with his pupil, even when his pupil was one of the most famous men in America, responded: "Now right there's the best piece of writing you ever did."

The natural tendency of Wallace students and Nashvillians in general was to address the headmaster as Doctor Wallace, a practice that caused him to bristle, for he did not have a doctorate. But it was almost as if people didn't believe him, for the habit of calling him Doctor persisted throughout his life. Even Grantland Rice, writing of Wallace in his memoirs and again paying him homage, referred to "my old prep-school master, Dr. C. B. Wallace."

The university that the Wallace School prepared most of its graduates for was Vanderbilt, not yet twenty-five years old in the mid 1890s (it would celebrate its first quarter-century amidst much fanfare in the fall of 1900, Rice's senior year) but already the dominant cultural and educational presence in the city of Nashville.

Vanderbilt University takes its name from Commodore Cornelius Vanderbilt of New York City, the shipping and railroad magnate and the wealthiest man in America during his time; but it has its roots in the Methodist Church. The Methodist Episcopal Church, South had secured a charter from the State of Tennessee for a university in Nashville that they proposed to call Central University. But in the Reconstruction economy the church could find no money to build or endow their proposed "institution of learning of the highest order." Bishop Holland McTyeire of Nashville, whose wife was cousin to Vanderbilt's wife, approached the Commodore in 1873 and secured a $1 million endowment. Vanderbilt never traveled to Nashville, never saw the campus that bore his name. His only statement of purpose—the words that are today inscribed on the base of his statue at the entrance to the university—came in a letter to McTyeire in December of 1875: "If it shall, though its influence, contribute, even in the smallest degree, to strengthening the ties which should exist between all geographical sections of our common country, I shall feel that it has accomplished one of the objects that led me to take an interest in it."

By the middle of the 1890s, Vanderbilt could reasonably lay claim to being the finest university in the South; it boasted the highest entrance requirements in the South, "accessible only to able students who enjoyed an exceptional secondary education." Such was the education Grantland Rice had received at the Wallace University School. Vanderbilt in fact had entered into an alliance of sorts with Wallace and other local prep schools. The prep schools agreed to tailor their curriculums to include liberal and rigorous instruction in Latin and Greek, history and geography, English, and mathematics—the foundations of the Vanderbilt academic curriculum. Vanderbilt in turn would waive the university entrance examinations for graduates of these select prep schools. It provided a nice conduit of quality students for Vanderbilt, not only from Wallace and other Nashville schools, but eventually from schools throughout the Southeast.

Vanderbilt's link with Wallace was but one of the social and cultural phenomena tugging Rice towards the school, making his selection of a college almost inevitable, or pre-ordained. There was, first of all, its presence—not just its location, but its *presence*—in Nashville. The citizens of Nashville had taken a proprietary interest in the university—it was an unquestioned jewel in the city that prided itself on being "the Athens of the South." They reveled in its splendid campus. Fifteen hundred trees planted by Bishop McTyeire in 1874 had flourished and grown, taking their place amidst the handful of majestic oaks that dated back to Revolutionary days. They were joined by some 450 different shrubs and plants, the gift of a nursery owner in Germantown, Pennsylvania. The flora blended with the gentle, sprawling lawns, and all of it was crisscrossed by a maze of meandering walkways and drives. The campus became known as "the Vanderbilt Arboretum." It invited leisurely and lingering visits; and its location at the terminus of the West End Avenue trolley line made it a convenient and popular destination for a family's Sunday outing. The lure for any young Nashville man to be a part of this, to be a Vanderbilt student, was surely great. Great too must have been the urgings of the parents who took their sons on Sunday excursions to visit the campus.

There were other, more considered reasons for a young Nashville man in the 1890s to choose Vanderbilt as his college. To begin with, there were the school's roots in the church. The Rices were Presbyterian, not Methodist. But Vanderbilt, aside from the Biblical Department, had never forced Methodism on its students, and while Methodists predictably dominated the early student body, over the first two decades the campus had become decidedly more homogenous—at least within the limits of Southern white Protestantism. The appeal of Vanderbilt's underpinnings in the church thus rested not on dogma but on values. "Vanderbilt's guiding purposes remained those of a church-related liberal arts college—the forming of Christian character," said historian Paul Conkin in his exhaustive and eloquent history of the institution.

Next, there was the matter of location. Not only was Vanderbilt in and of Nashville, it was of the South. America was still very fervently a country of regions in the 1890s; nowhere was this more true than in the South. Thus, Vanderbilt was, according to Conkin, not only "morally and doctrinally safe, but also geographically convenient and culturally familiar." For Grantland Rice, the geography was particularly convenient. The family's new home on Demonbraun Street (a block or so away from what is today known as Music Row—the recording studios and museums of the country music industry) was just a fifteen-minute walk from campus.

Rice was just sixteen years old when he came to the Vanderbilt campus as a freshman in the fall of 1897. It was an oppressively, unseasonably hot September in Nashville that year. Temperatures were in the upper nineties when classes began on Wednesday the twenty-first. Rice's discomfort—along with that of his seven hundred fellow students—was, without question, exacerbated by the vested, wool, four-button suit he wore over a heavily starched, long-sleeve shirt with a high, starched-paper collar—the *de rigueur* garb at turn-of-the-century Vanderbilt.

He was six feet tall and weighed 130 pounds; so thin he barely cast a shadow on the campus walks and lawns. Still, he came to Vanderbilt determined to play college football; and if he had had any doubts, they would have been chased by the exhortations of *The Hustler,* the student newspaper, which, in the first issue of the year left no doubt that the freshmen would be shirking their duty as citizens of the university if they did not do their part in bringing Vanderbilt greater glory through football.

> Upon the new men we would urge the necessity of at once allying yourself to the University and making yourself recognized as a factor in college life. Many a man who would never be heard of if he kept aloof from athletics comes to wield a powerful influence when he has shown that he is unselfishly working for the glory of his college. . . .
>
> Of course you will receive many hard knocks and rough handling, but this will make a better man of you and start you into life with a sound and vigorous constitution; the rewards will more than repay you for your trouble.

Rice received more than his share of "hard knocks and rough handling." During his college football career he would sustain a broken arm, a broken collar bone, a broken shoulder blade, and have four ribs torn from his spinal column. Only during his junior year would he survive the season injury-free. Still, he never gave thought to abandoning the game, even when it began to become clear that his slender frame was fragile, and susceptible to painful bruises and breaks, however much grit and heart he may have possessed. "Because football calls for courage, both the highest physical and mental condition, because of its ruggedness, I suppose I like football more than any other sport," said Rice years later, no doubt in agreement with *The Hustler*'s prediction that the rewards more than repaid him for his trouble.

In addition to enjoying the game so, he also played it with some finesse and success—when he was a whole man. He failed to make the varsity his freshman year but was the starting left end on the freshman team, and earned praise from *The Hustler* for his play in the team's only game, a 10–6 win over his old school, Wallace.

The 1897 season was a special one for the Vanderbilt students. The varsity went undefeated and unscored upon, though they were held to a scoreless tie by the University of Virginia. But the students treasured the season not so much for its success as for its very existence. For the school's Board of Trust had come very close to banning football and all intercollegiate athletics at their annual meeting in June; only an impassioned plea from Chancellor James Kirkland prevented a motion to abolish sports from coming before the Board, where it would have surely met with overwhelming approval and spelled doom for the intercollegiate teams. Support on the church-dominated board for sports had never been high; but in the spring of '97 it was at its lowest ebb ever in the wake of and ugly and embarrassing near-riot at the game between Vanderbilt and the University of Nashville in the fall of 1896.

The University of Nashville had become Vanderbilt's most bitter rival—their games drew crowds of up to five thousand people. It was an exhilarating time for the students, who wore buttons and carried flags of Vanderbilt old gold and black, and brought along megaphones to assist in the elaborate set of "yells" that were as well orchestrated as the football plays themselves. "Let the yell go up," urged *The Hustler*. "How are we to carry the day without great yelling?" The position of "yell leader" was much honored, usually bestowed upon the captain of another of Vanderbilt's intercollegiate teams. But there was a darker side to the crowds and the game in the 1890s. Vanderbilt's few women students seldom went to the games, because beyond the innocently boisterous student section there was rampant drunkenness and vulgarity. Fighting in the stands was common. And the behavior on the field was not much better. Cheap shots and spontaneous fights abounded; eligibility rules were lax; officiating was poor; in short, the game was a mess. In 1894, the Board of Trust informed the faculty that if they could not rein in the game, the Board would take action. Wanting to save the game for the students' sake, and prodded partly by the revenue that football was producing for the university, the faculty acted. They tightened eligibility requirements—only matriculated students could play, and nobody would be allowed to matriculate merely for the purpose of representing Vanderbilt in intercollegiate athletic competition. A committee would set the days and hours of all games so as to minimize the conflict with class work. Travel would be restricted so that no athlete was away from the university for more than four days a term; there would be no travel on Sunday. This tighter faculty control appeased the Board—until all hell broke loose at the Vanderbilt-Nashville game in '96.

It began the way all such incidents do. The players and the fans were all a little extra heated for this one—the rivalry was such an important one for both sides. The players early-on crossed the sometimes fine line between emotional aggressiveness and dirty play. They were prodded by fans on both sides of the field who were partisan to the point of hostility. With this atmosphere, it was not at all surprising that a Vanderbilt player, frustrated by what he perceived as dirty play, cold-cocked a Nashville player, precipitating an on-field fracas between the players. A very tenuous order was restored, and play resumed, whereupon a Nashville player promptly cold-cocked a Vanderbilt player in retaliation. This time the first punches brought not only both benches, but both bleachers as well. The fans wielded knives and canes and a few even brandished pistols, though no shots were fired. Several people were injured; and police and university officials from both schools, helpless before the mob, could only be thankful that nobody had been killed.

What infuriated the members of the Board of Trust, already of a mind that the "game detracted from the purposes of a true university," was the behavior of the student body. That the men of Vanderbilt, decent men of high moral character, possessed of deep-rooted Christian values, could be whipped into such a savage frenzy by the mere witnessing of a football contest, seemed to bear out the belief that football was inherently evil, that it had the capacity to "degrade . . . and injure morally to a noteworthy degree," as a Northern Methodist minister would put it that same year.

The incident attracted national attention, and as their annual spring meeting approached, there was little sentiment among members of the Board of Trust for saving intercollegiate competition. Had this been an isolated incident—unique to Vanderbilt—the mood might have been more sympathetic. But the Vanderbilt drama was unfolding against the backdrop of a sustained nationwide movement to clean-up or abolish the game of football. Ten to twenty men a year were being killed playing college football, and trustees and administrators all over the country were looking anew at the propriety of football on the campus. Columbia, Northwestern, Stanford and the University of California would all briefly drop football in the next decade; and in 1905, President Theodore Roosevelt, appalled by the naked brutality of the game, told college officials that unless they could clean up and civilize their game, he would outlaw it by presidential edict.

When the Vanderbilt Board of Trust met on June 14, 1897, Chancellor James Kirkland persuaded them not to consider the motion to abolish intercollegiate athletics, but to accept as a compromise the appointment of a five-man committee to study the situation, make recommendations, and report back to the Board at the 1898 meeting; in the meantime, football, and the other intercollegiate sports, would continue. The language of the committee's report was harsh:

> We believe that serious evils have resulted from such contests—interruption of study, substitution of interest in athletics for interest in scholarship and intellectual culture, excitement of bad passions, feuds, fights, bodily injuries, brutality, betting, desecration of the Lord's day. If these were necessary or invariable effects, there would be no room for discussion.

But the committee did find some redeeming social value in sports:

It is claimed in behalf of these contests that they are very popular among the students, are sustained not only by an extensive public sentiment, but also by the deliberate approval of many educators; help to develop physical health, strength, agility, courage, and endurance; and, by furnishing exercise for the exuberant animal spirits and activity of young life in an open way and under rule, prevent secret disorders and violent outbreaks to which institutions of learning have ever been liable.

The committee also found the standards adopted by the faculty committee three years before, as well as the standards of the Southern Inter-Collegiate Athletic Association—of which Vanderbilt was not only a member but a moving force—were sufficient to safeguard the integrity of Vanderbilt University, as long as they were adhered to. So, the committee's recommendation was that the Board adopt a very mild resolution, putting the onus for preventing "the evils incident to inter-collegiate athletic contests" on the chancellor and faculty. The Board approved the resolution at its 1898 meeting and the threat to the Vanderbilt athletic teams was ended.

Perhaps the bottom line on all of this for Grantland Rice is the fact that he came to a campus that was acutely—if recently—aware that participation was a privilege—a privilege to be cherished; and the best, indeed, the only way to ensure that this privilege was protected was for the Vanderbilt athletes and students to see to it that the integrity of their athletic program was protected, by adhering to the principles of honesty, fair play, and sportsmanship. The disappointment in losing is one thing; the disappointment in being shut out of the contest quite another.

It is folly to suggest that this single experience shaped and inspired Rice's passionate and unswerving espousal of these values in his forthcoming career. But it is equally foolish to dismiss its effect entirely, to believe that the possibility that athletics would be denied to him—which to Rice would have completely eviscerated the college experience—would have had no enduring effect.

Rice participated with a vengeance. In addition to playing football, he pledged the Phi Delta Theta fraternity, was elected secretary-treasurer of the Wallace School club, and served as captain of the class basketball team in the winter intramural competition. He did well in his class work. And he waited for February and the start of baseball season.

When practice began he impressed more with his hustle than with his bat or his skill afield; notices in *The Hustler* referred to him as an "active player" who "by very hard work . . . may make the team." He did, but he was playing behind the team captain, Dan Merritt at shortstop, and thus saw very little action. But he did get to travel with the varsity, making an early April trip to Philadelphia for a pair of games with Penn. (The Quakers won both games, 10–0 and 7–0.) And as he bided his time on the bench, he consoled himself with the fact that Dan Merritt was a senior; next spring the shortstop position would be his.

But first he had the perils of football and basketball to negotiate. He collected the first of his injuries at the start of the football season, missing out on any chance he might have had of making the varsity, though he did recover in time to play for the sophomore class team in their 10–0 loss to the freshmen in early December.

Basketball was a risky game on two fronts. First, there were the hazards of the game itself. The game that Dr. James Naismith had invented at Springfield College at the start of the decade was not played with a great deal of finesse in its early years. "The game was really a rough one," remembered a man who played in some of the free-for-alls that passed for basketball in Nashville in the 1890s. "Personal fouls were hardly known. Possession of the ball went to the man who was the stronger. Sometimes a player would be swung around by a man with a better purchase on the ball and be thrown sprawling across the floor."

But Rice was not a timid soul, and so perhaps the greater hazard he ran in electing to play basketball was the risk of incurring the wrath of the coaches and officials of the Vanderbilt Athletic Association. Basketball was not yet an intercollegiate sport at Vanderbilt (it would be in 1900); couple that with the fact that it was hazardous to the health of their potential football, baseball and track men, and the game was made anathema to the powers that be in Vanderbilt athletics. As the class intramural tournament approached in 1899, William Dudley, the president of the Vanderbilt Athletic Association (and the man for whom the school's current football stadium is named) appealed to the varsity athletes to eschew the game.

"The Executive Committee of the Athletic Association desire to express the hope that all students who expect to be candidates for the track, baseball and football teams of the University, in fact, all who have athletic ability, will avoid the game of basketball as played in our gymnasium," wrote Dudley in *The Hustler*. The gymnasium was not suited to the game, he argued, "and the practice of playing [basketball] in their gymnasium suits is considered unsafe by all experts"—a peculiar statement, perhaps, in light of where, and in what clothing, the game is played today, but not at all peculiar when considered in the light of the earlier testimony as to the game's gladiatorial atmosphere in those days. "Our athletes have had bitter experience with this game in the past," said Dudley; and he closed his plea with words that suggested that any varsity athlete who ignored his advice was not only a fool, but a callous citizen of his university as well:

> This game lost us the track championship of the Southern Intercollegiate Association last year, and it is hoped that this year we may have winning teams in the field. To bring this result about we must have the

hearty cooperation of every student, and no chances should be taken in a game like basketball in which there is nothing at stake.

But there was something at stake—class pride, a not-at-all inconsequential matter in campus life. The intramural tournament was immensely popular with the students; more than two hundred students paid ten cents each to crowd into the gymnasium to witness the games and a halftime gymnastics exhibition. At six feet, Rice was one of the tallest men in the sophomore class, his natural coordination and enthusiasm made him an obvious choice for the team—not to mention the fact that he had been captain of the "class five" the year before. His loyalties no doubt torn, Rice played—and played well, according to *The Hustler,* though he did not score in the sophomores' 8–1 win over the freshmen. (This was not a wide-open game they played.) Judging by the rosters printed in *The Hustler,* most of the other varsity athletes joined Rice in ignoring Dudley's suggestion that they avoid the intramural basketball tournament.

For Rice and his baseball teammates, the basketball tournament was also something to keep their minds off the miserable Nashville winter that delayed the start of baseball practice until the last days of February. Once practice did begin, it was plagued by a spate of rainy days and raw, windy weather. Still, on the few days they were able to get outside and practice, Rice showed well, exhibiting good range and a strong arm at shortstop; and by the time the season opened against the University of Nashville on Saturday, April 1st, the job was his.

Rice's play in his debut as a varsity starter—a 17-to-5 Vanderbilt win—might most charitably be described as uneven. He went hitless in five at-bats, though he reached base three times on a walk and two errors and drove in a Vanderbilt run with an infield out in the fifth inning. In the field he made three errors, but he also made what *The Hustler* called "the sensational play of the game" in the sixth when he made a diving catch of a pop fly behind third base after the ball had bounced out of the third baseman's hands.

The first game proved a harbinger of the season to come for the sophomore shortstop. He batted just .162, with six hits in thirty-seven at-bats, and he led the team in errors with twelve. But generally his defensive play impressed. He began or turned the pivot on a number of pretty double plays; he handled more chances than anyone else on the team; he led the team in assists, and his .771 fielding percentage was the second highest among the team's regular infielders. And it must be remembered that on the pebble-strewn fields of the 1890s, playing with the flat, padded-pancake gloves, there was no such thing as a routine ground ball; they were all an adventure. Rice played every inning of Vanderbilt's nine games, as the team went five and four, beating Cumberland once and Tennessee and Nashville twice each, and losing once to Cornell and three times to the University of Georgia. The pitching staff lacked depth—three of the four losses came by the scores of 12–11, 7–6, and 10–9.

His status as a varsity baseball starter did not diminish his determination to play football, and when school convened in the fall he was back on Dudley Field in his leather helmet and padded jersey—at his top college weight of 134 pounds—spending most of his time with the scrub team, punishing himself in a thankless daily scrimmage with the varsity. This year, for the first and only time in his college career, he survived the football season injury-free and saw enough game action to earn his varsity letter; he cherished it every bit as much as he did his four varsity-baseball Vs.

The football season ended with another ugly incident against the University of Nashville on Thanksgiving Day. With five minutes remaining in a closely fought, scoreless game, Vanderbilt center Felix Massey picked up a Nashville fumble on the Vanderbilt twelve-yard line, "and dashed down the field with a solid mass of gold and black interference behind him and sped on to a touchdown after the longest run ever made in a Varsity game on Dudley Field.

"In an instant," continued *The Hustler* breathlessly, "what might have been defeat was turned into certain victory, and the crowd went madly wild. It was a great run and will live in Vanderbilt football history."

It has indeed lived in Vanderbilt football history, though less for its own heroic merits than for the commotion it precipitated. Nashville argued that Massey had been down before he crossed the goal; when the referee ruled it a legal touchdown, fans from both schools poured onto the field to contribute their opinions, and before order was restored the game had to be called on account of darkness. The referee then ruled that though it had been a fair and legal touchdown, because the game was never concluded the game should be entered into the record books as a scoreless tie. Vanderbilt has never accepted the referee's post-game ruling. The game remains in the Vanderbilt football records as a 5–0 win, the crowning touch on a successful seven-and-two season. Though the students were chastised by both the administration and the editor of *The Hustler* ("Thursday this unsportsmanlike breach on the part of the students was a costly piece of thoughtlessness"), the incident in no way rivaled the near-riot of 1896, and this time there was no talk by the Board of Trust of abolishing intercollegiate athletics.

The first weeks of the twentieth century again brought unpleasant and atypical winter weather to Nashville, and Vanderbilt's preseason practice was again disrupted. It was not a good year for such troubles, for the first games of the '00 season were against the powerful University of Chicago team of Amos Alonzo Stagg.

Stagg's renown came from football; while a Yale undergraduate nearly twenty years earlier he had been named to Walter Camp's first All-America team. But since being named Chicago's first coach in 1892, his All America status as a player had become a mere footnote to his football résumé. Most of what was deemed dynamic and innovative in turn-or-the-century football had come from the fertile imagination of Amos Alonzo Stagg—sleight-of-hand such as the reverse, the man-in-motion, and backfield shifts; offensive weapons that opened up the game such as the lateral and the cross-body block. In eight short years he had built the Chicago program to the point where it shared billing with Michigan as the class of the West, and was more and more frequently being mentioned in the same breath as Yale, Princeton, Harvard and Penn as the class of the nation.

The football accomplishments tended to obscure Stagg's baseball credentials. During his Yale days he was considered one of the finest pitchers in America; he repeatedly spurned offers to play professional baseball in order to remain in school. As a coach, his Chicago nine enjoyed a success and a stature that was the equal of his Chicago eleven.

Stagg's visit to Vanderbilt during the last week in March would mark Grantland Rice's first up-close-and-personal brush with a major national sports figure. In years to come Rice and Stagg would come to know one another well, and they would share the dais as equals at hundreds of football dinners around the country. During this week in March of 1900, however, it is very likely that Rice was as star-struck as the rest of the campus by the presence of Mr. Stagg. There is nothing on record of their first meeting—or indeed, any certain evidence that they did meet that week. But it is difficult to believe that Rice would not have availed himself of the opportunity to at least walk up and shake Stagg's hand. For, as *The Hustler* noted, he had made a great impression:

> Of the conduct of the Chicago team, every man who witnessed the games has much to say, and the universal sentiment on that subject is that we have never had here a team composed of cleaner gentlemen.
>
> Mr. Stagg and the members of the team are to be congratulated not only for the fact that the team plays good ball but because they have a high appreciation of what sportsmanship means, and because their conduct both on and off the field was such as to win the admiration of every man who came in contact with them. . . . It is much to be regretted that we were unable to show them more attention while they were here, and if at any future time they should decide to visit us again, we will endeavor to show them the full meaning of "Southern Hospitality."

Vanderbilt had to be satisfied with the experience of spending time in Stagg's company, because his baseball team whacked them around pretty good in the three-game series—winning 18–3 in a rainstorm on March

30th; taking the measure of the Vanderbilts 11–5 in a snow squall on Saturday the 31st; and finally proving their adaptability to all weather by winning 22–7 under sunny skies on April 1st. Rice went one-for-ten over the three games, but again drew notices from *The Hustler* for his fielding.

Vanderbilt would have a disappointing season, finishing at seven and nine. Personally, Rice developed—albeit very slowly—into a more consistent hitter than he had been as a sophomore; and he emerged as a defensive shortstop of the first order. After a disastrous first six games, where he scratched out just two hits in twenty-four at-bats—an .083 pace—he went twelve for forty-two—a solid .285—over the rest of the season to finish with an average of .212. He led the team with three home runs.

And while he held his own at the plate, he turned heads in the field. After he handled eleven chances without an error in a game against Tennessee, the Knoxville *Journal and Tribune* took note of it, calling him "the star fielder of the team." Week after week *The Hustler* sang his praises: "Rice again played a brilliant game at short. Rice is playing a very fast game now and is taking in everything that comes his way." "Rice played his most brilliant game of the year at short. . . . His stop of Fisher's hot grounder near second was phenomenal." "He is generally regarded as the fastest and the best all-around shortstop Vanderbilt has ever had."

There were, however, moments to keep Rice humble. Against the University of Texas, he committed two errors in one inning, allowing two runs to score in a game Vanderbilt lost, 3–2. All told, while it was not the sort of season that sets a young man to dreaming of glory with the Giants or the White Stockings, it must certainly have left him eagerly awaiting his senior year. And it provided pleasant memories, even if they blurred a bit as the years passed. In writing his memoirs fifty years later, Rice remembered the best game he ever played coming against Tennessee in Knoxville. "I had fifteen assists, no errors, plus a home run and double. Vanderbilt won 4–3." He was no doubt remembering the three-game weekend series with Tennessee—the only time Vanderbilt played Tennessee in Rice's career—where he had those eleven assists in the first game, and a home run in the third game two days later—neither game a 4–3 Vanderbilt victory.

When fall arrived, Rice dutifully reported for football, his weight still hovering just above the 130–pound mark; and the game finally exacted a serious toll. Before the season started, Rice broke his right shoulder blade while playing for the scrubs in a scrimmage. His body was less resilient this time. Where he'd been able to bounce back from the broken arm, broken collar bone and the ribs he tore from his spinal column, this time, the approach of baseball practice found his shoulder still bothering him— he couldn't raise his arm to throw; he had to flip the ball underhanded.

This didn't prevent his teammates from voting him captain of the team at a meeting in January, a popular choice with the student newspaper. "His brilliant work has excited comment wherever the team has gone," said *The Hustler.* "He knows the game thoroughly, is very popular, and has all the qualities of a successful captain. . . . The election of Rice as captain is an honor well bestowed and much deserved."

Among the first matters the new Captain Rice needed to concern himself with was the very integrity and reputation of Vanderbilt University athletics, which had been besmirched in the February issue of *Outing,* a popular sporting news magazine of the day—and a magazine for which Rice would one day write. In an article alleging professionalism on the previous fall's football team, the magazine charged that two players "came out of the North at the eleventh hour" to play for Vanderbilt. The article further criticized the Southern Intercollegiate Athletic Conference for being lax in policing abuses in its member schools. "There is a crop of promises annually, but very little fulfillment," it said. The charges stung deep, and stung immediately at Vanderbilt, so sensitive to the rigid rules the faculty and chancellor had put into place following the football fiasco with the University of Nashville five years before.

Two well-attended mass meetings were held on campus the week the magazine hit the newsstand, and a series of resolutions were passed, denouncing the charges as "ungrounded and uncalled for" and "grossly libelous and false, made without a shadow of justice or any effort to ascertain their truth." The degree to which Rice may have been involved in the speechmaking at the meetings, or the drafting of the resolutions is unclear, but the catalysts for both were the key figures in the Vanderbilt Athletic Association—both student and administrative; and "Captain" Rice was now clearly one of those key figures.

The resolutions were sent to Casper Whitney, the editor of *Outing*; the Nashville newspapers got the carbons. The *Nashville American* came out editorially in support of Vanderbilt. "*Outing*'s random charges [of] professionalism at Vanderbilt, which has ever so strenuously stood for and maintained a high athletic standard, are so baseless as to be stupid," said the *American.* The Vanderbilt reaction also brought a letter from Casper Whitney who stopped short of apologizing, or retracting his remarks, but did say he meant them to be no reflection on Dr. Dudley—who was president of the S.I.A.A. as well as Vanderbilt athletic director—or on Vanderbilt. Only later, in a subsequent edition, did he publicly exonerate the school and its athletic program: "Vanderbilt's commendable career in Southern athletics is as familiar to me perhaps, as [to] the most enthusiastic of its supporters. I know it well as the Southern pioneer in sport for sport's sake—for long almost the only Southern college fighting semi-professionalism. There is no question of its having been, and of its being the leading disciple of amateur sport in the South."

The Vanderbilt athletes didn't let it rest at that. Feeling a need to make an additional statement they drafted and signed an official pledge, promising to subject themselves to the authority of the captain and coach of the team and "to abstain from all things of such nature as to injure our athletic condition, in particular to abstain from improper eating, late hours, smoking, drinking, immorality and any other form of dissipation." Rice was one of the authors of the pledge, and he never broke training in his four years at Vanderbilt—never once smoked or took a drink. Nevertheless he interceded on behalf of some of his friends on the baseball team and had a rider attached to the oath that allowed them "to except a limited amount of pipe and cigar smoking from their pledge."

The honor of the captaincy at the turn of the century carried with it a measure of responsibility and obligation that extended beyond the carrying of the lineup to home plate at the start of the game. With Vanderbilt coach William Guild—late of the Princeton varsity—part time, unpaid, and not scheduled to arrive until the middle of March, it fell to Rice to organize and supervise the first practices. He began working the pitchers and catchers in the dining hall of Westside Row, the campus dormitories, in early February; brought the whole team outside during the teasingly warm days at the end of the month; and no doubt acquainted them with the most dramatic of the new rules in effect for the 1901 season: from now on a foul ball would be counted as a strike, unless the batter already had two strikes on him.

None of this was making Rice's shoulder feel any better. He wasn't having any trouble hitting—his stroke at the plate was as crisp and as clean as it had ever been, and in batting practice he was rattling balls all over Dudley Field. But he still couldn't throw; in fact, so weak was his arm and so feeble his underhanded throws that after discussion with Coach Guild, it was agreed that he would move to second base, with its much easier throw to first.

Bedecked in their new black-and-gold-trimmed flannel "outfits," the Vanderbilt varsity opened the season with a three-game swing through Georgia during the first week of April. With Rice at second base they lost to Mercer, and split a pair of games with Georgia Tech. Rice acquitted himself well at second base, but must have looked a little out of place, for *The Hustler* reported that "he is a little weak on receiving thrown balls, but this is due to the fact that he is not accustomed to playing the position." But that discomfort apparently didn't last long. After a 25–0 win over Cumberland on April 8th, *The Hustler* reported that "Captain Rice makes second stronger than it has been in some years."

Vanderbilt now took a break from their collegiate schedule and played a six-game exhibition series against the Nashville club of the Southern League. The professionals thoroughly outclassed the college men, winning all six games; but the series was a tour de force for the Vanderbilt captain. In the first game of the series, a 9–5 Nashville win, Rice went four-for-five, including a bases-loaded double, and he fielded his position with aplomb, turning the pivot on two double plays, giving evidence that his weak arm was starting to come around. For the balance of the series, he continued to hit consistently and field flawlessly, commanding the attention of club management, and beginning to believe himself that perhaps he had a chance at a career in professional baseball when his Vanderbilt career ended in a few weeks.

His arm now stronger, Rice also returned to his old position at shortstop mid-way through the Nashville series. He was needed there—while he was coming around at second base, the four different men who trying their hand at shortstop were making errors that were costing Vanderbilt runs.

On May 2nd, an undefeated University of Alabama team came to Vanderbilt for the start of a three-game series. Vanderbilt kicked away a chance to win the opener, blowing a five-run lead and losing 11–10 in a game where the greatest excitement took place in the stands. Two Vanderbilt graduate students, alumni of the University of Alabama, came to the game and cheered loudly for Alabama, much to the consternation of the Vanderbilt students, who were so appalled by this seeming disloyalty that they angrily confronted the pair. Only an intercession on the part of the Vanderbilt players prevented what would have been a decidedly one-sided fistfight.

In the second game the next day, Rice hit a towering two-run home run in the top of the ninth (the home team did not yet always take the last at-bat) that extended a slender Vanderbilt lead. "When Captain Rice came to the bat the rooters were calling for a home run, for the Varsity's lead was uncomfortably small," reported *The Hustler* in a delightfully typical example of the era's sportswriting. "Then Captain Rice rose to the occasion and knocked the ball a mile. It was the longest hit of the season and the Captain was back at the bench a minute or so before the ball was run down and thrown back."

The 11–7 win evened the Vanderbilt season record (excluding the six games against the Nashville professionals) at three and three. But more important, it was the start of a season-ending, seven-game winning streak that allowed Vanderbilt to lay claim to being "Champion of the South." With "Captain" Rice hitting .304 out of the number two spot in the batting order and anchoring the defense with his steady, consistent, at times even brilliant play at shortstop; and with Vanderbilt finally getting the pitching they were missing during the last two years, they beat Alabama 10–3 in the third game of their series; then swept Central College 5–2, and 10–0; and finally concluded the season with three consecutive wins over Sewanee. With each win the crowds at Dudley Field swelled, spilling out of the wooden bleachers and lining the foul lines two and three deep.

Vanderbilt belonged to no formal league—the Southern Intercollegiate Athletic Association was an umbrella organization that set standards and guidelines and shared information. But the member schools played no common schedule; there were no post-season tournaments, and thus it was left to the successful schools to lay claim to the mythical "Southern Championship" the same way settlers in the West laid claim to their land. You said it was yours, and if nobody came along to dispute the claim, so it was. And so it was that nine-and-three Vanderbilt hailed itself "Champion of the South," in a bold headline on the front page of the May 30, 1901, *Hustler.*

"Captain" Rice, flushed with the success of a seven-game winning streak, a championship, and a personal season that justified the faith his teammates had shown in naming him captain, made plans to barnstorm throughout Tennessee, Arkansas, Mississippi and Alabama on a semi-pro team with some of the other Vanderbilt players. That is, if the Nashville Southern League club didn't call and give him a chance to play pro ball.

But first there was commencement, a time to reap honors earned over four years. His grades earned him a Phi Beta Kappa key, though he would not formally receive the key for another twenty years; Vanderbilt's chapter of the honor fraternity was just being formed. He prepared and read the class history during Class Day ceremonies, his "breezy" delivery being interrupted countless times by "peals of laughter and applause." Rice was honored by the student newspaper as one of thirteen members of the class of '01 "who have won honors in every department of college life and . . . have figured conspicuously in Vanderbilt affairs." He was cited for his baseball prowess and also as "one of the most popular men in his class as well as the university." On Wednesday morning, June 19, Henry Grant-land Rice—as the name appeared on his diploma—and his classmates—the first class in Vanderbilt history to graduate wearing cap and gown—marched from Wesley Hall, across Alumni Lawn to the chapel in University Hall to receive their degrees. Shortly after twelve noon, Grantland Rice, age twenty, stepped out of the darkness of the university chapel into the brilliance of the June sunshine, ready to begin his life away from the womb of family and school.

His immediate plans involved baseball. Throughout his career as a journalist, Rice hinted that he had had the chance to play professional baseball after college, but, as was his modest way, dismissed it with a smile and a wave, leaving the impression that it was an offer neither seriously tendered nor considered. The rumored offer, according to old-time Nashvillians and ex-teammates who shared their memories of Rice with other sports-writers down through the years, was apparently firm and came from the Nashville team in the Southern League. It makes sense. Rice had played well against the team in the exhibition series that spring. In addition, as the Vanderbilt captain, he was something of a local hero. It would have entailed very little risk on Nashville's part in signing him. He wouldn't have cost much; and he would have likely brought a few additional fans out to the park. Thus, even if didn't pan out, and the club was never able to sell his contract to a major league team, they would have at the very least broken even on their modest investment.

Bolling Rice and Henry Grantland said no. It was one thing to play baseball for Vanderbilt; it was quite another for their son and grandson to expect to make a living as a professional baseball player—a "ballist"—spending his life in the company of unlettered, uncouth, often unwashed scoundrels, living in seamy boarding houses and worse hotels, inviting consumption or worse. It was a squandering of one's youth, one's only opportunity to build a career, to make a life and family. It was something the sons of decent families—or graduates of schools like Vanderbilt—simply did not do in the summer of 1901. Rice acceded to his father's wishes—to go against them would have been completely out of character. But perhaps, too, Rice, and Rice alone knew the limits of his glass arm, suspected that a fling with pro ball would only be postponing the inevitable, and thus didn't press the matter with his father.

At any rate, his father apparently had no objection to Rice's cavorting with his college chums on the semi-pro circuit for a few weeks and shortly after graduation Rice left for West Tennessee and points south. He got as far as Memphis, about four weeks after he left, when his father wired him that he had found a job for him in a dry goods store; it was time to "come home to Nashville."

With a profound sense of melancholy but a clear sense of filial duty, Rice packed his bags, returned home, and reported for work the next Monday at J. S. Reeves & Co., on the Public Square in Nashville. He sensed immediately that this was not what he wanted; no matter how his father felt. There was a new newspaper, the *Nashville Daily News,* just getting started at the time, and Rice sounded out his father on how he'd feel about his son trying a career in journalism.

The most striking aspect of Rice's Vanderbilt career is the absence of any involvement with the campus publications. Neither the listings on the mastheads of *The Hustler,* the student newspaper, and *The Observer,* the student literary magazine, nor Rice's own listing of his campus activities in the yearbook and elsewhere, show any connection with journalism. Nor was he even the class poet. He was among the chronicled, not the chroniclers. His only brush with undergraduate journalism came when he served as one of five associate editors on *The Commencement Courier,* a four-page tabloid published daily during graduation week. A month after graduation, however, his father thought exploring a career in journalism was a fine idea. "My dad figured that inasmuch as I hadn't gone in for engineering, law or medicine at college—but had done creditably well in the arts—I might try my hand at journalism," said Rice. So with his father's blessing, Rice went down to the *Nashville News* and applied for the job of sports editor.

"There's no such job," Buford Goodwin, the editor, told him. "If you want to write sports and cover the State Capitol and the Courthouse and the Custom House along with it, we'll hire you."

Rice quickly agreed. The added jobs seemed a small enough price to pay for the chance to write sports. The salary was five dollars a week, and though he could not know it, his career path was set.

4

Apprentice Sporting Writer

*T*he *Pilgrims and Puritans* and a dozen or so generations of their descendants were not a terribly playful lot of people. On the nascent endeavor of American sports journalism, this had two effects—ramifications at once contrary and complementary. There were precious few athletic contests to write about, so "sportswriting" as we recognize it today did not take on a recognizable form until after the Civil War. Those few early athletic contests there were, however, easily fell into the category of "news" and thus warranted space in the sparse early American newspapers. The newspapers of Revolutionary days, for example, would write of the frequent horse races that would accompany the social gatherings of a community's leading citizens.

After the Revolution, American sport evolved on two widely separated parallel planes. On the higher plane, the upper classes indulged in horse racing, sailing and rowing races. (The jockeys and oarsmen for these competitions were frequently the servants and slaves of the owners.) Among the lower classes, boxing, dog- and cock-fighting provided diversion from a dreary life. Only the pastimes of the upper crust commanded any attention in the newspapers, and with the advent of the popular press in the 1830s they became a regular source of news. Even during the Civil War there would be the regular snippet of sports news, there amidst the war news, ads for patent medicines, and classified ads offering rewards for the return of stolen horses—the results of an Independence Day regatta in Boston, New York or Philadelphia, perhaps, with prose that in no way presaged the color and energy that sportswriting would one day come to enjoy: "The regatta of the Palisade boat club was held yesterday morning. The races were one mile down the river to Ludlow's Pier."

Sport burgeoned after the Civil War, fueled by the immense popularity of base ball—it was two words until the 1880s. The game provided recreation in the bivouacs and prison camps of both the Union and Confederate armies. It cut across class lines, and after the war it cut across geographic lines. Town would play town, city would play distant city, and the fiercely loyal community citizens would look to their newspapers to keep them abreast of the fortunes of their teams. For the first time, publishers came to see sports news as a source of circulation and by the 1870s, sports news—particularly base ball news—was a daily feature of most major metropolitan newspapers, just like the market report and the police blotter. But it was not yet a big part of the paper. In the *Boston Globe* of the 1870s, when the local team—anchored by the legendary Albert Spalding—played, the story would generally warrant two or three column inches. When the Bostons were idle, the out-of-town games would get just a line score and maybe a sentence or two. While the sports news could sometimes fill a column or more, it seldom ran more than five or six column inches in the eight- to twelve-page paper.

But the coverage quickly expanded, as did the range of sports that commanded attention. By the 1880s football, boxing, croquet, yacht racing homing-pigeon racing, bicycle racing and lacrosse had joined baseball and horse racing as sporting news staples. Instead of being scattered throughout the paper as they had been, these stories were put together on a single page—sometimes even the back page of the paper—generally under the headline "Sporting News." And by the mid-1880s it was not uncommon for the accounts of important baseball games to run a column or more. The style was the languid, leisurely, chronological narrative typical of much nineteenth century journalism, with salient information—in this case the final score—often lost in the morass of secondary detail, as in this 1884 story from the *New York Tribune:*

The New Yorks Again Victorious

About 1500 spectators witnessed the game of baseball at the Polo Grounds yesterday afternoon between the New York and Philadelphia league clubs. As on the previous day, the home club outplayed their opponents at every point and won easily. The local club had its reserve battery in and several of the best players, owing to sickness, did not play. Up to the ninth inning the Philadelphia nine had not scored a run, but then hits by Purcell, Farrar and McClellan and errors by Richardson and Dorgan allowed the visiting players to score three unearned runs. With the exception of the last inning the New-York nine played in fine form. Loughran is still poor in throwing to bases, Ward and Hankinson did the best batting and base running while Mulvey excelled in fielding. The score was:

The reader then had to wade through a box score that included runs, hits, put-outs, assists and errors to find our that New York had prevailed by a score of 9–3.

Occasionally, however, there were splashes of elegance and fun in the sporting news. Witness this passage from the *Brooklyn Eagle* from the same year, 1884, a peculiar and pleasing mix of Victorian formality and locker-room jargon:

> The Brooklyn team signalized their return home from the West on Monday by allowing their New York rivals, the Metropolitans, to "Chicago" them by the score of 4 to 0, and that not by superior pitching or batting, but through costly errors in two innings out of the nine and by inferior baserunning. The way it was done was as follows: The Metropolitans went to the bat first and after Nelson's retirement at first base Brady hit to Walker, who partially stopped the ball, Cassidy running in to field it. He failed to handle it in turn, however, and a life was given. Then Ester made a base hit and sent Brady to second. Roseman was then well fielded out by Geer, but on the latter's muff of Orr's ball a life was given, and as Brady had reached third on a passed ball, he ran home on the error. A wild throw then let Ester in, after which Geer gave Troy a life, and another passed ball sent him around, but before another run was scored Cassidy threw Relp out from right field. This inning's play, with its two runs given on two single hits, virtually gave the game to the Mets.

From the first, sportswriting was something of a stepchild of the city room. In a pattern that would reach its zenith during Grantland Rice's career, writers of talent and conscience would soon abandon the arena for other venues of art and life. There were exceptions, of course, but a goodly number of those who remained were not the sort of people likely to be welcome through the front door of polite society. Famed *New York Herald Tribune* editor Stanley Walker, in a probing and witty essay on sportswriting in his 1934 book *City Editor,* called them "unlettered chroniclers (muggs) writing for an audience of their own kind." According to historian Richard D. Mandell, the "overwhelming mass of sports journalism [was] slovenly, childish, and venal. . . . Typically sports writers were poorly paid and were often tools of the promoters. They got passes to events and the company of sweaty masculine men. A sports businessman of the 1890s is purported to have said, 'Sports writers—you can buy them with a steak.' "

Though their talent may have been humble, the sportswriters of the late nineteenth century were a proud lot, and, in their own twisted way, a creative one as well. Allowed more latitude in their writing than their brethren on the city desk—a comment on the triviality of sports and their place in larger society—sportswriters engaged in a frenzied competition to out-write one another, in the process giving birth to a memorable, if lamentable, argot that is a much a relic of the game as Cap Anson's bat or Al Spalding's glove. "[W]riters . . . got the idea that all sports should be written in a bizarre patois," said Stanley Walker, "and that to use good English was a sissy trick."

Thus, according to Walker, "a baseball became 'the old apple,' 'the horsehide pellet,' 'the elusive spheroid.' In football the ball became the 'pigskin' or even the 'oblate spheroid.' Home runs were 'circuit clouts,' 'four-ply wallops,' 'four-masters,' and 'four-baggers.' Baseball parks were 'ball yards,' or 'orchards.' A base hit was a 'bingle.' When a man struck out he 'whiffed the ozone.' A bat was 'the ash,' 'the willow,' 'the war club' or 'the bludgeon.' A left-handed pitcher was a 'port-side hurler' or a 'southpaw,' although for some reason right-handed pitchers never were 'northpaws.' A pitcher's throwing arm was his 'salary wing.' The manager was a 'mentor,' 'wizard,' 'miracle man,' 'generalissimo,' or 'master mind.' " Walker called this the "Dada school of sportswriting."

"Most of these nicknames and phrases were rubbed smooth by constant usage," he wrote. "Sports writers became dippy trying to think of new ones, which were almost invariably worse [Is is possible that something could be worse than four-ply wallop?], until the sports pages were so much maudlin balderdash, an esoteric jargon which did not even have the authentic ring of American slang. It was purely synthetic; no one but the writers would understand what it meant."

Understand it or not, however, readers devoured it, because mixed in with the "maudlin balderdash" were the scores and results the fans craved. A big baseball game or prizefight would routinely warrant an "extra"; and, as was the case with war news or election returns, the first paper on the street with the news of an important athletic contest could sell thousands of extra copies. Chalkboards in the windows of the newspaper offices would inform those too impatient to wait for the "extra" of the inning-by-inning or round-by-round progress—a custom maintained until the late 1940s, when the saturation of radio and the introduction of television made it irrelevant. Saloons and pool halls saw the commercial benefits of up-to-the-minute sports news, and installed telegraph wires to keep their clientele informed, entertained and in place—a nineteenth-century version of the wide-screen TV.

Generally, for much of the 1890s, the height of the sensation-mongering "yellow" journalism period, the quantity—and the quality—of sports news was in inverse proportion to the class of people the paper was trying to attract as readers. By the turn of the century, however, even the high-minded papers had recognized the circulation value inherent in a good sports page, though coverage did remain reflective of class. In New York, for example, the *Times* and the *Tribune* gave far more attention to country club sports, golf and tennis, than did the populist *Sun* or the two protagonists in the yellow journalism wars—Joseph Pulitzer's *World* and William Randolph Hearst's *Journal*—who concentrated their sports coverage on the earthier matters of baseball and boxing.

By the time Grantland Rice wandered into the city room of the *Nashville Daily News* to begin his life in a field that for all intents and purposes was scarcely older than he was, there was hardly a newspaper in the land that did not have a man covering the sports beat. There were some isolated small papers where such a staff position was deemed either unnecessary or a too much an expensive luxury—witness Buford Goodwin's reply: "There's no such job," when Rice inquired about the position of sporting editor on the *Daily News.* But even on the small papers, editors and publishers were quickly learning—as Buford Goodwin would have found, what with Nashville vying for the pennant that summer in the inaugural season of the Southern League—that in this new century a newspaper simply could not survive without a sports page, and a good man to run it.

By mid-afternoon on Friday, July 19, 1901, a sizable crowd had gathered at the corner of Church and Cherry streets in downtown Nashville. The were awaiting the arrival of the *Daily News,* Nashville's newest newspaper, and had in fact been waiting all week, while pressmen inside the newspaper's offices at 316 Church Street scrambled to assemble the recently arrived Mergenthaler typesetting equipment and get the presses in working order. Shortly after three o'clock, the first copies of Volume One, Number One, came off the presses. With breathless hyperbole, and a knack for shameless self-promotion that it took some newspapers years to polish, the *Daily News* described the scene at its own debut:

> The first copies from the press were eagerly caught up by those in the press room, and though the newsboys stormed and fought upon the outside it was several minutes before an effort was made to supply them with copies.
> When the first newsboy received his batch of papers he experienced considerable difficulty in fighting his way to the outskirts of eager youngsters that besieged the window. He was not fully free from the crowd before he was selling papers like he had never sold them before in all his life. His breath was taken away by the rapidity of the sales, and before he had scarcely turned into Church Street and before he had broken the tense atmosphere was a single cry of "Daily News," he was

standing with a bewildered expression on his face, empty hands, and a pocket full of pennies.

This was the experience of many of the newsboys. The sale was unheard of and it was almost half an hour before a newsboy could get out of the neighborhood of *The Daily News* office with a single paper of his stock left.

The citizens of Nashville seemed to realize that there was little hope of securing copies of the paper at any distance from the office for some time after the press started. Many came in person to the business office and bought one, two, three, and many more copies.

The article also claimed that there was a "Bright Future Predicted," though it neglected to mention that it was the *Daily News* itself doing the predicting.

The *Daily News* was created by Colonel Jere Baxter, Nashville businessman, former state senator and president of the Tennessee Central Railroad. Baxter needed a newspaper, because he was then engaged in a campaign to have the City of Nashville invest one million dollars in a rail line he was planing to build. The question was set to come to a vote in Nashville on August 8, 1901; and Baxter needed a newspaper of his own because the city's two existing newspapers, the *American* and the *Banner,* were published by men involved with Nashville's other railroads and thus not favorably disposed to Baxter's plans. So the *Daily News* became his mouthpiece and served him well. The bond issue passed and Baxter got his rail line.

The *Daily News* was a lusty paper; it continually and sometimes viciously railed at the enemies of Jere Baxter, and complemented this dubious crusading with news coverage that specialized in the accounts of lynchings, disasters, and the plaintive suicides of jilted maidens. Its editor-in-chief was Van Leer Polk, a relative of President James K. Polk. The editorial staff was comprised of three editors and seven reporters.

Grantland Rice's work on the paper was spirited, alive, and undisciplined; eloquent and nearly illiterate; original and formulaic; positively brilliant and downright awful. In short, it was inconsistent; it was plagued by the sins and shortcomings common to nearly all green reporters. While it showed flashes of promise and creativity at every turn, it was also inclined to be a shade brash and arrogant, woefully overwritten, and frequently sacrificing material that was relevant for the sake of that which was clever. In writing his memoirs in the 1950s, Rice selected a story from his third week on the job to illustrate his *Daily News* days. The piece pretty well illustrates everything that was good and bad about Rice's apprentice work:

> Baker was an easy mark
> Pounded hard over park

> Selma's infield is a peach,
> But Nashville now is out of reach
> All of the boys go out to dine
> And some of them get full of wine.

After their long successful trip the locals opened up against Selma yesterday afternoon at Athletic Park, and when the shades of night had settled on the land the difference that separated the two teams had been increased by some dozen points.

Throughout the whole morning a dark, lead-colored sky overhung the city, and a steady rain dripped and drizzled, only stopping in time to call the game, but leaving the field soft and slow. . . .

The verse actually appeared as a bank of headlines at the top of the story, which ran on page one. The story was Rice's first chance to do any reporting. Nashville had been on an extended road trip since the *Daily News* opened for business, and Rice had been desk-bound. Combine the cabin fever with his youthful exuberance and the hyperkinetic style of the day, it's surprising that the piece is as restrained as it is.

Rice's comment after reprinting the piece in his memoirs was " . . . wonder what the score was!" A particularly apt question, for not only did he not mention the score up near the top of the story, he never mentioned it at all—not once in twelve inches of narrative. The reader had to comb the box score to learn that the locals had won by a count of 11–6. Still, the piece is breezy and descriptive, and if you're not particularly fussy about details like the final score, a fun and altogether serviceable piece of journalism.

Twenty-four hours after writing this article, Rice fashioned the first of the countless journalistic milestones of his long career. This one is far less celebrated than the others; in fact it has been entirely forgotten—as it should be, for in writing on the second game of the Nashville-Selma series on August 13, 1901, Grantland Rice produced what is undoubtedly the worst "lead" in the history of American sportswriting:

> Did you ever hear of the battles of Gettysburg, Bull Run or Water-loo? Of how Napolean crossed the Alps on a mule and Washington the Delaware on a piece of floating ice? Well, all these were mere skirmishes compared with the struggle that took place yesterday at Athletic Park. ★

★In 1988, when Dan Jenkins set out in his novel *Fast Copy* to parody the labored, overwrought efforts of sportswriters inspired by Grantland Rice, he produced a lead that is strikingly evocative of Rice's actual 1901 lead. The novel is set in 1930s Texas, and the sportswriter is a character named Clarence "Big 'Un" Darly, who writes a column called "Sportanic Eruptions:" "You know how it was at the Somme when the little corporals limped out of trenches with their purple guts in their hands and asked the Kaiser if that was the best he could do, but if you wanted to see a real war, you should have been at Clay Field last night when our brave. . . ."

And so it went with his writing that first summer—up and down, like a .500 ball club that look like champs in beating the league leaders on a Wednesday, only to look foolish in losing a laugher to the cellar-dwellers on Thursday. He was much more consistent in his news judgment. His first week on the job he wrote an insightful piece on the status of the Southern League, then in its first year of play, and pronounced it firm. When he gained a bit more confidence, he urged the league to name a Nashville businessman its first president in an unsigned but reasoned piece of advocacy. When the Nashville baseball team was out of town he filled the sports columns with news of a Nashville-bred champion trotter and preview articles on the Vanderbilt and University of Nashville football seasons. When the baseball season was over—beginning the career-long practice that would culminate in his being selected by *Collier's* magazine to select the football All Americans in the 1920s—he wrote an article offering up his choices for an All Southern League team. And while his writing may have been uneven, it was certainly distinctive. There was a Darwinian reference in one story—"That man is descended from monkey is believed by many. Doctor Darwin published several books on this interesting subject some years ago"—a fact that in itself is probably unworthy of notice aside from the fact that Tennessee was the state that had a law on the books that forbade the teaching of such a theory, as John Scopes and the nation would learn some two decades hence. Rice was also confident enough in his opinions to brook no criticism when he felt it unwarranted. A Birmingham, Alabama, man wrote a letter to the *Sporting News* alleging that the reason the Nashville ball club was so unpopular in other Southern League cities was "caused by the brazen manner in which the Nashville papers sneeringly referred to the other teams. . . ." In his rejoinder, Rice was able, rather deftly, to call his antagonist a jackass:

> It is difficult to see how a baseball writer can could maintain his self respect and keep from scoffing at the aggregation which represented the proud city which [the writer] calls his home. They were about as fast as a lot of ham strung snails with bunions on their feet, and played like a lot of colonial dames. Birmingham never won a game in Nashville and blooming few out of Nashville. . . .
> In his concluding remarks [the writer] advises [Nashville manager Newt] Fisher to bridle the baseball writers of this city. We balk. The baseball writers of Birmingham may be the kind of animals that are bridled but we refuse to back up. Long ears are fitting accompaniments of the bridle, and the article from the sunny south smacks of extended hearing appendages.

Rice got his first by-lines in October when he traveled to Chattanooga to cover the year-end meeting of the Southern League board of directors. The meeting was anything but routine. The pennant was in dis-

pute, both Nashville and Little Rock were laying claim to the gonfalon—
to use a 1901 term. The heart of the dispute was a melee-interrupted series
between Nashville and Little Rock in September, in which two wins were
awarded on forfeit to Nashville after umpires were unable to restore order,
forfeits the home Little Rock club contested. Also in dispute were two
games between Little Rock and Selma, Selma protesting the two losses
claiming that Little Rock had used an illegal player in one and an illegal
ball in the other. The week before the meeting was scheduled, Rice summed
it up with his still-unchecked capacity for overstatement. "The task which
will confront this selection is of such monumental magnitude that the rock
of Gibraltar sinks into the insignificance of a broad side pebble in comparison. The various boards of international arbitration, treaty committees,
etc., have easy sailing compared to what these people will have to undergo."

Nashville was awarded the pennant, and Rice's by-line underwent a
rapid evolution during the weekend meetings. On the first day his dispatch from Chattanooga was signed "H. G. Rice." On the second day
and forever after his writing was signed "Grantland Rice."

Rice was the only *Daily News* staff writer to receive a by-line in those
first months of the paper, and while managing editor Buford Goodwin
and city editor Edward Martin were no doubt pleased with the industry
and initiative he was showing in sports, they did not relieve him of his
other responsibilities. He began each day down at the produce market—
"EGGS BRING CENT APIECE . . . The gloomy weather today has
caused the local market to be somewhat less active than is usual but in all
the larger houses there was good business reported."

He also continued to cover the state capitol, though the legislature
was not in session during the summer and fall and that eased Rice's reporting burden somewhat. What political news there was in the *Daily News*
is ably done; but it is impossible to tell if Rice actually wrote any of it, for
he had entered into a conspiracy with Louis Brownlow of the *Banner.*
Brownlow had been a boyhood chum of Rice's; they had both hung around
the soda fountain at De Moville's Drugstore. Like Rice, Brownlow was
wearing many hats as a reporter, and he found sports as loathsome as Rice
found politics. So, when the opportunity arose, they traded off. Rice would
go to the ball game and file an extra story for Brownlow; Brownlow
would cover for Rice at the political affairs. Their respective editors were
either none the wiser or unconcerned about the bartering that went on.

Rice would occasionally draw an extra assignment from Goodwin or
Martin. The one he remembered best was the society ball Martin sent him
to one evening.

"That's a women's beat," Rice protested.

"We're all out of society writers," replied Martin. "Tonight you're
it."

So Rice reported that he went back to the newspaper's morgue and found the write-up of some similar event in the *New York Tribune,* and proceeded virtually to transcribe the *Tribune* article, inserting the Nashville names. "I had attributed all the current New York styles and descriptions of the fashionable gowns worn at that New York dance to the madames and belles of Nashville's upper crust," he said. "All seemed happy with that column because Martin received nary a kick and congratulated me on a job well covered."

In the fall Rice received a job offer, and while the source of the offer was a likely one, the job was odd. The offer came from Herman Suter, whom Rice had known as the Sewanee football coach. He was headed to Washington to edit something called *Forester Magazine.* "I didn't know a Christmas tree from a Northern Blue Horned Spruce," Rice remembered. "But when the call came from Suter, I joined him."

Why? Rice never said. Perhaps it was because he had never in his life been out of Nashville and at the age of twenty-one he might have been nervous about when he would get another chance. But a better guess at why he jumped at the chance to write about trees was the perilous financial condition of the *Nashville Daily News.* It was a troubled paper throughout its short life. (It lasted just three and a half years, publishing its last issue in February 1905.) From the first the advertising was scant and the paper was notorious for missing payrolls. Rice's salary of five dollars a week was not much; it bespoke a great love for newspapering. But nobody's love is so great that it can withstand the test of pay envelopes with nothing in them.

On Thanksgiving Day he covered the Vanderbilt-Sewanee football game, getting a by-line in the next day's paper, and then left for Washington. The experience was not what he had hoped. Soon after arriving in Washington he was felled by appendicitis. After a five-week stint in a Washington hospital, Rice's mother arrived from Nashville to take him home, so that he might complete his recuperation under the watchful eye of family.

Rice returned to the *Daily News* after he was well, but the lengthy recovery from appendicitis had done nothing to cure either Rice's wanderlust or the *Daily News*'s financial ills, and in November he accepted a job as sporting editor of the *Atlanta Journal* for $12.50 a week. He did his first work for his new employers on Thanksgiving Day, covering the Vanderbilt-Sewanee football game. The next day November 28, 1902, twenty-five Nashville newsmen from the *Daily News,* the *Banner* and the *American* gathered at the Duncan Hotel to honor Rice and Louis Brownlow, the *Banner* writer with whom Rice exchanged assignments, who was also leaving Nashville, for a job with the *Louisville Courier-Journal.* His personal and professional goodbyes said, Rice then left his hometown in early December, in search of his destiny, for the second time in less than a year.

He was to find it in Atlanta, for it was in Atlanta that Grantland Rice came of age as a journalist. In Atlanta he experienced for the first time the nomadic life of a sportswriter, and for the first time built the close personal relationships with athletes, coaches and "moguls" that would enrich his life and shape his work. In Atlanta he worked with men whose talent was equal to his own, and as they worked and lived and drank together they forged a camaraderie cum competition that advanced the talents and the art of each. And, it was in Atlanta that Grantland Rice fell in love with the woman who would share his life for forty-eight years.

The *Constitution,* not the *Journal,* was Atlanta's pre-eminent newspaper when Rice came to town in the waning days of 1902. Under the legendary Henry Grady, with his editorials during Reconstruction calling for a "New South" centered on industry, business, and a diversified agriculture, the *Constitution* had become not only the pre-eminent newspaper in Atlanta, but the most important and respected editorial voice in all of the South. Grady died in 1889 at the age of thirty-nine; still, thirteen years after his death the luster of his newspaper remained.

All of which made the *Atlanta Journal* the perfect paper for Rice. Founded in 1883, fifteen years after the *Constitution,* the *Journal* made no effort to match the *Constitution*'s editorial advocacy on behalf of the South's identity and place in the American economy and culture—to do so would have been to cast a whisper against the thunder. Rather they chose to be the foil to the older, richer paper, eschewing intellectual debate, as it were, and opting to fight in the streets, concentrating their energies on such staples of the newspaper game as crime, local politics—and sports.

Rice put the stamp of his own puckish personality on the sports page of the *Journal* right away. The banner over the sports page never read simply "Sports" or "Sporting News" as was customary. Rice's flags read "A Trip to Sportville, via Many Scenic Routes" or "A Sporting Feed from Soup to Cigars" or "One Day's Fishing from the Streams of Sport" or "Some of the Sparklers in the Sporting Milky Way" or "Driving in Record Time Across the Sporting Speedway—Chauffeur: Grantland Rice." There was a different goofy headline practically every day. In the columns of sports news, he continued to struggle—unsuccessfully for the most part— with his tendency to overwrite. When an Atlanta Colonels' ballplayer died suddenly of pneumonia, Rice reported that "The Great Umpire of the game of life . . . has called him out on strikes. . . . A white shroud today has supplanted his uniform and the hands that flagged many a hot drive in life are crossed upon his chest."

But here was also much in Rice's sports page that was noble. It was a comprehensive page, with box scores and game stories from the major leagues, the Southern League, the new South Atlantic, or "Sally" League, as well as from college and prep games. It was also an innovative page. He wrote of the off-field hobbies, interests and personalities of the members of the Atlanta club, humanizing them in a way that ballplayers were not in those days. When Rice traveled home to Nashville with the Atlanta club in May of 1903 he took the ballplayers on a tour of the Hermitage and of the Belle Meade horse farm, and delivered to *Journal* readers what modern-day readers will recognize as a pretty sturdy and distinctive travel article. After another trip he rated the cities in the Southern League according to their appeal for a visitor—Atlanta and Nashville, perhaps not surprisingly, rated tops on Rice's list; Montgomery ("it's a case of four days shalt thou labor and on the fifth wish thyself dead"), Little Rock ("scrapping with pop-eyed mosquitoes") and Shreveport ("nothing to see or do save keep posted on railroad schedules") fared the worst.

He was not a man to let himself be constrained by the historical limits of what constituted sports news, or what was in the purview of a sports reporter. He studied up on the arcane language in the major league and Southern League constitutions and by-laws, and wrote stories on how the lackadaisical enforcement of roster rules in the still-new American League would effect Southern League teams. He would be in touch, via telegram, with American League owner Charles Comiskey on impending roster moves involving Southern League players, and wrote stories anticipating pending moves. By the close of his second season in Atlanta, Colonels manager Abner Powell was seeking his counsel on rules interpretations; by the time he left for Cleveland in 1906 he was recognized throughout the South as the leading authority on the Southern League rules. All the while his voice continued to emerge. He advised a young Southern League umpire in print to stand up to the bullying he was taking from certain players. "Use just a little more backbone." The verse on the sports page also became a little more frequent, though at this point he had not yet begun to write on deeper themes—most of his Atlanta verse was merely a device for telling his story of the game he was covering.

Other papers throughout the South took note of his writing, and his reputation grew. It is thus not surprising that when an ambitious and savvy eighteen-year-old minor league rookie went looking for some publicity that would speed his way to the majors, he figured Grantland Rice was the writer whose reach and influence would serve his interests best.

Ty Cobb sent Rice an unsigned telegram before he left for spring training in 1904: "Tyrus Raymond Cobb, the dashing young star from Royston, has just started spring training with Anniston [of the Southeastern League]. He is a terrific hitter and faster than a deer. At the age of eighteen he is undoubtedly a phenom." Rice was in the midst of a post-deadline poker game when the telegram arrived and sent back a note to the Royston Western Union office: "After this the mails are fast enough for Cobb."

But as the baseball season passed Rice heard more about Cobb. Postcards from every port in the South arrived at the *Journal* offices.

"Keep your eye on Ty Cobb. He is one of the finest hitters I have ever seen."

"Watch Cobb of Anniston. He is sure to be a sensation."

"Have you seen Ty Cobb play ball yet? He is the fastest mover I've seen in baseball."

The cards were all signed by men Rice did not know—people named Brown and Smith and Jackson and Holmes. Finally Rice wrote a column about this kid from Georgia who was "the darling of the fans." After Cobb moved up to Augusta of the Sally League, Rice went over to see him play. "I've been hearing about you," said Rice by way of introduction when he found Cobb in the Augusta dugout. "My name is Rice. I write baseball for the *Journal*.

"Is that so?" said Cobb. "I've heard of you too."

A year later Cobb was in Detroit, and by the time Rice reached New York some years later the Cobb story was a part of his legend. Profiles of Rice would invariably feature something on his "discovery" of Ty Cobb.

It was not until after World War II that Cobb admitted to Rice that he had been the one to send all those postcards. In retrospect, thought Rice, he probably should have known; for the shameless was consistent with the enormous drive that had carried him to his remarkable career. "Self-confidence is the hallmark of a champion," said Rice, "any champion. Not only had Cobb had that amazing cheek and flare at eighteen—more important, he knew how to use it, something few can handle at any stage of life."

Upon arriving in Atlanta, Rice had taken rooms at the Aragon Hotel, sharing them with a young man from Illinois whom he had met in the *Journal* newsroom. At twenty-four, Don Marquis was two years older than Rice; he was writing editorials and covering the theater for the *Journal*. Like Rice, he was blessed with enormous talent, and destined to become one of the most popular, widely-read, and highest-paid journalists in the country. Between 1912 and 1925 he was one of New York's most popular newspaper columnists, first with the New York *Sun* and later with the *Tribune*. He would go on to write more than thirty books and four plays, in the process creating some of the most durable and popular characters in Jazz Age popular culture—characters like archy the cockroach. archy's name was never capitalized because, as Marquis told his readers, archy typed his own stuff, late at night on Marquis's typewriter, and archy had no way of working the shift key for the uppercase letters; in fact, to work the typewriter at all he had to dive headfirst onto the proper key.

The humor in Marquis's writing did not come from gimmicks, though, but rather from a biting and bittersweet satire on a broad range of American society, from matters such as intellectual pretension—he called Greenwich Village a "cultural bordello"—and the societal affectations of the flappers to the self-righteousness of the prohibitionists. Marquis gave voice to feelings on prohibition through his most famous creation, a chap named Clem Hawley, known as The Old Soak, an appealing sot who said he "believed in old-time religion, calomel and straight whiskey." The Old Soak was an effective wag, railing against the unpopular Eighteenth Amendment; he became not only the most enduring of Marquis's column voices but the title character in a successful Broadway play as well. But there was a darker side to The Old Soak—heavily veiled by his wit and charm, but there nevertheless for anyone who cared to probe. Marquis once described The Old Soak as "one whose devotion to alcoholic fellowship and endearing generosity of spirit superseded the practical concerns of work and family support."

There was a lot of The Old Soak in Don Marquis. Whatever darker side, whatever fears he may have harbored were lost in the wash of wit, ebullience and studied insouciance that he presented to the world in both his writing and his personality. During his Atlanta years he made it a point to tell those who asked that his goal was to succeed in order that he might one day retire and lead a lazy life. In truth he was a tireless worker, who strived for years in obscurity in New York before finding himself suddenly famous in the years before the Great War. "When the tide finally turns in New York," he said, "it turns swiftly and with a rush."

Having lived with him for three years, Rice knew how hard Marquis worked for his success. He knew too Marquis's inner torment, and knew how it helped to shape Marquis's art. "How Marquis could write!" Rice wrote,

An unaffected genius, at times he was a black brooder, but his physical and mental courage were magnificent.

> . . . There I stood at the gate of God,
> Drunk but unafraid.

That closing line of one of his verses mirrored Don's scorn for any human soul lacking the courage of its convictions.

But mostly, Rice knew and loved Marquis the blithe spirit. On Christmas Eve, 1902, shortly after arriving in town, Rice came upon Marquis in the *Journal* offices, "high as two kites and lathered with red ink and oil from a battered old hand letter press."

"I'm putting out page one of my Christmas issue—the way Hearst would do it," Rice reported Marquis roaring in greeting. And when Rice inspected his roommate's work he found red, four-inch letters running atop the mock front page, reading "CHRIST IS BORN!"

Marquis and Rice used to while away the hours in the bar at the Aragon Hotel, playing cards, drinking, talking about writing and newspapers, oftentimes scribbling verse on scrap paper and napkins when the Muse struck. The were often joined at the Aragon by Frank Stanton, columnist and poet from the *Constitution*. After Stanton died it was said that he produced work "inspired by the sublime simplicities of the genuine Christian life, however slow he may have been to claim such a life for himself." It is a judgment with which Grantland Rice would no doubt concur. Stanton's column and verse were noted and loved for their gentleness and unabashed optimism, and the tenor of his work would be found echoed in Rice's own writing; but when Rice first met the man it was in the bar in the Aragon, and he certainly wasn't living the life he may have been espousing in his column.

"After a rough night, Stanton might wander into the old Aragon Hotel and offer to 'write a poem for thee' in exchange for three fingers," said Rice. "He turned out more classic lines for bartenders than most mortals have composed for publishers."

However troubled Stanton may have occasionally been by drink, as Rice also maintained, "the inbred sweetness of the man never deserted him." He was named the first poet laureate of Georgia two years before his death in 1927; but—as Rice would do in future years when someone took to analyzing his verse—Stanton was given to dismissing discussion of his poems as art. For in his own mind, he was naught but a newspaper columnist, and his verse simply the daily offerings of another journalist. "Still," he acknowledged once to an interviewer, "if a feller goes fishing every day of his life, he is bound to get a nibble now and then."

The lines to Stanton's "Sweetes' Li'l Feller," more popularly known as "Mighty Lak' a Rose," were published as a song and became one of the most popular lullabies of the early twentieth century; but the lines for which he's perhaps best remembered—illustrative also because they so sum up the philosophy embodied in his writing—is a four-line ditty entitled "The World."

> This world that we're a-live' in
> Is might hard to beat;
> You git a thorn with every rose,
> But ain't the roses sweet.

Through Stanton, Rice and Marquis came to know Joel Chandler Harris, the doyen of Atlanta journalism, now in retirement from his duties as associate editor of the *Constitution* and devoting full time to his fiction—the Uncle Remus stories that had won him international acclaim. Harris had been the man who brought Stanton to the *Constitution*, saving him from a life as an itinerant printer. He was also the man who laid out what

Stanton's column should be. "We all, up here, are just plain home-folks," Harris told Stanton, "kind of like children that crawled in under the tent of the big circus, and when we look back to where you and lots of us came from, we are apt to feel that we're a long ways from home. Make your column home-like." Stanton did, and while its tenor and tone are badly dated today, in its day it was unique, fresh, and one of the best-loved columns in the land.

Stanton was more than twenty years Rice's senior, and the two men seldom saw one another after Rice left Atlanta in 1905. Still, of all the influences on Rice's work during these Atlanta years, that of Frank Stanton was unquestionably the most profound; the similarities are too distinct to be dismissed as coincidence. Rice never heard Harris's advice to Stanton, yet it was a philosophy Rice applied to his own work, particularly when he worked for the *Nashville Tennessean* later in the decade. Rice too was noted for writing that was homey, nostalgic, a little wide-eyed, and possessed of a subtle wit and great tenderness. A half-century after sharing a midnight glass in the Aragon bar, Rice's fondness for the humble poet, who taught him more than either man ever realized, remained undiminished. "By all odds the finest poet I ever met, he was like a mocking bird singing in a Georgia oak which happened to be flooded with moonlight."

Sometime around the summer of 1904, the bar at the Aragon lost some of its appeal for Rice, for he had begun to spend his evenings in the company of Katherine Hollis.

When she met Grant Rice, Kate Hollis was twenty-two, a tall, dark-haired beauty with one of the busiest social calendars in both her home-town and the big city of Atlanta. She lived in Americus, a small town about 120 miles south of Atlanta, with her widowed mother and three of her five sisters; she traveled to Atlanta frequently to visit friends and partake of the big city's much richer social life.

A proper young woman, she was no doubt a bit unsettled by her first glimpse of Rice. She was at an Atlanta amusement park, together with a friend and their dates, riding sidesaddle on the merry-go-round when she spied the slender, blond-haired Rice standing rakishly off to the side, boldly staring at her and smiling. They were properly introduced later and she agreed to let Rice come to call. (The amusement-park date was apparently not with one of her more serious suitors.)

But she apparently had second thoughts, for when Rice arrived the next day, Kate Hollis was gone. Undaunted, the smitten Rice scratched out a poem and left it for her:

> Since this hyper-torrid weather leads to sunstroke and "dys-
> pepsy"—
> To heat prostration, nettle-rash, neurosis, epilepsy—

Allow me to refresh you with a bite or so to eat—
(Tho' when I went to meet you there was no one there to meet).

Still, if the fates are kind to me and things pan out alright—
I hope to run across you out on Kimball Street tonight—
I have but one suggestion—that is—"will you won't" repeat
Your morning "dodge" and leave me there with "no one there to
    meet."

Kate was quickly as smitten with the handsome young sportswriter as he was with her, and her memories of the early courtship summon images of sepia tones and soft ragtime: "A few days later, he came calling for me again," Kate remembered, "this time with a rented horse and buggy—and took me for a ride. I recall he wasn't very sure of himself . . . or the horse. It started to rain. There was a clap of thunder and the horse bolted and started running—away! Granny couldn't stop him, and the next thing both of us were pitched out of the cart. The horse galloped clear out of sight. We 'hitched' a ride back to Atlanta on a milk wagon."

In letters home to her mother, Kate provided a day-by-day itinerary of how thoroughly Rice was dominating her time and thoughts:

Listen to how many times I have seen Grantland Rice: Saturday for lunch; a ballgame afterwards until dark; Sunday afternoon a long walk all the afternoon; Monday night box party at the theater. This afternoon (Tuesday) the ball game at 3 until dark. Tonight he is coming out. In the morning he leaves or I could have seen him more. I like him more each time I see him. People are even calling me "Mrs. Rice."

Evidently, though her infatuation with Rice was apparent and complete—"Have an engagement with G. Rice every night next week that I am here. What joy!!! Will it last?" read one letter home—Kate did not immediately jettison her other beaux, as a later letter home tells of one she is keeping on the string. "Willis is in a rage because I haven't written him everyday. Wrote me that I need not bother myself to do so again. He is crazy. I have written him twice. Must write to my Grantland now, so goodbye."

By the end of the summer, 1905, Rice was determined to propose marriage—his problem was how to convince Kate—and her skeptical mother—that he could be a good provider and a worthy husband on the pittance he was being paid by the *Journal*. Instead of trying, he went looking for a better job. In October he went north to cover the World Series between John McGraw's Giants and Connie Mack's Philadelphia Athletics, intending to return home with more than just a handful of stories and memories from his first World Series.

After Christy Mathewson shut out the Athletics on four hits in the first game on October 9, Rice wired the *Journal* that the Giants would win the Series and that Mathewson would pitch three games and shut out the Athletics in each one. That's exactly what happened; with the Series tied at one game apiece, Mathewson pitched on two days' rest in game three and threw another four-hitter; he came back in game five on just one day's rest and allowed but six hits, closing out the Series for the Giants. His line score for the Series—played over a span of just six days—read: 3 wins, no losses; an earned run average of 0.00; 27 innings pitched; 14 hits; 1 walk; 18 strikeouts. Word of Rice's soothsaying anent Mathewson's heroics spread throughout the newspaper business, cranking up his already considerable renown a notch as he prepared to look for work.

He went to Cleveland, where a new afternoon paper to be called the *News* was being put together from the pieces of a couple of existing but struggling afternoon dailies. The Cleveland newspapers took notice of his visit—not hinting or even suspecting that he was trolling for work. To those undoubtedly curious as to why he had chosen Cleveland for his post-World Series vacation, there was the fact that the Cleveland ball club took spring training in Atlanta, and that Rice thus had a lot of friends on the shores of Lake Erie. "Mr. Rice endeared himself to the members of the Cleveland club and the newspaper writers with the team," reported the article noting Rice's Cleveland visit, "by his unfailing courtesy and his earnest endeavors to make the stay of the Cleveland people in the Southern city a pleasant one."

There were some of the *News* staffers who felt the sporting editor's job should go not to Rice, but to a man named Bill Phelon, whose roots were Midwestern, not Southern. Rice was apparently confident that he would be able to win over the disaffected staffers, and he was certainly impressed with the money, for he accepted the job the moment it was offered; the salary was fifty dollars a week. "That was real money and for a fellow with marriage on his mind—money never hurt." He returned to Atlanta by way of Americus and proposed to Katherine Hollis.

Grantland Rice's proposal, and Kate Hollis's inclination to accept, touched off a furor in the Hollis home. Kate's family had never approved of her romance with Rice; he was known around the home, somewhat derisively, as "that ballplayer," and seen as a man without prospects—a stark contrast to some of Kate's other suitors; she had had a lingering flirtation with a successful young Atlanta banker which the family had been hoping would flower.

Florence Davenport Hollis, Kate's mother, was a strong-willed woman who, as a teenager, had nursed the Confederate wounded and gawked at the Yankee prisoners in Andersonville, and who later held a family of eight young children together after the death of her husband in 1893. She was a woman of strong opinions and strong designs on how her daughters' lives should be lived. While the particulars of what happened after Rice proposed to Kate are lost, it is clear that Florence Hollis forbade her daughter to marry him, for what does survive is a plaintive letter from Grantland Rice to his future mother-in-law:

Dear Mrs. Hollis:

I received Miss Jenny's [Kate's older sister] letter today and one from Kate telling me how everything stood now.

Mere words are so useless and futile in what I want to say that I am at a loss how to tell you what I feel; what I have been through since I've been here, with the climax Friday when I saw it was my duty to go out of Kate's life. All seems like a strange dream but since it is over, I want to come to you tonight and ask you, not to take Kate out of your life, but to let me share it with you and those who love her and whom she loves in turn. I ask you this not only because I love her so, and she loves me, but because I have a life that I know for her sake I can make worthwhile to offer her, and because in my heart I know I can make her happy.

Promises are so easy to make and break, I know, and words by themselves mean so little, that I feel it would all be in vain to tell you how I feel.

I can only say if all the love a heart and soul and every thought only for her can make her happy, then she will never know a second's pain.

To try to tell you how I love her would be an impossible feat for me, but I will give my life to her—my heart and soul. If I thought that anything would happen to her under my care to make her unhappy, the torture would be unendurable, but with God's help, I will guard her and shield her from that so that neither she nor anyone else will regret the choice she has made.

She may not have the great things that others could have given her, but what is lacking there will be more than made up by a love and life that is only for her and will be to the end. And I could never be happy beyond the end if I knew we would be separated even then.

If you could only see my heart bared to you, I know then you would be content.

I know better than anyone in the world the sacrifice it will involve for I, too, thought she was lost to me and I know the struggle it cost my very soul to even think of giving her up.

You are her mother and I know what it means to you before anyone else in the world, but even while asking you for Kate, I want to ask you to feel, if you can, to me as if I belonged to you and needed your love and care.

Have faith in me until I fail in anything concerned in Kate's happiness just once and I can promise you that I will keep the faith until the end.

While I can make no pretensions to any greatness in any way, I can tell you in all truthfulness that I can offer Kate a life that no man or woman can ever say has done them a wrong.

I have taken part in nothing, believe me, that either she nor any member of the family need ever be ashamed.

In return for her, I will give her my life and live it only for her sake as she is my all in all. I can take care of her in all comfort now and some day with the strength of her love to help me on, I feel in my heart that I can do even more.

Forgive me for taking her away if you can, but when you give her to me you know beyond all doubt that she will still be loved and cared for beyond all time.

Sincerely,
Grantland Rice

Whether it was Rice's tender baring of his soul, or the apparent parallel beseeching of her daughter, Mother Hollis—as Rice was wont to call her—capitulated and gave her blessing. Rice went north to start his career on the *Cleveland News* and Kate returned to Americus to plan the wedding. The date was set for April 11, 1906, a Wednesday.

Rice was in Georgia for a month prior to the wedding, covering Cleveland's spring training in Atlanta. He worked out with the ball club on March 17, and found that five years removed from the playing fields of Vanderbilt had left him unaccustomed to the demands the sport placed on his body:

> I scampered like a young schoolboy from center field to
> right.
> And, like some raw recruit, I tried for everything in sight.
> I scooped up grounders high and low in great exceeding
> glee,
> Nor had one thought of what the morrow had in store for
> me.
>
> But when from out of my downy couch I tried in vain to
> rise,
> I found a string of charley horses stabbed me in my thighs.
> My joints creaked like some rusty hinge, my feet were full
> of lead,
> And I had to get three bellboys just to roll me out of bed.

The wedding was timed to coincide with the end of spring training, when Rice could steal a couple of days off as the club was breaking camp and starting north.

Grantland Rice and Kate Hollis were married at the First Methodist Church in Americus at 7:30 in the evening on April 11, in what one newspaper clipping called "an event of large social interest in several states, both young people being prominent and widely known. Miss Hollis, as one of the beauties and belles of the Empire State of the South, and Mr. Rice as a writer, whose career as a newspaper man, though covering only a few years, has been notably brilliant."

The newspaper article went on to say that the twilight wedding had Americus "astir socially, from end to end of the darkening streets," and that the bride and groom "made a bright picture of youth and loveliness at the altar." Rice's brother John came down from Nashville to serve as best man; Kate's sister Elizabeth was maid of honor. The reception was held at the Hollis home on Taylor Street; "the beautiful old home was en fete for the crowning event in the life of the much-loved daughter and was a beautiful setting for a southern wedding," according to the newspaper account. The home was decorated with a profusion of pink and white—roses, carnations, ribbons, tablecloths, candies, even the lamp shades.

Grant and Kate left the party at eleven. Their wedding night was spent aboard a Pullman coach, on a train wending its way north to Louisville, where Rice was scheduled to rejoin the Cleveland baseball team; their honeymoon was spent following the team back to Cleveland. When the newlyweds caught up to the team in Louisville, Nap Lajoie, the Cleveland manager, presented Kate with what Rice called "a huge barrel of china . . . , enough chinaware to stock a hotel" as a wedding present.

The Rices' stay in Cleveland lasted just about a year, long enough for Kate to give birth to a daughter, named Florence Davenport Rice, after her grandmother. "Floncy" had her father's blond hair and blue eyes, and Franklin P. Adams would later call her "the prettiest girl ever born in Cleveland who moved to Nashville at the age of one month."

Rice's life at the *News* centered on the Cleveland baseball club. They were not yet called the Indians—they would not be until 1914. When the writers felt a need to apply a nickname they most commonly referred to the team as the "Naps," after player-manager Napoleon Lajoie. All but forgotten today, Lajoie was one of the genuine titans of the game, every bit the equal of Cobb as the best of the dead-ball era. His .422 batting average in 1901 still stands as an American League record, and likely will for all time. In the season to come, 1906, he'd lead the league in hits with 214, and his .355 batting average was just three points off the league-leading .358 posted by George Stone of the St. Louis Browns. Behind Lajoie the Indians shot out to a big lead in 1906, but by early summer injuries—including a spike wound to Lajoie that didn't keep him out of the lineup but did ever so slightly blunt his brilliance—had started to expose Cleveland as pretenders and Rice predicted in his column—correctly—that the White Sox and Cubs would win their respective pennants and meet in an all-Chicago World Series.

In these early days of sportswriting, when eloquence or a distinctive style mattered not a whit to most readers, a reporter would make his reputation on how he presented the "dope." The dope was purported to be expert analysis and inside information. At least that was how it was touted in the newspapers. In truth it was nothing more than a reporter's passing on the rumors and guesses he'd gleaned about the ball park and then making some self-important prediction about what it all meant. Rice was particularly good at doling out and making sense of the dope. He was a particularly well-sourced reporter; the ball players and managers knew him, liked him and took him into their confidence. Rice was also an indefatigable reporter; he was constantly milling about the dugouts and hotel lobbies—talking, listening, fitting the pieces into the big picture. He was also a fortunate reporter, in that his popularity and prominence prompted sports people—players, managers, front-office types—to seek him out. In Cleveland, as it had been in Atlanta and would be later in Nashville, it was the rare out-of-town sports figure who would come to town and not feel obliged to call on Rice.

He was also smarter than the average reporter, able to offer up a hypothesis based on a correlation that others may have missed. So his dope—the same rumors and guesses that everyone else was offering carried the crack of authority, which, when matched with his graceful prose, made it all seem to make sense. When a reader read Rice, it was with the conviction that he was getting the real dope—and often he was. Rice would make a living writing such a column long after changes in both sports and newspapers would make it something of an anachronism.★

It was his verse, however, that was really beginning to distinguish Rice's work; and a couple of his best-known and most frequently reprinted poems dated from his days in Cleveland.

They both star "Mighty Casey," Ernest Lawrence Thayer's eternal symbol of cockiness humbled, who had left "no joy in Mudville" when he struck out to end the game with "Jimmy safe at second, and Flynn a-hugging third" and the Mudvilles by two runs back in 1888. That was the year "Casey at the Bat" first appeared in the *San Francisco Examiner,* published by Thayer's Harvard classmate and fellow *Lampoon* staffer, William Randolph Hearst. The poem was signed "Phin," apparently an abbreviation of Thayer's college nickname, Phinny, and the *nom de plume* under which he contributed occasional ballads to the *Examiner.* While the poem became almost instantly renowned, Thayer remained virtually unknown. It was William DeWolf Hopper who was the embodiment of

★In the years after Rice's death, as his contemporaries and imitators gradually retired or died off, the column of "dope" became all but extinct; and it's off-season cousin, the hot-stove-league column, was pushed off the sports page by burgeoning interest in basketball and other sports. In the 1980s, however, it underwent something of a renaissance. In major metropolitan papers there are once again columns—often in the Sunday edition—of rumor, speculation, notes and miscellany that are strikingly evocative of the style so common in Rice's day.

Mighty Casey. DeWolf Hopper was a Broadway actor who gave a recitation of the poem one evening as intermission entertainment for the theater's special guests, the members of the New York Giants and the Chicago White Stockings. According to the next morning's *New York Times,*
the recitation was "uproariously received," and it swiftly became DeWolf
Hopper's signature piece; he claimed in his memoirs to have recited it
more than ten thousand times. It was also widely republished after Hopper's popularization—seldom with any credit to Ernest Lawrence Thayer—
and taken up by elocutionists less celebrated than Hopper; bringing it to
virtually every corner and crevice of the American consciousness.

   Rice may not have been the first man to attempt a sequel to "Casey
at the Bat," though his is the first of over twenty-five sequels and parodies
that researcher Martin Gardner uncovered for his book, *The Annotated Casey
at the Bat.* Gardner also considers "Casey's Revenge" the best of the many
sequels that gave Casey a chance to redeem himself with another at-bat.
"Casey's Revenge" takes place a week after the strikeout, and opens with
Casey a demoralized man and the Mudville fans bitter and vengeful:

. . . But now his nerve had vanished—for when he heard them hoot,
He "fanned" or "popped out" daily, like some minor league recruit. . . .

"Back to the woods with Casey," was the cry from Rooters' Row,
"Get someone who can hit the ball and let that big dub go."

   Inevitably, facing the same pitcher who fanned him in Thayer's original, Casey hits a prodigious bases-loaded home run and "Mudville hearts
are happy now." The poem was widely reprinted, but, like Thayer, Rice
generally suffered the indignity of having his name omitted; and sometimes the verse was even credited to somebody else. A 1907 quarterly
magazine called *The Speaker* and at least two later anthologies credit the
poem to a James Wilson. Gardner offers the theory that Wilson may have
been a pseudonym Rice used, but there's no evidence that he ever wrote
under any name other than his own; it is more likely that James Wilson
was a name that some editor simply invented—not knowing or having
any way of finding out who the real author was.

   The second poem in Rice's Casey series—also widely reprinted
throughout his career—came later that summer, and "Mudville's Fate"
seems to hold up a little better than "Casey's Revenge." There is something artistically unsatisfying about Casey as conquering hero. As poet
Donald Hall wrote: "None of the triumphant sequels will do. . . . Casey
must strike out: Casey's failure makes the poem. . . . We do not want
Gods or kings—that's why we crossed the ocean west—but human beings,
fallible like us."

   "Mudville's Fate" is melancholy—if a little glib—evoking, as it does, a wistfulness for vanished innocence and abandoned playgrounds, and heroes since revealed to have feet of clay:

I wandered back to Mudville, Tom, where you and I were boys,
And where we drew in days gone by our fill of childish joys;
Alas! the town's deserted now, and only rank weeds grow
Where mighty Casey fanned the air just twenty years ago.

Remember Billy Woodson's place, where, in the evening's shade,
The bunch of us would gather and discuss the home runs Casey made?
Dog fennel now grows thick around that "joint" we used to know,
Before old Casey whiffed the breeze some twenty years ago.

The grandstand, too, has been torn down, no bleachers met my gaze
Where you and I were wont to sit in happy bygone days;
The peanuts which we fumbled there have sprouted in a row
Where mighty Casey swung in vain just twenty years ago.

Oh how we used to cheer him, Tom, each time he came to bat!
And how we held our breath in awe when on the plate he spat;
And when he landed on the ball, how loud we yelped! But Oh
How loud we cursed when he struck out some twenty years ago!

The diamond is a corn patch now; the outfield's overgrown
With pumpkin vines and weedy plots; the rooters all have flown—
They couldn't bear to live on there, for nothing was the same
Where they had been so happy once before that fatal game.

The village band disbanded soon; the mayor, too, resigned.
The council even jumped its graft, and in seclusion pined;
The marshal caught the next train out, and those we used to know
Began to leave in flocks and droves some twenty years ago.

For after Casey fanned that day the citizens all left,
And one by one they sought new land, heartbroken and bereft;
The joyous shout no more rang out of children at their play;
The village blacksmith closed his shop; the druggist moved away.

Alas for Mudville's vanished pomp when might Casey reigned!
Her grandeur has departed now, her glory's long since waned.
Her place upon the map is lost, and no one seems to care
A whit about the old town now since Casey biffed the air.

Rice was hardly through with Casey. That same summer he wrote thirty-two lines of verse—his entire game story—about an actual Cleveland rally that fell short:

But here in poor old Cleveland, all the population's sick—
Remember what old Casey did? Well, so did Elmer Flick.

In response to letters asking for more about Casey, Rice put him in a football game that fall, playing for "Yarvard" in their big game against "Hale." That verse was not reprinted very often—and justly so.

Twice more in his career, Rice returned to the irresistible Thayer hero. In Nashville a year later, Rice wrote a verse called "The Man Who Played with Anson on the Old Chicago Team," a take-off on another popular nineteenth-century poem, Eugene Field's "The Man Who Worked with Dana on the Noo York Sun." It was also as much a tribute to the Cap Anson White Stockings of the 1880s as it was a Casey adventure. And finally, in 1926, after reprinting "Casey's Revenge" in his column, Rice received a note signed with the initials L. F. K. It read in full: "Dear Sir: I have just read your 'Casey's Revenge,' which I understand is a sequel to 'Casey at the Bat.' I had never heard of this poem before. Where can I get a copy." Rice absolutely couldn't believe it:

> I knew a cove who'd never heard of Washington and Lee,
> Of Ceaser and Napoleon from the ancient jamboree,
> But bli'me, there are queerer things than anything like that,
> For here's a cove who never heard of "Casey at the Bat"!

He went on for fourteen stanzas, alternating his own incredulity with lines from Thayer. And, while "Casey's Revenge," "Mudville's Fate" and "He Never Heard of Casey" would occasionally be reprinted over the next thirty years, Rice was finally done with him.

And by the spring of 1907 Rice was done with Cleveland, too.

5

Cumberland's Calling: Hometown Sporting Writer

Rice could have gone anywhere he wanted in 1907. Good literate newspapermen were rare and had their pick of the country's opportunities. He could have set out for the West and joined Jack London and W. W. Naughton, writing sports in San Francisco. He could have gone east to New York, the mecca of American newspapering, but strangely without a brand-name sportswriter, save for perhaps the old cowboy Bat Masterson, whose notoriety did not spring from his prose.

The most logical career move from Cleveland would have been west to Chicago. Chicago rivaled New York as a newspaper town, and as a sportswriting town it was without equal. Hugh Fullerton of the *Examiner* stood at the head of the class. He had made his reputation by successfully doping the 1906 World Series. While the smart dope was picking the mighty Cubs, Fullerton's lonely voice picked the White Sox, the "hitless wonders," to win it four games to two. That was exactly what happened and Fullerton was catapulted to a celebrity that some jealous critics unjustly claimed he coasted on for the rest of his career.

Rice's favorite in the pantheon of Chicago sportswriting was Hugh E. Keough, the columnist of the *Tribune*. Keough took a charitable view of the world, and, a man after Rice's own heart, sprinkled his column with verse and witty aphorisms. "Broad, liberal and forgiving toward hu-

83

man frailties," said Fullerton of Keough after the latter's death, "understanding and giving quick sympathy, he was tolerant of everything save sham and hypocrisy."

For sheer talent, however, Keough and Fullerton had to bow to Charley Dryden of the *Tribune*. As droll as Keough was sentimental, Dryden contributed some of the most enduring phrases in American sport. Watching the futile efforts of Washington Senators he opined that "Washington was first in war, first in peace, and last in the American League," and seventy years of hapless Senators teams kept the phrase fresh. Of White Sox star Ed Walsh he noted that "he is the only man who can strut while sitting down." Scarce are the vainglorious twentieth-century figures who have not had the same descriptive applied to them, by writers who never knew they were plagiarizing Charley Dryden.

Dryden had a novelist's eye for detail and irony, and Rice admired him for his felicity with the language as much as he did Keough for his sentiment. He would frequently cite the work of both men in his own column.

With two major league teams—two of the best major league teams—and nine newspapers, Chicago could handle all the baseball-writing talent that came its way. (Later that same year of 1907 the *Inter-Ocean* would hire a young writer from Michigan by the name of Ring Lardner.) As a sportswriting town it must have been all but irresistible, but there is no evidence to suggest that Rice looked there or anywhere else when it came time to leave the *News*. Nor is there any evidence to suggest that he and Kate were unhappy in Cleveland or that he was professionally restless. He was simply faced with an offer he couldn't refuse—seventy dollars a week and the chance to return home.

Nashville in 1907 was a city of some 85,000 people. It had been passed in size by both Atlanta and Memphis over the preceding fifteen years, though it clung to its identity as "the Athens of the South," and Nashvillians of the time claimed that Nashville, Boston and Richmond were the American cities where the English language was spoken with the purest accent. While it is all but impossible to imagine how anyone found symbiosis between the Boston- and Nashville-accented dialects of the mother tongue, it is nevertheless a revealing suggestion of the Zeitgeist of Grantland Rice's hometown in 1907.

For all of its cosmopolitan airs and flavor, however, Nashville in 1907 was still very much a close-knit small town. The organizers of the Tennessee State Fair that September set out with a goal to invite anyone who had ever lived in Tennessee back for a grand reunion at the fair. They sent out 26,000 invitations, and some 10,000 erstwhile Tennesseans returned for the party. There is no question that this "small townness" and the womb of family and the familiar were a large part of what lured Rice back to Nashville:

"Once again among your people"—is there any phrase as sweet,
As you wander down the stretches of a distant, alien beat—
Where the lights of Home are shining and the sun is dripping gold,
On the pathways of your childhood and the lanes you knew of old?

In an era when generations of one family commonly remained together under one roof, the Rices quite naturally moved into his parents' home on West Broad Street. On moving day, Rice came in for a berating from Kate as he carried the baby up the porch steps. "It seems I was carrying Floncy under one arm like a football," said Rice. "She wasn't much larger."

In early May, Rice began work on the *Nashville Tennessean,* the new newspaper that had brought him home—the paper he would help to make the most successful in the city over the next four years. His presence in the paper was enormous. He wrote and edited the entire sports section, a section dominated by his signed column. He also contributed a signed column of verse and comment on the editorial page; his were, in fact, the only by-lines to appear in the *Tennessean* during those first months and years.

Herman Suter, the old Sewanee football coach who had lured Rice away from Nashville to Washington and his appendicitis-abbreviated stay on *Forester* magazine, was the editor of the *Tennessean* and the man who decided that Rice's unique talent and value to a new paper were worth the very un-Nashville salary of seventy dollars a week. The money was Luke Lea's, the publisher of the paper, a Nashville businessman with heavy interests in real estate, the Nashville phone company, and Democratic party politics. The *Tennessean* was well capitalized; Lea and Suter assembled a large and seasoned editorial, production and sales staff, and from the first the *Tennessean* was a successful newspaper. It quickly rivaled both the *Banner* and the *American* in advertising as well as news coverage; and within three years it had forced the *American*—its morning competition—into a merger.

It published its first issue on Sunday morning, May 12, 1907, with front-page stories on the birth of a male heir to the Spanish throne and speculation on whether Theodore Roosevelt would seek a third term as President. Rice's position as the paper's most prominent voice was evident from the signed, thirty-line and oh-so-sweet poem in the middle of page one, entitled "God's Country":

Above that land He painted first of all
His sky a deeper and a richer blue—

He gave each bird a sweeter, clearer call,
    A rarer note than Eden ever knew,
Each blade of grass He painted deeper green—
    And to each flower greater fragrance fell—
He poured more gold into the sunlight's sheen
    And when He finished murmured, "It is well."
And so "God's Country" knew its birth and He
    Thus marked it on his world map—"TENNESSEE."

Rice's arrival back in Nashville was a celebrated affair. He had become something of a noted figure throughout the South for his work on the *Atlanta Journal,* and a number of Southern papers took note of both his leaving to go to Cleveland and his return to the South a year and a half later. Rice's poetic inclinations sometimes elicited tributes in kind, as in this one from the *Bowen Blade:*

No longer North or South we roam,
But now complacently sit at home.
From a Nashville paper we take our "dope,"
For the *Tennessean*'s fulfilled our hope;
And its sporting page isn't bad a bit
For Grantland Rice is editing it.

If Rice was at all disappointed at having to watch Southern League baseball again after a year in the majors it was not reflected in his copy. It was just as spirited, just as alive, just as distinctive, just as much fun to read as the stuff Nashvillians remembered from the old *Daily News.* But so too was it more disciplined, more polished, and possessed of a very definite point of view—the fan's point of view. Rice was fun to read because he had mastered that elusive quality of universality in his prose. Readers who possessed not a whit of Rice's insight or eloquence identified with his stuff, read it and said, "Yeah, that's the way I feel. Wish I had said that. This guy really knows the game." Here is an example from June of that first summer on the *Tennessean,* a pleasant, simple, whimsical piece he entitled "The Fan's Revery":

When I have finished up my shift upon this world below
When they have borne me to my rest with music sad and slow—
When I have reached the Pearly Gates—have climbed the Golden Stair—
I wonder if I'll get a chance to see a ball game there?

I don't mean any Angel game where everybody's good—
I wanter see the two teams meet and fight it out for blood—
And when the umpire hands us one that puts us on the blink—
I wanter rise up in the stands and tell him what I think.

I wanter take my same old place on Rooter's Row up there—
And help the other fellers put the pitcher in the air—
To holler "Slide, you lobster, slide," and at each dumb play, make
Some few remarks that I'm afraid might not exactly take. . . .

I'd like to go to heaven and yet it brings me tears
To think I'll never see a game through many million years—
The six months rest from fall to spring seems awful long to me—
So tell me then how I could last through an eternity?

When I have been "called out" below—my time "at bat" is done—
When I "have slid across" the Styx—"have scored my final run"
If I but knew beyond the skies I still could be a "fan"—
Sometimes I have a hunch that I would be a better man.

With the Nashville baseball team tumbling ever lower in the Southern League standings (they would eventually finish last), Rice's column by the middle of August began featuring teasing little references to the upcoming Vanderbilt football season. From the evidence in his work, of all his assignments in Nashville, Rice derived sheer pleasure from his alma mater's football team. There were first the pleasant afternoons on the Vanderbilt campus, a more comfortable working environment than Sulphur Springs Bottom on a 100-degree August day. Then there was the fact that Vanderbilt played fine football, the best in the South. It was unbeaten by other Southern schools for a seven-year stretch between the Sewanee game of 1902 and the Sewanee game of 1909. During that stretch Vanderbilt acquitted itself well against Fielding Yost's Michigan team, Jim Thorpe's Carlisle team, other powerhouses such as Ohio State and Navy, and even held mighty Yale, the unchallenged colossus of college football, to a scoreless tie in 1910.

Last, football at Vanderbilt was fun for Rice because of his friendship with Vanderbilt coach Dan McGugin. A Michigan man, and a lawyer, McGugin was twenty-eight years old in 1907 and in his fourth season as the Vanderbilt coach. He was still commuting to Nashville from his law practice in Michigan; he stayed in Nashville only during the football months. But at the end of the 1907 season he would make a commitment to Vanderbilt and Nashville; he would move to the city, accept a position on the Vanderbilt faculty, and remain as the Vanderbilt football coach for another twenty-six seasons, stepping aside after the 1934 season with a thirty-year record of 197–55–19.

The first time McGugin and Rice met they stayed up till dawn talking football. They would travel to out-of-town and prep school games together, and when Rice would cover a football game other than Vanderbilt's he would file a scouting report to McGugin in addition to his story for the *Tennessean*. McGugin would freely share with Rice his hopes and doubts on particular games or players, and Rice would then share them with the *Tennessean* readers. McGugin would remain a close friend and one of Rice's contacts with Nashville after Rice left for New York in 1910.

As close to McGugin as he was, and as partisan towards Vanderbilt football as he was, Rice was more than a little bit miffed at the apathy he perceived among the Vanderbilt students during the opening game of the 1907 season on October 5th. The next Monday he apologized to his readers for becoming unpleasant in his column—"It is much nicer to say pleasant things of people than to come forward and let the anvil chorus loose, but when the time arrives to speak out plain in meeting and lay bare the cold bleak facts as they are, we haven't yet, and don't intend to be, very far from the job." And in this column this day he was very plain:

> We feel perfectly satisfied that we are sticking closely to the path when we say that the bulk of the Vanderbilt student body have, or at least are showing, less college spirit than any other institution in the South. . . .
>
> By college spirit we do not mean the wild and rowdy outbreak of a bunch of Indians, but loyal support to the team, attendance at the games and concerted cheering and enthusiasm.
>
> . . . Three hundred Vanderbilt students or more sat in the eastern bleachers Saturday afternoon and let Captain Blake and his squad come upon the field . . . WITHOUT A CHEER OR A RIPPLE OF APPLAUSE! Just at the time the team needed encouragement most—needed most to know that a loyal student body was back of them—the bulk of said student body EITHER QUIT COLD or else was TOO INERT AND INEXCUSABLY INDIFFERENT to care. The main support offered the team all the afternoon was by the alumni present and sympathizers in the city WHO NEVER WENT TO VANDERBILT!
>
> If you can beat this for a down right shame anywhere on the collegiate map, we would like to know the location of such a point.

The tirade had a dramatic effect. Two mornings later a mass meeting was held in the Vanderbilt chapel for the purpose of learning and coordinating new yells, and 300 students attended. That afternoon 450 Vanderbilt students lined the field for the team's last practice before they entrained for Annapolis and a Saturday date with the Midshipmen.

Rice went along on the trip north to Annapolis, and was something of a traveling secretary as well as reporter; he arranged for practice fields in both Atlanta and Washington. Loosed from the chains of the desk at the *Tennessean* for the first time since he arrived, Rice enjoyed himself immensely, and his in-transit dispatches back to the *Tennessean* reflect the zest and adventure of a two-day train ride with a corps of spirited and emotional young men. He reported on the skull sessions that McGugin conducted aboard the train, and of the impromptu workouts the team went through when the train stopped for water and coal at various wayside towns in Georgia and the Carolinas. "At every chance McGugin would lead his people out into the open where they were viewed with mingled wonder and astonishment by the peasantry. . . . It's a safe bet that the Vanderbilt war cry has penetrated places on this trip which heretofore have known no greater vocal cataclysm than the penetrating call of the kind-faced cow or the sombre chant of the Sunday morning choir."

Rice rekindled old friendships in Atlanta, but resisted the temptation to consecrate the reunions with a toast. "The call of the demon rum failed to entice us from the straight and narrow path which must be trod by all football teams and those who accompany them. Tough luck, too. . . . "

In Annapolis, Vanderbilt stunned Navy and came away with a 6–6 tie, an altogether improbable stalemate; Navy had been favored by as much as thirty points. To that point, no Southern team had ever even played a Northern school with Navy's renown. It was widely held that no Southern school could. Rice wrote 3000 words for the Sunday paper; the piece began on page one, and was datelined "ON THE NAVY FIELD, ANNAPOLIS, MD."

"Aye, there was a game for you my countrymen, worth traveling the continent to see." At the large rally on campus to welcome the team home on Tuesday, among those coming in for thank-yous and huzzahs from the likes of Tennessee governor Malcolm Patterson and Vanderbilt chancellor James Kirkland was one Grantland Rice of the *Tennessean,* "whose brilliant pen and graphic accounts of the game have done so much to make the Vanderbilt team as famous as it is."

Today, such incest between a newspaper reporter and the story he was covering would be a grand embarrassment to a newspaper, but in 1907 it was altogether commonplace and a source of pride for the newspaper which could boast of a man so well-connected that he was literally on the inside. Rice didn't make anything of it himself; he remained the unassuming witness in his column and never mentioned his own involvement. But the *Tennessean* was not so reticent. Whenever they had occasion to mention the important place their scrivener held in the affairs of his alma mater, they would do so without hesitation or embarrassment—and Rice was on the dais at an awful lot of Vanderbilt athletic functions during his years on the *Tennessean.*

During this fall he also began his habit of regularly filling his Saturday morning column with predictions for that day's college football games. It would be a signature piece throughout his career. He had done nothing to diminish his reputation as a sage since arriving in Nashville. In early September, with ten games left in the Southern League season and Memphis leading Atlanta by three games, Rice said that the dope and the schedule favored Atlanta and that they would win. They did.

Now, on October 19, he opened his column by predicting that Sewanee would beat Auburn 18–0; Georgia Tech would beat Tennessee 12–

0; and Harvard would beat Navy 12–0. "Don't bet over a million dollars on these tips," he cautioned; but, while the final score in each game was closer than he suggested, he did have all three winners right. It was an auspicious beginning to a forty-five year column habit.

On December 1st, after fashioning another 3000 words on the Vanderbilt-Sewanee game, Rice left Nashville for two weeks of quail hunting in the Georgia mountains, "where the shooting was good, if the killing was bad." It was a respite well-earned. Since he had started on the *Tennessean* in May he had worked seven days a week, writing his column for both the sports page and the editorial page, as well as editing the sports page, which on Sundays included a four-page "pink" section, consisting of "Sport Lore, Dope & Comment Up-to-Date." He often doubled as theater critic, filing reviews of the never-ending string of troupes that would play Nashville in one-night stands. He lived a killer schedule, working at least twelve and often as many as sixteen hours a day. In the *Tennessean* news room on the northwest corner of Church and Spruce streets, Rice worked at a small typewriter desk at a window overlooking Spruce Street and the Tulane Hotel. Beneath his window, streetcars would rattle westward on Church, others would turn off Church to begin their northward rumble up Spruce. The clatter of horses' hoofs pulling steel-wheeled wagons over the granite-block paving added to the cacophony, an urban symphony mixing with the shouts and typewriters of the newsroom.

Rice began his working day at eight in the morning, dispatching first his editorial-page column of verse and comment called "Tennessee 'Uns." Before and after an early lunch he'd write some sports. By mid-afternoon (during the baseball season) he was out at Sulphur Dell for the ball game that generally began at four or four-fifteen. Baseball was a faster game than we know it today and nine innings would generally take but an hour and a half, seldom more than two. The game finished, he'd rush back to the *Tennessean*, write his game story, assemble the scores and standings from the major leagues and the other Southern League games, and put the sports page to bed. If there were no show to review, his day was done, twelve hours after it had begun. On show nights he'd grab a quick dinner at the Tulane and be off again, returning to his desk sometime after ten, and, with the clamor outside his window having receded somewhat, finish the last of his day's toils before catching a trolley back to his West Nashville home around midnight.

A newspaper is a demanding mistress, consuming the passion, the energy, the spirit of her minions. She is a voracious and dispassionate harridan, caring only that her needs are satisfied, and demanding satisfaction anew each morning—no time for anything until the mistress is serviced, and as soon as the task is completed it is time to begin it again; tomorrow is just a blur of whirling twenty-four-hour cycles stretching out to infinity. Time away from the task was rare, and something to be cherished for Rice, but insofar as he could he found time for himself and his family and, during those years in Nashville, lived a semblance of a normal, workaday life. He kept a small vegetable garden in the yard behind his Broad Street flat, and on mornings when the Muse had afforded him a jump on his "Tennessee 'Uns" column, or when he just couldn't resist, he lingered at home to play with his young daughter.

There are precious few glimpses into Rice's private life in his published work; he would draw on his own life and family and experiences far less than other newspaper columnists of his reach and output. Often this autobiographical element is simply a matter of survival for columnists; writers often need every little morsel of their life simply to fill the space. But Rice was able to keep his private life private, and what glimpses he did afford were generally in the area of emotions rather than anecdotes, though one charming little forty-line verse on one morning spent with Floncy offers the reader a little bit of both:

> How can I think of songs to sing
> > With you, wee Elf, a-cooing there?
> How can I scribble anything
> > While you are tugging at my hair?
> You'd better let your Dad alone—
> > Some day you'll want a dress or bonnet,
> Which you can never get, my own,
> > By chewing up his salaried sonnet.
>
> I cannot sing when you are near—
> > No lilting song can I indite,
> For you are so much sweeter dear,
> > Than any song your Dad can write;
> Turn loose my ear and I will try
> > To grind out something à la mode;
> YOU LITTLE SUCKER—on the sly
> > You've eaten up my Sunday ode.
>
> Now honest, Babe, I'd like to play
> > With you forever, ever more;
> But I must spin a roundelay
> > To keep the old wolf from the door;
> The old wolf's waiting for us there—
> > No use for you to try and "scat" him—
> Just turn a-loose your daddy's hair
> > And let him throw a poem at him.

Rice took up golf during these years, and took it up with the same intensity and seriousness of purpose that marked his career as an athlete at

Vanderbilt. The game would give him much, both personally and professionally, though no more than he would give in return. But that was still in the future. For now, golf was simply a diversion to serve as a counterpoint to the pressures of the office and to satisfy the needs of the erstwhile athlete who "hankered for some participant sport" that might fit in with my strange hours.

Still, golf and the press box weren't enough. Rice remained a sports junkie and during these years it was not always enough to be a witness to the game; he had to be a part of it. Through his friendship with Dan McGugin, and with his considerable reputation for understanding the rules and nuances of football, he became one of the busiest referees in the South. He officiated at Sewanee games, University of Tennessee games, even some Vanderbilt games—including the 1908 game between the Commodores and the University of Michigan in Ann Arbor—usually serving as head linesman or umpire. He was paid fifteen dollars a game, and he spent the money on Kate. Kate also enjoyed the social whirl that Vanderbilt football offered. "I went to the first football game Saturday and enjoyed it so," she said in a letter to her mother in 1908. "It was a bum game but everybody was there. Grantland umpired it and got $15.00, so I will get me a new hat this week; otherwise, I couldn't have it for a while."

Rice did not, however, let his passion for the fray compromise his good sense. When the Vanderbilt varsity played the alumni in 1909, the battle-scarred old end from the '99 team demurred, opting to referee.

In the spring of 1908, Rice coached the Vanderbilt baseball team. He took on the task in early March, and kept the news out of the newspaper. *Tennessean* readers learned only that "arrangements have been made for graduate and alumni coaching as Tom Davis was not able to return." Coaching college baseball in the first decade of the century was not a big deal; the role Rice served was as much overseer and chaperone as it was strategist and tutor. He did not celebrate his role anywhere in his column; not once that season did he mention the name of the Commodore coach. His accounts of the games were straightforward, three- or four-paragraph jobs absent any stylistic flourish; they were perhaps not even his, but the work of a Vanderbilt student. In his column he did now and again make reference to the team. For those readers aware of his role, the remarks in the column could be read as quotes from the coach. For those ignorant of his role as coach, the remarks could be read as simply those of a well-sourced reporter, passing on the "dope":

> From the Commodore point of view, the surprise of the day was the showing of Beasley, a big freshman sent in at the close. You can put it down that this fellow will bear watching. He is a bit raw now, but after the S. W. P. U. series he will be taken in hand and worked for the

main games of the year and it wouldn't be any great surprise to see him nose out one of the early regulars.

The Commodore squad will have to improve considerably to head off Clarksville's crack squad. Only a few men on the team are doing any batting, and the advent of so many new people means practically a green team for the first few games. Alden had been shifted in from center to second, and Costen removed to his old stand in the outfield. This switch balled up play a bit Tuesday, but with a few more work-outs it should bolster up the squad.

The chief weakness of the Commodore squad now in the inability of the slabmen to fire within twenty rods of the plate.

Under Coach Rice, the Commodore baseball team finished the season at 11–8–1.

Not only was Rice close to the athletic department at Vanderbilt, he was a big part of the alumni association as well. In the spring of 1908, that same spring he coached the baseball team, Rice composed a poem in honor of the alumni reunion in June. A poet laureate of sorts for Nashville, and especially for Vanderbilt, he was used to churning out verse on demand for such occasions, most of it forgotten as soon as presented, the program on which it was printed either tossed away on the way home or stashed in the attic with other keepsakes of dubious value. This poem would have no doubt been consigned a similar fate were it not for two lines in the penultimate stanza. For these two lines comprise the phrase that has immortalized Grantland Rice. The phrase that is still being chiseled into the concrete of high school fields and gymnasiums across the land. The phrase that has been offered as solace by every parent of a little-leaguer disconsolate over a loss. The phrase that has been bastardized, parodied, twisted and ridiculed as we have evolved into a society that too often perceives triumph as the only negotiable currency. But a phrase that has nevertheless endured, even thrived as the embodiment of that which is noble in sport. Carl Yastrzemski, at his induction into baseball's Hall of Fame in 1989, cited the line, and its author, and said that of all he had achieved, he was most proud of the fact that he could look back and say that his career was a testament to the ideals and values in Grantland Rice's phrase. One hundred years from now, two hundred years from now, for as long as sport in played in America, the phrase will remain, for however tattered and frayed it might become, there is a gentleness, a lesson, a relevance in the words that is timeless.

The phrase, of course, is: it is "not that you won or lost—but how you played the Game." The poem was titled "Alumnus Football," and a few days after the alumni reunion, Rice reprinted it, together with the note: "Manufactured for the Vanderbilt Alumni gathering, 1908, where it first happened." Here is the full poem, as it appeared in the *Tennessean* of June 16, 1908:

Bill Jones had been the shining star upon his college team;
His tackling was ferocious and his bucking was a dream.
When husky William tucked the ball beneath his brawny arm,
They had a special man to ring the ambulance alarm.

Bill hit the line and ran the ends like some mad bull amuck;
The other side would shiver when they saw him start to buck;
And when a rival tackler tried to block his dashing pace,
His first thought was a train of cars has waltzed across his face.

Bill had the speed, Bill had the weight—the nerve to never yield;
From goal to goal he whizzed along while fragments strewed the field—
And there had been a standing bet, which no one tried to call,
That he could gain his distance through a ten-foot granite wall.

When he wound up his college course, each student's heart was sore;
They wept to think that husky Bill would hit the line no more.
Not so with William—in his dreams he saw the Field of Fame,
Where he would buck to glory in the swirl of Life's big game.

Sweet are the dreams of campus life—the world that lies beyond
Gleams ever to our inmost gaze with visions fair and fond;
We see our fondest hopes achieved—and on with striving soul
We buck the line and run the ends until we've reached our goal.

So, with his sheepskin tucked beneath his brawny arm one day,
Bill put on steam and dashed into the thickest of the fray;
With eyes ablaze he sprinted where the laureled highway led—
When Bill woke up his scalp hung loose and knots adorned his head.

He tried to run the Ends of Life, when lo' with vicious toss,
A bill collector tackled him and threw him for a loss;
And when he switched his course again and crashed into the line,
The massive guard named Failure did a two-step on his spine.

Bill tried to punt out of the rut, but 'ere he turned the trick.
Right tackle Competition fumbled through and blocked the kick;
And when he tackled at Success in one long vicious bound,
The fullback Disappointment steered his features in the ground.

But one day, when across the Field of Fame the Goal seemed dim,
The wise old coach Experience come up and spoke to him;
"Old boy," said he, "the main point now before you win your bout.
Is keep on bucking Failure till you've worn that lobster out!

"Cut out this work around the ends—go in there low and hard-
Just put your eyes upon the goal and start there yard by yard;
And more than all—when you are thrown, or tumbled with a crack—
Don't lie there whining—hustle up—and keep on coming back—

"Keep coming back for all they've got and take it with a grin,
When Disappointment trips you up or Failure barks your shin;
Keep coming back—and if at last you lose the game of Right—
Let those who whipped you know at least they, too, have had a fight—

"Keep coming back—and though the world may romp across your spine—
Let every game's end find you still upon the battling line;
For when the one Great Scorer comes to write against your name,
He marks—not that you won or lost—but how you played the Game."

★　★　★

Such is Alumnus Football on the white-chalked field of Life—
You find the bread-line hard to buck, while sorrow crowns the strife.
But in the fight for name and fame among the world-wide clan,
"There goes the victor," sinks to naught before "There goes a man."

That Rice himself did not appreciate the power his words would have is suggested from its subordinate placement in the piece. He had been playing with variations on the thought for some time before "Alumnus Football." One year before he had written a sports page verse called "Out on the Line" that read

It isn't so much—"Did you make a hit"—but "How did you swing at the
　　ball?"
Did you go up to the bat with your nerve all gone and never half try at all?

It isn't so much—"Did you win the game"—but "How did you play old
　　scout?"
Did you give 'em a fight to the bitter end and scrap till the last was out?

And just twelve days before the appearance of "Alumnus Football," Rice published a poem titled "Two Creeds," which expressed virtually the same thought without the football or baseball metaphor:

> But to your Soul—which pays the cost—
> It matters not who Won or Lost;
> It only matters how you fought
> Or strove or struggled—worked or wrought;

"Alumnus Football" lay dormant for a couple of years. But after he arrived in New York in 1911, Rice returned to it and did some fiddling. He made a lot of small changes with the language, making it a lot brighter in some places, making it merely different in others. He added two stanzas, one of them between stanzas eight and nine of the original, introducing "Envy" and "Greed" as two more defenders in Bill's path. The second addition came just before the how-you-played stanza—where old coach Experience continues his advice: "You'll find the road is long and rough,

with soft spots far apart,/ Where only those can make the grade who have
the Uphill Heart." But the only notable change in the 1919 version comes
at the end; Rice discarded the last stanza and let "how you played the
Game" stand as the close.

The reworked version appeared in Rice's column in the fall of that
year, and once he was syndicated in 1915, nearly every year thereafter;
and "how you played the Game" passed quickly and forevermore into the
language.

By the end of the first year at the *Tennessean,* Rice's life had settled
into a cycle that would persist for the duration of his stay there. In the
spring and summer he'd cover Southern League baseball; in the fall his life
would center on college football—especially Vanderbilt football. After the
Vanderbilt-Sewanee game he would depart for the Georgia mountains and
vacation. On his return he'd plunge into the hot-stove-league dope, which
would sustain him and his readers until the start of spring training in March.
The rest of the panorama of American sport—major league baseball, prize-
fighting, track and field, crew, the London Olympics of 1908—he would
cover from a distance, through correspondence and exchanges with out-
of-town newspapers.

In the spring of 1908 he played a part in giving the Nashville baseball
team a new nickname. In February, Rice sat down with Will Ewing of the
American and Richard Yancey of the *Banner* to choose a name for the
Nashville baseball team, which throughout its seven-year history had never
had a nickname and had been indiscriminately called dozens of different
things by players, fans and sportswriters, including the Fishermen, the
Finnites, the Dobbers (these last two after Nashville managers) and the
Boosters. The three sportswriters came up with a slate of three names—
"Volunteers", "Lime Rocks," and "Rocks"—and asked the fans to vote.
Volunteers was the runaway winner, and according to Rice would stand
as "the Nashville sobriquet for all time to come." He could not, in 1908,
have anticipated Nashville's later identity with country music, a symbiosis
that would prompt the team to change its name from the Volunteers to
the "Sounds" in the 1970s.

But as much a part of the fabric of his community this would become, it was, like the naming of Sulphur Dell, a trivial *journalistic* matter;
of far more consequence was his reporting on improprieties in intercollegiate athletics, scandals that are strikingly evocative—are indeed the
antecedents—of the seaminess so pervasive in big-time college sports
today.

In the fall of 1907, Rice's contacts in Atlanta fed him the information
that the University of Georgia had paid three former varsity athletes from
Georgetown and one from Syracuse to play for Georgia in their game
against Georgia Tech. According to Rice, each man had been paid $150,

plus expenses, for their Saturday afternoon toils for Georgia, and confessed as much on their train ride back north. In reporting the story, Rice also unloaded a cannon shot on Dr. Edward Holmes of Mercer College, the S.I.A.A. overseer for Georgia. "Nor can Prof. Holmes . . . be depended upon to set things right. Dr. Holmes is honest and well meaning, but most evidently incapable for his post." Holmes responded to Rice's charges with an indignant letter to the *Atlanta Georgian,* defending his actions (and lack thereof) and charging that some of Rice's charges went "wide of the mark." Holmes's subsequent actions would seem to bear out Rice's charge that he was "incapable for his post," however, for, like Pontius Pilate, he washed his hands of the matter and dumped the whole mess in the lap of S.I.A.A. president William Dudley. For the record, Dudley censured Georgia, and Georgia fired its football coach.

One year later, Rice reported that the Louisiana State University football team, all of a sudden undefeated that autumn after years as a southern also-ran, was also engaged in the scandal of choice for that era—playing ineligible and non-matriculated players. This information had come to Rice from a man named Chet Clarke, coach of the Rose Polytech team, as he spoke with Rice in the *Tennessean* office after Rose Poly had lost to Vanderbilt. Clarke claimed that he had been approached by L.S.U. and offered a monthly stipend if he would come to Baton Rouge and play football for the Tigers, and that he had played against two of the current L.S.U. players when they had previously played for Wabash College and Butler College. Clarke then wrote the L.S.U. president outlining his charges; a copy of this letter also appeared in Rice's column in the *Tennessean.* Within a month Tulane had leveled similar charges against L.S.U.; the president of L.S.U. denied the charges on behalf of the university, and every paper in the South was running the story, with Rice and the *Tennessean* consistently out in front on it. The commotion led ultimately to a formal investigation of L.S.U. by the S.I.A.A., an investigation that found that "there have been during the past two years grave irregularities in athletics at the L.S.U." The S.I.A.A. lacked any disciplinary power in those days, it could merely make the recommendation that the coach at whose feet they laid the blame for "most, if not all" of these irregularities "should therefore be disqualified for serving as coach or as athletic director in any institution belonging to the S.I.A.A."

Rice's reporting on these stories was reported in other Southern newspapers, with Rice duly cited and credited, solidifying his position of first citizen of southern sportswriting. In his hometown, meanwhile, he was becoming an outright celebrity. By the fall of 1910, a Nashville haberdasher was using Rice's name to sell his wares. "Grantland Rice Says," read the copy, with these opening words set in forty-eight-point boldface type, "Sewanee had the advantage in kicking. He should tell them to buy their clothes from Joe Morse & Co. Then they won't have cause to kick.

Ours are always correct. Just notice the good dressers at the Thanksgiving Day game. They are wearing clothes with this label."

Also swelling Rice's Nashville renown in 1910 was his first book. *Base-Ball Ballads,* a small paperback containing a collection of fifty-seven pieces of verse from Rice's *Tennessean* and *Cleveland News* work, was published by the *Tennessean* in time for the start of the baseball season. The collection contained the "Casey" poems and other oft-repeated favorites from his column. It met with much enthusiasm in Nashville, and later, after Rice had attained his great fame, the tiny volume would become something of a collector's item. But it attracted little attention beyond Nashville when published in 1910. Rice's reputation as a leading purveyor of popular American verse was still nearly a decade away. This despite the fact that by this point in his career he had already written some of his most enduring and most popular verse.

Most of the verse appeared first, not in the sports pages, but rather on the *Tennessean* editorial page, in the daily potpourri that Rice offered up under the rubric "Tennessee 'Uns." The column virtually always contained a verse—or two—and generally offered a paragraph or two of prose— sometimes nonsensical fiction, and sometimes silly, sometimes trenchant comment on politics, fashion, sports, the weather, human foibles or anything else that passed in review before Rice's nimble and fecund mind.

The column is as much of an anachronism today—as much a period piece—as bustles, petticoats and high button shoes. But in Edwardian America, prized was the writer who could generate a daily verse on matters either topical or sentimental. For one thing there was a civility, a gentility about these columns that provided a nice counterpoint to the boisterousness of the front page and the shrillness of the editorial page.

Frank Stanton's column in the *Atlanta Constitution* was the prototype for "Tennessee 'Uns;" at least that was the column with which Rice was most familiar. But the same sort of thing was being done in a lot of newspapers, and Rice, a fan and student of the form as well as one its foremost practitioners, was familiar with them all—Franklin P. Adams's (F. P. A.) "Always in Good Humor" in the *New York Mail;* Bert Leston Thayer's (B. L. T.) "Line 'o Type or Two" in the *Chicago Tribune;* old friend Don Marquis in the *New York Sun;* Edgar A. Guest in the *Detroit News;* Judd Mortimer Lewis in the *Houston Post;* John Wells in the *Buffalo News;* Eugene Field of the *Chicago Daily News*—forgotten men, all of them, today. The spirit of their sort of column might live today in the works of an Andy Rooney or an Art Buchwald, but there is no similarity of form or style. It might be said that if Buchwald and Rooney had lived at the turn of the century they would have likely been versifiers and paragraphers; if Thayer, Adams, Marquis et al. were alive today, they would probably be writing in the Buchwald-Rooney vein.

The form thrived before the Great War but dried up during the twenties, and by the start of the Depression it was all but extinct. Rice himself, lamenting its passing, provided perhaps the keenest insight into why it passed. "This is a much more serious age than the old days ever were," he said. "There was a lightheartedness that the world knew before the first World War that has never been worn since. . . . There has been war or the shadow of war for the past forty years and the dark shadow hasn't ever been absent from the scene in that time. Most of the true singers have had little heart with which to sing."

Be it modesty, or lack of time, or both, Rice never pointed to one aspect of his career and said: "This. This is my best work and this is the work of which I'm most proud." But had he done so it would have surely been his verse to which he pointed. Of the millions of words of prose he wrote during his half-century in the breach, he saved not a word of it—threw it all away. But he lovingly saved every line of verse he ever wrote, whether profound or banal, enduring or ephemeral, and pasted them in scrapbook after scrapbook until his collection grew to be ten scrapbooks large, each some two inches thick, each page with clippings covering every square inch of the page and hanging over the sides.

Of the six to seven thousand pieces of verse that Rice wrote and published during his career, he wrote some fifteen hundred of them—or very nearly one-quarter of the total—during the fewer than four years that he worked for the *Tennessean*. Years later, erstwhile colleagues were still somewhat aghast at his felicity with rhyme. "Grantland could write a poem on any subject almost as quick as you could write it down," said one. Damon Runyon once said that Rice was the only man he knew who could write verse as quickly and freely as he could write straight prose. During his *Tennessean* years Rice would produce a minimum of twenty-four lines of verse a day. Forty lines was more the norm and frequently there were days when he would write sixty to eighty lines; it tumbled from his typewriter like rain from the springtime sky.

"How or why I ever fell into the habit of breaking up my columns with verse I don't know," he said. "But rhythm and rhyme seem to come naturally, perhaps as a reflection of the meter I had enjoyed scanning in Latin poets."

In an interview in 1948, Rice allowed that he found writing verse a little bit tougher than Runyon and his anonymous colleague at the *Tennessean* suspected, but not much. "Sometimes I can write one in twenty minutes and sometimes it takes three or four days," he said. "You know how it is. It all depends on how I get started, and sometimes I can't get started off right for a day or two. Sometimes not for three or four days. Once I'm started, it's easy."

Rice as poet is not held in great esteem. It comes in part, no doubt, from Rice's having been unabashedly a sportswriter, and sportswriting, as everybody knew, was the ghetto of literature. How seriously are we supposed to take the scribbling of a sportswriter, we ask; especially when so much of his verse was so palpably bad—sentimental and saccharin and all? "[I]t rarely rose more than a razor thin cut above the rhyming roadside signs of Burma Shave," said one modern day critic of Rice's verse. Rice himself never made any highfalutin claims for his verse—beginning with the fact that he insisted on calling it verse, never poetry. It wasn't humility—Rice was proud of what he did. He was simply aware that no matter how loudly some of his fans may have sung his praises, what he was doing in his daily newspaper column bore but faint resemblance to the work of Homer and Virgil and Shakespeare and Milton, or Byron or Shelly or Keats, or Kipling or Longfellow, or any of the other poets he read so regularly and admired so passionately. He was a poet of the people—a "near poet" he called himself—along the lines of James Whitcomb Riley, the popular Indiana poet of the day and the man to whom Rice's fans most often compared him. And all of this was fine with Rice:

> The average life is all I care to lead—
> The average salary for the average need;
>
> The average dream—the average clothes to wear;
> The average joy nor more than average care;
>
> The average struggle, neither more nor less.
> To reach the goal of average success;
>
> Yet, with it all, to strive along the way
> To raise the average, bit by bit, each day.

A "Tennessee 'Uns" column from September 1907, called "The Story of Two Poems"—neither of them Rice's—is most revealing of Rice's attitude toward verse. The column begins by noting that Ambrose Bierce, writer, critic and literary titan of the era, had called a recent poem by one George Sterling one of the greatest poems ever published. "Of his work I have the temerity to think that both in subject and art it nicks the rock as high as anything of the generation of Tennyson and, a good deal higher than anything of the generation of Kipling," said Bierce of Sterling's poem, titled "A Wine of Wizardry."

Rice quoted a bit of the Sterling poem:

> Dull fires of dusky jewels that have bound
> The brows of naked Ashtaroth around;
> Or hushed at fall of some disastrous night,
>   When sunset, like a crimson throat to hell,
> Is cavernous, she marks the seaward flight,

Of homing pigeons dark upon the west. . . .

Halls—
In which dead Merlin's age hath split
A vital squat whose scarlet venom crawls
To ciphers bright and terrible. . . .

Ere the tomb-throned echoing have ceased,
The blue-eyed vampire, sated at her feast,
     Smiles bloodily against a leprous moon.

By comparison, Rice then told his readers that six years before, James Whitcomb Riley had been asked to submit the name of the best poem he had ever read. Riley's favorite was something called "Brave Love." Rice could not remember the author's name.

He'd nothing but his violin,
     I'd nothing but my song;
Yet we were wed when skies were blue,
     And summer days were long. . . .
We sometimes supped on dewberries
     Or slept among the hay;
But oft the farmers' wives at eve
     Came out to hear us play
The rare old tunes, the dear old tunes;
     We could not starve for long.
While my man had his violin
     And I my sweet love song. . . .

"If it is possible for any two things to be absolutely and hopelessly opposite, the above two contributions must be listed in this class," said Rice by way of analysis.

"One is a monument of complexity and grandeur, built from the mystic. The other is a monument of simplicity and lyrical beauty, built from the heart.

"As to which of the two should rank as the greatest poem, there is no doubt at least as to how the great mass of the public would judge— and Riley's choice would not come out second best."

Given these feelings on poetry, Rice quite naturally set out in his own writing to produce verse that would speak to "the great mass of the public" and didn't worry about whether it was acclaimed high art:

When you ponder with acidity
Upon the keen avidity
Which critics, in stupidity,

Have shown in roasting you—
Never let the burning ember
Turn your June into December—
Just remember, friend, remember
    That they roasted Byron too.

When the critics, railing curses,
Take a wallop at my verses,
No wet tear my eye immerses,
    But I laugh with merry glee
When I think that Burns was roasted,
Shakespeare as an upstart posted,
Keats and Wordsworth badly toasted,
    And I mutter, "Why not me?"

Let the critics slip their tether,
Light on us like sloppy weather—
All us poets, linked together,
    Simply scribble all the more—
If Bill Shakespeare's stuff was hooted,
Wordsworth as a bard refuted,
If old Burns and Keats were booted,
    Tell me—why should I get sore?

Not that Rice was deluged with criticism. Most of the criticism came posthumously, and most of that from people unfamiliar with anything beyond "how you played the Game" and maybe the derivative Casey poems and a verse or two of sing-songy patter describing a ball game or paying obsequious homage to some no-account jock. In his time, however, and particularly in the years surrounding the First World War, Rice was one of the most popular of American poets—not only one of the most widely read, but one of the most widely praised as well. Irvin S. Cobb said that Rice was the successor to James Whitcomb Riley as "our most typical writer of homely, gentle American verse." When Rice's second book of verse, *Songs of the Stalwart,* was published in 1917 it was widely praised. the *Boston Evening Transcript,* a genteel newspaper that had built a national reputation for its literary criticism, fairly gushed:

> One would almost say that his art, simple, glowing and precise as it is, was altogether too fine for some of his themes and subjects; but this is not so because nothing is too common for the spirit of poetry to light and reveal. This book is literature, an honor to the man who made it, and a delight to the reader who receives it. And Mr. Rice stands quite alone in his achievement.

Another anonymous critic of the World War I era called Rice's verse "truly American."

[O]ne side of Rice's work . . . expresses our sentimental side, the side that looks on to something, a something that is a returning to the things we have known, seen in a finer, clearer light than when we first knew them. . . .

The sporting side of the American, Mr. Rice writes about better than anyone else—there is no question about that. For his baseball poems are good baseball stuff, and at the same time they give the national game a broader aspect for they hook it up with life. This, after all, is what makes a poem or any other piece of writing worth while. . . . All the varying baseball moods have been caught and portrayed by a method that is as faithful as the camera, and back of them is the spirit of the artist that relates them to life's broader things.

Most of the broader things and themes in Rice's verse were melancholy; he wrote of defeat, of death, of loves and innocence lost, of dreams unfulfilled, and of wanderers and expatriates pining for home. Yet through it all there was always a glimmer of hope, an ever-so-slender sliver of light brightening the dull pallor. It is difficult to select a single poem to illustrate this canon of sad-voiced verse and call it the best, or the best remembered, or the most enduring, or the most illustrative. But the following example from March 1909, titled "The Return," is fairly typical:

> I felt that I was shadowed through the day
> > By some gray ghost of memory that came
> Upon the winds from Life's forgotten way
> > Until, as in a dream, one spoke my name,
> And called to me to wait for him, and so,
> > In turning back to solve the mystery,
> Borne back to me from Long and Long Ago
> > I saw the Man I Dreamed that I would Be.
>
> And he was all that I was not, and still
> > I knew him—every dream-line of his face—
> The crest he held upon Fame's laureled hill—
> > The crown he wore from triumph in the race—
> And through the hurt—the heartache and regret
> > At change so vast, I turned with spirit sore—
> "Why have you come to one who would forget?
> > To one who would remember you no more?"
>
> And then he spoke as one would soothe a child.
> > "Long, long ago, beneath the orchard trees
> A barefoot lad—a dreamer undefiled
> > By Life and Strife or Sin's dark mysteries—
> A lad with Soul of Honor—stalwart heart,
> > Who with the Sword of Right would cleave to Fame,
> Dreaming of Fields upon the battle-chart,
> > Sent me ahead to wait until he came."

"I left him there, asleep beneath the wide,
    Kind sky that God had placed above his rest,
And so I passed upon the dream-spun tide—
    Soul of his soul and spirit of his breast—
Passed from his Boyland into Manhood grown,
    Just as he made me—dreamed that I would be—
But through the years I've waited here alone—
    Weary for him that would not come to me."

"Weary for him, borne from his course today,
    Far from his dream-road in the World of Men,
And so, before he might forget the way
    I've come to lead him back again."

At least once or twice a week, Rice would produce verse with this tone, verse that would prompt a pause, a moment's reflection, and then a second and perhaps a third reading. But there was quite a bit more in the "Tennessee 'Uns" column, most of it far less somber and serious. There was first the whimsical and sentimental verse:

Wouldn't you be willing not to have a Christmas call
Not to get a penny's worth of anything at all,
To know that every ragged kid you see along the street
Would wake on Christmas morning and find the dream was sweet?

One of the regular prose feature of "Tennessee 'Uns" was something called "Daily Dope from the Styx," equal parts nonsense, and subtle and not-so-subtle comment, written in the form of an eclectic array of dispatches filed by a "stuff correspondent" on the scene in the netherworld:

STYXVILLE—At the annual track meet given here today several records were broken. Sampson put the sixteen-pound shot 297 yards, while Hercules won the hammer throw with two miles and ninety-eight yards to his credit—both eclipsing the old standards. Mercury won easily in the one-hundred-yard dash, setting a mark of three seconds flat. . . .

Shakespeare and Beethoven are now at work on a comic opera farce which is sure to make a big hit. In fact the critics who have read the "book" and heard the "score" are of the opinion that it will rank with anything George M. Cohan ever wrote. This is of course extravagant praise but the following lyric is bound to be whistled up and down the Styx. It is entitled "Although Your Teeth Are False, Sweetheart, My Love Is Always True."
    The chorus runs:

"Although you teeth are false, my dear,
    My love is always true;

> Although your hair comes off each night,
>   I'm bug house over you;
> I'll always love you dearly, dear—
>   Just as a lover should.
> Your face and form may both be bad,
>   Just so your checks are good."

Shakespeare is writing the book and Beethoven the music.

Another regular feature in "Tennessee 'Uns" played variations on a comedy staple—Rice responding to his mail. Sometimes the alleged letters from readers were obviously created by Rice to set-up a punch line. Sometimes they were inquiries ingenuously submitted by readers in search of information. And sometimes it was hard to tell:

Dear Tennessee 'Uns:
  If a base runner going to third should bump into the third baseman, knock the ball out of his hands and punch him in the eye, would he be safe?

  FANATIC

  A: Not very safe, unless he was bigger than the third baseman.

A large portion of Rice's mail concerned his felicity with verse and rhyme. How do you do it?, he was asked. Where does the inspiration come from? How can I get started as a poet? From time to time, Rice would oblige his readers with a little tutorial:

Dear Tennessee 'Uns—
  What is the best method to follow in writing verse.

  FLOSSY

  Pick out all the words you can think of that rhyme and then fill out the rest of the sentence with anything that happens to be handy. For example, select such words as "day—way," "sky—pie" and proceed as follows:

> The man came down the hill one day—
> And then he went upon his way—
> His soul was longing for the sky—
> His stomach for a piece of pie.

It's easy enough when you once get the habit.

"Tennessee 'Uns" was also replete with topical comment—sometimes in verse, sometimes in prose. Rice particularly enjoyed tweaking the obscene wealth of John D. Rockefeller and the bluster of Theodore Roosevelt. On the local political scene—a very acrimonious and tempestuous political scene at the time—he commented by professing boredom with it all and needling the earnestness and lack of humor of those who were obsessed by it. The political battle was joined in Tennessee in those years over the issue of prohibition. The *Tennessean* was aligned clearly on the side of the "drys." Each day its editorial page self-righteously proclaimed above the lead editorial that "The *Tennessean* does not print liquor advertisements," and the editorials would rail against the evils of the "whiskey shop" and about how the liquor itself was corrupting morals and family life, and the liquor industry's money was corrupting Tennessee government, particularly the administration of Governor Malcolm Patterson.

Rice's sympathies in the matter ran directly counter to those of his employer, and it is testament to the editorial latitude he enjoyed that on the very page that the editors of the *Tennessean* were excoriating the evils of liquor, Rice would frequently celebrate its pleasures. After the state did go dry on July 1, 1909, Rice would gleefully take note of how the prohibition was less than absolute ("Nashville may not be a wide-open town but you can squeeze through in several spots if you edge in sideways and hold your breath.") and once even went so far as to provide instruction on how to procure an illegal drink, in a thinly veiled "Tennessee 'Uns" item headed "Sherlocking":

> "Watson," commented the great detective, "They are selling whiskey in this town again, despite the law."
> "From what do you make this deduction?" gasped the astonished doctor.
> "Entering a certain ex-barroom a few hours ago I called for an imported ginger ale and winked at the barkeep. What I got leads me to the conclusion that my deduction given above is absolutely correct."
> "Marvelous—marvelous," whispered the bewildered doctor, who almost threw a fit over his friend's keen and searching intellect.

By the summer of 1910, the introspective verse in "Tennessee 'Uns" gave way more and more to the glib, and the "Tennessee 'Uns" column itself became less of a fixture in the paper; it was cut back from six days a week to four. And in the still-frequent paeans to life in Tennessee there was a new element, the acknowledgment that despite his choice to have returned four years before, perhaps his destiny did yet lay beyond the bluegrass of Middle Tennessee:

> The Hills of Fame still beckon where the Paths of Glory lead,
> Crowned with the Victor's Laurel in the Gleam of Valiant Deed.

Viewed from the perspective of these eighty years, there is evidence of a man exhausted by nearly four years at a frenetic pace that should have been creative suicide, a man ready for a change. That fall, following the Vanderbilt-Sewanee game, he took his customary vacation. But this time, instead of going to Georgia in search of quail, he went to New York in search of a job. His reputation within the newspaper business was such that he could have knocked on any door and found a sympathetic editor willing to listen. But he was drawn to the offices of the *Evening Mail,* downtown at Broadway and Fulton Street. The *Mail* stood in the shadow of City Hall and the Brooklyn Bridge, as did most of New York's other papers, clustered together on nearby Park Row. He had a sponsor of sorts at the *Mail,* Franklin Adams, the man known to his readers—and virtually all of New York read him—as F. P. A., the country's best known practitioner of newspaper versifying and paragraphing. Rice's work had appeared frequently in F. P. A.'s column, "Always in Good Humor," as frequently as Adams's work had appeared in "Tennessee 'Uns." Adams had brought Rice to the attention of *Mail* editor Henry Stoddard, who had written offering him a job. When Rice presented himself in New York, Stoddard told him: "Rice, I've been reading your verse. I never met a sportswriter worth fifty dollars a week, but in your case I'll risk it!"

Fifty dollars a week represented a considerable cut in pay from his Nashville salary. Nevertheless, he accepted, drawn partially by the challenge of the big city, partially by the fact that the job at the *Mail* entailed merely writing a six-day-a-week column—no desk work in the sports department, no theater reviews, no humor column for the editorial page. The salary may have been just 70 percent of what he was making in Nashville, but the job would entail only about one-third the work. And he also no doubt presciently realized that in New York his opportunities would be constrained only by the limits of his own substantial imagination and industry. As much as he loved Nashville, he had to have realized that he had gone as far as he could there, and to stay any longer would cause him to begin slipping back.

Because this deal was consummated while he was away from his desk at the *Tennessean,* and because he was set to start at the *Mail* immediately, there was no public leave-taking in the pages of the *Tennessean*—rather sadly, he never said goodbye to his readers in print. The *Tennessean* itself paid him a rare tribute when it took notice of his departure in an editorial on January 7, 1911.

> Considering Mr. Rice's great natural ability as a newspaper writer, and that it is combined with an unusual capacity for work, it is not surprising that his efforts have attracted the attention of New York publishers.
>
> His ability to write serious verse of a high order and his cleverness as a paragrapher have attracted wide attention to his work and to this paper. The rollicking humor, the pathos and the sweet sentiment of his

poems are only the well-expressed personality of Grantland Rice himself. . . .

He is thoroughly deserving of the success he has won, and we take
no small measure of pride in the fact that this paper has been the vehicle
through which his work for nearly four years has been given daily to a
large and steadily increasing list of readers.

He would not, in truth, be gone from Nashville for long. Within four
years the *New York Tribune* would begin syndicating his column, and the
Nashville Banner would be one of the first newspapers to subscribe.

But even in the interim his presence in Nashville would loom large.
That first summer after he had departed for New York, a prospective
sportswriter presented himself to Marmaduke B. Morton, the managing
editor of the *Banner*. As the supplicant began laying out his qualifications,
Morton, a gangly Kentuckian once called "saturnine and profanely em-
phatic" by the elegant Atlanta columnist Ralph McGill, interrupted, thrust
his corncob pipe into his would-be sportswriter's face and said: "Listen
young man, I don't care whether you can tell a baseball bat from a bull
fiddle. This goddamn Grantland Rice has changed sports writing. What I
want to know is can you write rhymes? If you can, you're hired. If you
can't, you aren't."

Big City Sporting Writer

*I*t **has been said** that modern journalism has experienced three eras, and in 1911, when Grantland Rice arrived in New York, it was savoring pieces of all three.

First came the era of the great publisher—Horace Greeley and James Gordon Bennett, Joseph Pulitzer and William Randolph Hearst, their dominant nineteenth-century newspapers an extension of their forceful personalities. The second era was that of the great reporter, which dawned during the Civil War and reached it apex at the turn of the century in the person of the rakish Richard Harding Davis, a man with the presence and stature of a network anchorman, a man who conveyed a certain importance to a story by his simple presence. The third era, the era that persists to this day, is that of the great story—riveting events either profound or trivial that fix a city's attention and render a newspaper not a luxury but a lifeline. The "arrival" of this era is probably best embodied by the sinking of the *Titanic* in 1912, but earlier stories—the assassination of William McKinley, the San Francisco earthquake—proved a harbinger of its destiny.

When Grantland Rice stepped out of New York's sparkling new Pennsylvania Station in the waning days of 1910, and headed south towards Park Row, the three eras of modern journalism were experiencing a re-

109

markable confluence. Joseph Pulitzer, though he had less than a year to live, and though blindness and shattered nerves forced him to keep his distance from the *World*'s elegant headquarters on Park Row, was still very much the publisher of the *World,* and the paper retained the visionary, voice-of-the-people identity upon which it had risen to prominence during the 1880s.

Likewise, James Gordon Bennett the younger was still the boss at the *Herald.* The son of the man who had founded the *Herald* and had perfected modern journalism back in the 1830s, Bennett was, like Pulitzer, an absentee proprietor. His exile had been prompted by what stands as perhaps the nineteenth century's most noted social indiscretion. One New Year's Eve in the 1870s, Bennett, already the publisher of the *Herald* but still Manhattan's most reckless blithe spirit, urinated into the ballroom fireplace of his fiancée's home in the presence of a house full of shocked revelers. The subsequent abuse and resulting embarrassment drove him to Paris, where for more than forty years he had continued to rule the *Herald* by cable, by telephone, and by threat of one day sweeping into the building unannounced to vent his capricious wrath upon the *Herald* staff.

These two titans from the era of the great publisher were joined by Adolph Ochs of the *Times* and William Randolph Hearst of the *American* and the *Journal,* though neither cast the shadow of Pulitzer or Bennett. Ochs was a man who entrusted the editorial operation of the sedate *Times* to the newsroom professionals he hired, and Hearst's unsuccessful forays into the political world had removed him from the day-to-day operation of his papers; it made him more distant and less real to the laborers in his vineyard, at the same time it made him larger than life.

The publisher's stature during these first years of the new century was shared by the great reporters. Richard Harding Davis himself was still a presence on Park Row. Though he spent most of his time on his farm up in Mount Kisco, fashioning the stories and plays that made him among the most widely read and critically celebrated authors of his day, he remained—in his own mind as well as in the public consciousness—principally a journalist. He was still on call for big stories—particularly wars and armed skirmishes. Rugged, handsome, impeccably tailored and mannered, Richard Harding Davis was journalism's first star—its first sex symbol really.

In the first decade of the new century Davis counted himself among the most famous men in America, and stood as the most admired man in American journalism, a figure not so much to be envied as to be held in awe. It was a time, said one writer, when "young boys no longer dreamed of becoming president: they dreamed of becoming Richard Harding Davis." Whether dining at Delmonico's, or standing conspicuously in the open surveying a battlefield, or seeking out a cub reporter to compliment him on a story well done, Davis was a model of glamour, valor, and decorum.

All along Park Row, and on Newspaper Rows throughout the land, journalists took a vicarious pride in Richard Harding Davis—in his bearing, his status, his talent. For too many of them recognized in themselves those qualities not of Davis himself, but those that he had ascribed to young Gallegher, the copy-boy hero of one of his most famous stories. "All Gallegher knew had been learnt on the streets; not a very good school in itself, but one that turns out very knowing scholars. . . . He could not tell you who the Pilgrim Fathers were, nor could he name the thirteen original states, but he knew all the officers of the twenty-second police district by name, and he could distinguish the clang of a fire-engine's gong from that of a patrol-wagon or an ambulance fully two blocks distant."

Journalism was not a profession with a tremendous self-image in the early twentieth century. So many of those with talent would look to get out—to write books or to write for the magazines, or the theater. Or they'd leave for press agentry or advertising or their father-in-law's dry goods business. And for those who remained—for whatever their reasons—for everyone who took a pride in the Gallegher-like street-smarts that allowed them to do their job, there were several who nipped at the flask in their desk drawer and hung their heads in gloomy resignation and consent whenever a colleague would lurch forth with a new harangue on the social value of reporters. The best of these harangues come from Ben Hecht and Charles MacArthur, Chicago journalists during these years. They say it through the character of Hildy Johnson, the reporter in their immortal play, *The Front Page:*

> Journalists! Peeking through keyholes! Running after fire engines like a lot of coach dogs! Waking people up in the middle of the night to ask them what they think of companionate marriage. Stealing pictures off old ladies of their daughters that get raped in Oak Park. A lot of lousy, daffy, buttinskies, swelling around with holes in their pants, borrowing nickels from office boys! And for what? So a million hired girls and motormen's wives'll know what's going on.

Richard Harding Davis was a antidote for this. A kinship with Richard Harding Davis, no matter how remote or vicarious, was a source of pride. The age that was in flower as Grantland Rice walked down Broadway—past the Park Row headquarters of the *World,* the *Tribune* and the *Journal*—towards the newsroom of the *Evening Mail,* was the age that proved the beginning of journalism's gradual but inexorable passage towards professionalism. And Grantland Rice, in his own way, was as much a symbol of this new era as Davis himself. Here was a man who, like Davis, was college-educated, erudite, admired by readers, respected by sources, and both admired and respected by colleagues. And here was a man, like Davis, who, for all of the fame and opportunity that came his way, continued throughout his career to derive satisfaction, fulfillment and pride from his duties and identity as a journalist.

There were fourteen daily newspapers in New York City at the start of 1911, twelve of them on and around Park Row—only the *Times* and the *Herald* were quartered uptown. Still atop the circulation heap were Pulitzer's morning and evening *Worlds*, Hearst's evening *Journal* and morning *American,* and the bright and tireless morning and evening editions of the *Sun.* Atop the heap in matters of influence and journalistic prestige were the morning *World,* the comprehensive and scrupulous *Times,* the polite and earnest *Tribune,* and the socially correct *Herald.*

The *Evening Mail* rested atop the slag heap of Gotham journalism, a second division comprised of the *Mail,* the *Telegraph,* the *Press,* the *Globe* and the *Post.* All but the *Post* would be gone within a decade and a half. In 1911, however, in addition to its new sports columnist, the *Mail* could boast of two other men destined to be among the most famous newspapermen of their generation—Franklin P. Adams and Rube Goldberg. In 1911, their fame and talent, like Rice's, were still ascendant, however, and *Evening Mail* struggled to maintain a viable circulation in competition with the wealthier, noisier, and tradition-steeped newspapers that surrounded it. Rice's presence was to be a prominent one in the paper. A good sports section, first with the scores and details of the afternoon games, was essential for a paper hoping to be competitive in the afternoon market, and the *Mail* trumpeted the Southerner's arrival loudly.

"Reader's constant and fickle, we have a pleasant surprise for you," ran the notice on the *Mail*'s sports page the week before Rice began his toils.

> This paper has secured the services of Mr. Grantland Rice, technically known as Old Grant Rice of Nashville. It is Mr. Rice's habit to write entertainingly and authoritatively, in prose and verse, in winter and in baseball season, by and large, hither and thither, on topics pertaining to all kinds of sport, including wrestling. If you like sport and intelligent, amusing comment on it, you'll like Mr. Rice's daily grindings; if you don't like sport you'll like his department anyway. Mr. Rice will open Monday.

The *Mail*'s first citizen at Rice's arrival and throughout his tenure on the paper was Franklin Pierce Adams—the *Mail*'s most popular feature was F.P.A.'s column, "Always in Good Humor." It wasn't, of course—always in good humor, that is. It was prickly and puckish and unfettered by convention and that was why it was one of the most popular columns in New York. A slight, almost frail man of twenty-nine, with thinning black hair—eventually to be offset by a rich, full mustache—Adams had come to the *Mail* from Chicago in 1904, and virtually all he retained of his Midwestern heritage was a passion for his alma mater's football team, the University of Michigan. In every other respect he became the quintessential New Yorker. According to his biographer he "became integral to the genesis of the New York City myth," the myth "that New York City is the center of all things glamorous, intellectual, sportive, and generally worthwhile." His column was a sort of literary Schwab's Drugstore—a place where those with aspirations of stardom in the world of New York letters came to be discovered.

Each day, F.P.A. filled no more than a quarter to a third of his column with his own verse and wit. The balance would come from contributors. Sometimes he would lift things from the nation's other versifiers and paragraphers that would reach him through the newspaper exchange—such was the source of Grantland Rice's frequent contributions to his space. But most of it came not from other papers but from a remarkable stable of contributors who competed for a coveted cameo in the column.

Two Nobel laureates, Sinclair Lewis and Eugene O'Neill, contributed original material to F.P.A.'s column. So did Ring Lardner, Edna St. Vincent Millay, Robert Sherwood, E. B. White, Robert Benchley, James Thurber, John O'Hara, Ira Gershwin, Edna Ferber, George S. Kaufman, Moss Hart, Alexander Woollcott and Dorothy Parker, who told people that F.P.A. had raised her from a couplet. Like Parker, many of these eventual immortals were published first in F.P.A.'s column.

Adams would spend the better part of his day sorting through the mountain of material that came his way; and he paid his contributors all the same for their brilliance—nothing. Each year he gave a gold watch to the contributor with the most appearances. When asked why he rewarded quantity instead of quality, he responded it was because every contribution he accepted was a gem. And so it was—his contributors sent nothing but their best stuff to Adams, for so much more than money was at stake. An appearance in F.P.A.'s space provided not only cachet, but a credential that would turn other editors' heads—the clipping itself was negotiable currency.

Adams himself was a man of brilliant, biting wit, and his own contributions to the proceedings of the column were still its soul. "Dozens of things he said were quoted all the time in smart circles throughout the country," noted one writer. Things like: "Nothing is so responsible for the good old days than a bad memory," or "ninety-two percent of the stuff told to you in confidence you couldn't get anybody else to listen to anyway."

Adams was also a big sports fan; he would often accompany Rice up to the Polo Grounds of a summer's afternoon. He did not often poach on Rice's territory in his column; still on one foray into the sports world, during the summer before Rice arrived in New York, Adams turned a phrase that has proven as durable as anything any sportswriter ever fashioned:

> These are the saddest of possible words:
>   "Tinker to Evers to Chance."
> Trio of bear cubs and fleeter than birds,
>   Tinker and Evers and Chance.
> Ruthlessly pricking our gonfalon bubble,
> Making a Giant hit into a double—
> Words that are heavy with nothing but trouble:
>   "Tinker to Evers to Chance."

When Adams's little ditty appeared in July of 1910, Rice immediately reprinted it in the *Tennessean,* as did numerous other paragraphers in their columns. Many, including Rice, appended their own little verses in the Adams meter and cadence and playing on the same theme. They dutifully sent them back to Adams via the newspaper exchange and he dutifully printed them and the game had a summer's long run. (The Tinker-to-Evers-to-Chance Cubs, fittingly, won the National League pennant, but were beaten four games to one by the Athletics in the World Series.)

Rice's variations upon Adams theme have been rightfully forgotten, as have all the others. But their existence, back there on the fragile yellowed clippings of forgotten newspapers give evidence of the early awareness of the sturdiness of the Adams lines. Shortly after his arrival at the *Mail,* Rice fell into conversation with managing editor Theophilus England Niles about Tinker-to-Evers-to-Chance verse. "Frank may write a better piece of verse," said Niles to Rice. "But this is the one he will be remembered by." Rice was quick to agree, and history has borne them out.

Adams took note of Rice's arrival at the *Mail* in his column, and it is worthy of note today because in doing so Adams also made reference to one of his contributors, a gangly and obscure young man from Pittsburgh, destined, as so many of Adams's contributors were, for bigger things. Some contributors wrote under pseudonyms. Others, in emulation of their patron, signed their pieces with just their initials. One of the most familiar sets of initials were those of G.S.K.—George S. Kaufman, soon to be recognized as Broadway's most dynamic comedic voice. On the day Rice began his toils on the *Mail,* G.S.K., a regular contributor for nearly two years, was piqued at having had his recent submissions ignored. " 'If I don't get something in the column pretty soon,' writes G.S.K. 'I'm going on strike.'

"Have a care!" wrote Adams. "Old Grant Rice is going to start another league on the sport page, and if we can agree on the purchase price we may ask for waivers."

Old Grant Rice's different league was never in competition with F.P.A. He abandoned most of his political and social commentary and much of what made up the old "Tennessee 'Uns" column, left the territory to F.P.A., and tended to sports.

On the sports page, Rice shared the spotlight with the cartoonist and boxing writer who signed his work R. L. Goldberg. Goldberg, a twenty-seven-year-old San Franciscan who had been at the *Mail* for four years, would soon change his by-line to Rube Goldberg; and like those of Rice and Adams, it would become one of the most famous and best-loved signatures in American journalism. Moreover, the name would soon enter the language, where it persists to this day; "a Rube Goldberg" is the name for any jerry-built contraption fashioned by a well-intentioned hand seeking to plug a leak, build a better mouse trap, or facilitate some ordinary task by means of bailing wire, chewing gum and common sense run amok.

Goldberg had been trained as an engineer at the University of California, but he chafed under the rigors and formulas of the discipline; he was particularly frustrated by the complicated machinery and tools that drove and serviced the engineering world. Long after abandoning engineering for newspapering, he would exact his revenge by creating the character of Professor Lucifer Gorgonzola Butts, whose forte was a penchant and capacity for labyrinthine and lunatic inventions such as the "simple bookmark." The simple bookmark was set into operation by a reader picking up his reading glasses. This released a flock of moths who ate through a woolen sock out of which dropped a tear-gas bomb which caused a small dog to weep into a sponge whose added weight set into operation a magic lantern which cast upon the book's cover the likeness of a man who has stolen the wife of an angry dwarf. The dwarf plunges a dagger though the picture and into the book, stopping when he strikes a pet flea who had jumped between the pages to sleep when the book had been set down. The flea says "Ouch!" and kicks open the book to the right page.

This zaniness would make Goldberg wealthy as well as famous, but both—and Professor Butts as well—were still a few years away in 1911, and he was still complementing his drawings—which ran daily in the center of the *Mail* sports page—with coverage of selected athletic events, particularly boxing.

On the morning Rice reported for work at the *Mail*, Goldberg greeted him warmly and introduced himself; and the pair made their way quickly over to Frank Adams's desk. It was the beginning of two more beautiful friendships for Rice. As their paths diverged and their respective fames grew, the bond formed that first day remained.

Adams had anticipated that Rice would need a room and had arranged for temporary quarters for him in his building, a block from Columbia University at 616 West 116th Street. With Kate and Floncy still in Nashville, Rice scouted about for permanent lodgings, settling on a large flat at

450 Riverside Drive, just one block north and west of his temporary digs in F.P.A.'s building. The Morningside Heights neighborhood had a certain journalistic cachet in these pre-war days. Rice shared his building with such colleagues as *Tribune* editor Walter Trumbull, music critic Deems Taylor, and writers Max Foster and Irvin S. Cobb. Foster and Cobb in particular would became fast friends. Nearby lived not only F.P.A., but the young *Tribune* sportswriter and recent Harvard student Heywood Broun, and the new kid just in from Denver, name of Damon Runyon, covering sports for the Hearst papers. Runyon and Broun in the years ahead would be amongst a core of journalists and friends who would, as Rice put it, "wander by" the Rice flat in the post-midnight, post-deadline hour, to partake of some hard liquor and gentle conversation.

Kate and Floncy arrived from Nashville in mid-February, and with them arrived just a measure of melancholy for Rice, as he grew homesick for Nashville and those he had left—particularly, he said, his mother. But he set such thoughts aside:

> But we have set our soul ahead
> Upon a certain goal ahead
> And nothing left from hell to sky shall ever turn us back.

He discovered in the first spring in New York, much to his consternation if not his surprise, that baseball began not when the teams arrived for spring training, as it did in Nashville, but when they returned—a full six weeks later. It was the first spring training he had missed during his career; it would also be the last. Like a stallion pawing the paddock dirt, Rice restlessly bided his time during his first weeks on the *Mail,* until the Giants and Yankees returned north, growing perhaps a trifle irritable at dispensing the dope from a distance. "The box score doesn't absorb much space," he wrote, "but it has all the advance dope ever written crowded into the hatchway."

As Rice and the others on the city's sports pages made ready for the season ahead, they received a bit of chiding from F.P.A. on their growing penchant for wearing out a clever phrase and reaching too often for the familiar cliche:

> Pushers of the pencil,
>   And ye please to lamp,
> Throw away the stencil,
>   Stow the rubber stamp.
> Say "He swiped a stinger,"
>   But, the coming year,
> Chop the "portside flinger,"
>   Can "the leathery sphere"!. . . .

Talk of "labored hurler,"
  Write "the horsehide pill,"
Speak of "southpaw twirler,"
  Mention "mound" or "hill,"
Call a pennant bringer,
  A "gonfalonier,"
But jar the "portside flinger,"
  Tin "the leathery sphere."

Pushers of the pencil,
  Scorers of the hit,
Is my dope prehensible?
  Do ye gather it?
Then, o slangy slingers,
  Ere the season's here—
Nix on "portside flingers"!
  Nix on "leathery sphere"!

Rice would too, from time to time, brood about the state of his art. The rather remarkable argot of the sports page had been under siege for some years. As early as 1907 a *Collier's* article concluded: "Baseball slang is out of date." Rice took note of its "passing" in his column in the *Cleveland News.*

And so today, with tear ducts turned on full force, we record the demise of "baseball slang" as she is called. . . .
Any gent who will cut in with slang in the wake of the spiel launched against it has parallelopipedons in his attic, and his fountain pen is full of soup. He has mud on his skylight and the dent in his brain pan is deeper than a cistern.
Baseball slang, farewell. The populace no longer esteem it any great bliss to glue their lamps upon you. . . .
It's tough, but it's all in the game. The old Greeks used to wag a classic tongue, but it's out of date now, listed with the dead spiel of the Latin age.

But baseball slang proved remarkably sturdy, and during the halcyon days—which continued for at least a quarter century after *Collier's* proclaimed it dead—Rice was capable of laying it on as thick as anyone. In sizing up the Southern League season for his *Tennessean* readers in the spring of 1908, he told them that "baseballically speaking" to "keep your weather orb peeled on those Pelicans" who would be featuring a staff of "crack slabmen" that summer.
Newspaper readers cherish nothing so much as the comfortable or the familiar in their papers. Let a paper today undergo a re-design or drop a popular feature like Ann Landers or "Peanuts" and readers will howl and

write angry letters at having their routine disrupted. It was thus in Rice's day; so much so that as he read the dispatches that flowed from the training camps and noted that the hyperbole was absent, he was moved to ask in his column: "Are the scribes slowing up?"

> What has become of the old-fashioned war correspondent who, within ten days had some veteran or recruit inventing a double shoot or a tango twist or a curve that broke seven ways at the same time?
> There hasn't been a zig-zag or a spiral curve exploited and the training season is nearly ten days old. Is the old hop on the fast one drifting away from the fountain pen? Is the typewriter losing its old-time smoke?

The call for a banishing of baseball slang from the sports pages was not coming from the hundreds of thousands of sports-page readers, but rather from a core of militant purists in the matter of the care and feeding of the English language. Their numbers small, the argot they decried persisted, readers and writers both comfortable and satisfied with the familiar. And as late as 1923, Rice was still taking puckish note of a newspaper editorial that proclaimed: "The movement to eliminate slang from the sporting page has gained remarkable headway in the last year or two."

> I'll take mine in the simple buzz
> When Noah Webster led the batting,
> He had these slangsters full of fuzz
> When it came down to big league chatting.
>
> I'm off this slang all ninety ways,
> I'll take mine as straight as Hagen's brassie;
> And when some blinkin' blighter brays
> You'll never hear me call him "classy" . . .

Nevertheless, though it would take most of the rest of Rice's career before he and the more innovative of his colleagues chased completely the slang and the hyperbole from their columns, sportswriting was in the early throes of transition that spring of 1911. The change was coming in on the typewriters of what has been called the "foremost freshman class that ever broke into the New York press box." In addition to Rice, New York rookies that spring included Rice's uptown neighbors and friends, Broun and Runyon, and a small, slender, baby-faced lad from Philadelphia by the name of Fred Lieb, writing for the New York *Press*. Today, Rice, Broun, Runyon and Lieb are all in the writers' section of the baseball Hall of Fame in Cooperstown, New York.

What these men (and a few others, like Sam Crane of the *Evening Journal*, a star for the Giants during the 1880s and 1890s before turning to newspapering) were bringing to readers was not just information, but understanding. American fans were vastly less sophisticated on the nuances of baseball in 1911 than they are today, when, due to the writings of people like Peter Gammons, Tom Boswell, Roger Angell and others, and armed with statistics and facts assembled by the Bill Jameses and Bill Eliases of the field, an avid fan might know as much about the players and the game as some general managers—or believe he does anyway. In 1911, fans and newspaper readers knew who was in first place and who was hitting what, and that was about all. Rice and the others brought another dimension to the sports page, the "why" behind the "what." He saw more than what was simply happening at the moment. When Ty Cobb held out for the seemingly obscene salary of $15,000 in 1913, Rice said he was worth it, strictly from a business standpoint, and then provided a sturdy argument in support. When the Athletics and the Red Sox—the class of the American League—came to New York to play the Yankees they would ordinarily draw crowds in the 4000 to 5000 range, said Rice. When Cobb and the Tigers came to town, they would routinely draw crowds of 10,000 to 12,000. If only three hundred extra fans a game came out to see Cobb and only Cobb, reasoned Rice—a rather modest estimate considering the evidence—that would mean an extra 45,000 fans over the course of the season, or—based on the then-standard fifty-cent admission—an additional $22,500 in the Detroit till.

In that same column defending Cobb's claim to $15,000 a year, Rice probed much deeper into the business side of baseball. Asked whether he felt Cobb should be dealt to a team in a larger city, a more profitable club, better able to meet his salary demands, Rice answered with a resounding nay. "We don't think he should. If [Detroit owner Frank] Navin was forced to sell or trade Cobb to a larger city where he could get his price, it would only be a matter of a year or two before New York, Chicago and the larger cities could gather in all the stars," said Rice, eerily presaging words that would be written in abundance in the 1970s as free agency for major league players finally loomed on the horizon.

> Detroit can't afford to pay as much as New York can; for Detroit hasn't the people to pay the extra tax at the gate. If New York or Chicago corralled the star hold-outs, the pennant race would be a joke by July, interest would vanish, the attendance would wane and the big concrete stadiums would soon be as thickly populated as the western desert of the Florida Everglades.

From there, Rice grew positively prescient:

> Every one knows the reserve clause in baseball will not stand the test of American law. And yet without the reserve clause how long could

Washington hold Johnson? How long could Brooklyn hold Rucker? How long could Detroit have held Cobb?

If it was abolished there would simply be a private agreement among club owners not to raise the ante [emphasis added]. The smaller towns must be protected, the average ball player must be protected, for the general good of the game.

Congress might start forty investigations, in so far as the reserve clause was concerned, without getting anywhere.

Such sophistication in newspaper sports columns was revolutionary in 1911. Indeed, it would not become commonplace fare for another three-quarters of a century.

Rice shared the baseball beat at the *Mail* with Harry Schumacher. Schumacher was the *Mail*'s traveling writer; Rice the homebody. Schumacher covered John McGraw and the Giants at their Marlin, Texas, camp, and when the season began, he would travel with whichever team was on the road while Rice would set up shop in the northern reaches of Manhattan—at Eighth Avenue and 155th Street, the Polo Grounds, when the Giants were in town; at 168th Street and Broadway, variously called Hilltop Park or American League Park, when the Yankees were at home.

On the infrequent occasions when Rice would take to the road—he went on about two trips a season during his years on the *Mail*—it would invariably be with the Giants. In these days of John McGraw, the Giants owned the town. The Yankees were decidedly second class—both on the field and in the affections of fans and writers; the Dodgers had yet to be discovered by any but the denizens of their own cozy borough. The Giants, however, were New York personified—brazen, boisterous, fun-loving, arrogant, cocksure and glamorous. They were winners, and they were stars. National League champions in 1904, world champions in 1905, they had been bridesmaids to the Cubs in most of the ensuing seasons, but in the spring of 1911 they are about to embark on a ride that would bring three consecutive National League pennants. And that success, in turn, would make them the best-known, best-loved (in New York), and most-despised (everywhere else) collection of athletes the world had yet known. "The whole nation had become Giant-conscious," remembered the wife of manager John McGraw, "and fans couldn't get enough news of them."

"Goddamn! It's great to be young and a Giant," said Larry Doyle, the second baseman and team captain, whose ingenuous exuberance earned him the nickname "Laughing Larry." Though his teammates never expressed the thought in exactly the same way—sentiment would have been out of character; if this collection of Giants was playing today, they would be called a team with an "attitude"—Doyle's teammates shared his feelings. Each day they plied their trade before the richest and most famous people in New York. The Polo Grounds box seats would be filled with Broadway stars such as George Dillingham, Mabel Hite, Louis Mann, John McCormick and DeWolfe Hopper, the actor who had popularized Thayer's "Casey at the Bat." Former heavyweight champion Gentleman Jim Corbett was a regular too, as were jockeys and owners from the horse racing world. But the swagger and grittiness of the Giants gave them their largest following amongst working-class New Yorkers. "The Giants represented the New York of the brass cuspidor—" said writer Harry Golden, "that old New York which was still a man's world before the advent of the League of Women Voters; the days of swinging doors, of sawdust on the barroom floor. . . ."

There was a larger-than-life quality to these Giants. Rube Marguard, the pitcher who would win nineteen straight in 1914, also made headlines for dating actress Bloosom Seeley, and for being named in an alienation of affection suit filed by Seeley's husband.

Pitcher Bugs Raymond, who ultimately drank himself out of baseball and into an early grave, used to leave the Giants bullpen in the distant reaches of the Polo Grounds and while away some of the idle moments during the game in the saloons along Eighth Avenue near the ball park, trading the baseball he'd taken to warm up with for a few fingers of rye.

Mike Donlin, called "Turkey Mike" for the way he strutted on the field, was "one of the great lounge lizards of all time, the undisputed nightlife champ of either league," according to Giants' historian Noel Hynd. Four times during his fifteen-year career, in 1906, 1907, 1909–10, and 1913, Donlin walked away from baseball to pursue a vaudeville career. When his playing career was over, he ended up in Hollywood as a bit player in silent films. When his mind was on baseball, however, he approached brilliance. He hit a lifetime .333, and his scrapiness and his strut made him a favorite of the fans.

In contrast to these blithe spirits, and the most able and most charismatic Giant of them all, was the elegant Christy Mathewson. Mathewson was the antithesis of the typical turn-of-the-century ballplayer, generally rough-hewn, poorly educated, earthy, vulgar and naive. Mathewson, by contrast, was a graduate of Bucknell, well-mannered and introspective, the author of a treatise on pitching called *Pitching in a Pinch* that is read to this day. He was a strapping six feet, two inches, 195 pounds, an exceedingly handsome man, with soft features and a boyish complexion and a tuft of light brown hair generally protruding from beneath the bill of his cap. "What a grand guy he was," was the succinct and nothing-more-need-be-said appraisal of teammate Rube Marguard. Christy Mathewson was everybody's favorite.

According to Grantland Rice, he was also "just a little bit better at all games than anyone else. He played chess and checkers and poker better, for example, and usually drew in most of the pots." The game he played better than all others, of course, was baseball. He'd won twenty games for a last place team during his rookie year of 1901. He'd won twenty-two with a sore arm in 1906; thirty-seven in 1908, when he also led the league with an earned run overage of 1.43, and could have easily won forty-four or forty-five that year with just a little bit of luck. And of course, his exploits in the 1905 World Series, three shut-outs in six days, represent heroics the likes of which baseball will never, ever, see again.

Beneath Mathewson's mannerly demeanor—he never swore and never questioned an umpire's decision—there lived a competitiveness that was the equal of any in the game. "Always have an alibi," he said once in a statement that must have surely shocked his contemporaries.

> An alibi is sound and needed in all competition. I mean in the high-up brackets. One of the foundations of success in sport is confidence in yourself. You can't afford to admit that any opponent is better than you are. So, if you lose to him there must be a reason—a bad break. You must have an alibi to show why you lost. If you haven't one, you must fake one. Your self-confidence must be maintained.
>
> Always have that alibi. But keep it to yourself. That's where it belongs. Don't spread it around. Lose gracefully in the open. To yourself, lose bitterly—but learn! You can learn little from victory. You can learn everything from defeat.

One wonders how Mathewson learned as much as he did; he did not lose often.

But for all the enormous appeal of the supporting cast, for all the *duende* of Mathewson, these were forever "McGraw's Giants." The team was molded in the scrap-iron, never-give-quarter image of its manager, and no presence on the field—either individual or collective—ever overshadowed the presence of the man in the dugout.

"His very walk across the field in a hostile town was a challenge to the multitude," said Rice of McGraw. Often, McGraw's challenges were far less subtle. In Cincinnati he once offered to fight everyone in the stands. He would bait opposing players from the dugout with scatological harangues so original and so absolute in their vulgarity that they would prompt a blanching even among seasoned ballplayers who were certain they had heard it all before. Umpires would banish him from a game; undaunted, he would change into his street clothes and continue his verbal abuse from a box seat behind the Giants dugout. In 1905, he humiliated Pirates' owner Barney Dreyfuss outside the Polo Grounds, shouting down to him from the porch of the Giants clubhouse in front of several hundred fans, calling him a welsher on gambling debts, and accusing him of holding a number of the league's umpires in his pocket. He did it, of course, in the distinctive McGraw vernacular—a stream of blue epithets. When Dreyfuss brought a formal complaint to National League offices, McGraw, with the loud assistance of the New York newspapers, bullied the league into dismissing the charges. It did not help Dreyfuss's case that he was unable to persuade even a single one of the several hundred witnesses to testify on his behalf.

Most of these incidents spring simply from McGraw's natural bellicosity; but McGraw was a schemer and something of a psychologist as well. "He courted dislike in various cities," said Rice, "for he was shrewd enough to realize that additional fans would come out to see the 'hated Giants' beaten."

But if organized baseball beyond New York saw McGraw as the Antichrist, in New York he was the Messiah, for he had turned a moribund franchise into champions and put a sparkle and glitter into the city's signature sporting attraction. McGraw was a mere slip of a boy when he came out of the semi-pro leagues of western New York to join the Baltimore Orioles near the end of the 1890 season, not quite five feet, seven inches tall, scarcely 120 pounds. He filled out to a sturdy 155 while starring for the Orioles during the nineties, but since taking over the reins of the Giants in 1902, his playing days over, his fireplug body had sagged, softened and spread, and by 1911 his players had taken to calling him "The Little Round Man"—behind his back, of course. He professed to have an interest in nothing but winning. "In playing or managing, the game of ball is only fun for me when I'm out in front and winning," he said. "I don't give a hill of beans for the rest of the game." It was a pronouncement consistent with his image, and good for his image. It was also disingenuous, for he was a man devoted to his players and devoted to teaching. He'd often keep rookies, destined but not yet ready for the starting line-up, next to him on the bench, explaining the game and his decisions and philosophy as the season passed before them, convinced they would learn more about baseball sitting at his hand then they would playing under lesser tutelage in the minors. With Mathewson, McGraw's devotion was nothing short of paternal. The two men and their wives went so far as to share a seven-room flat on Columbus Avenue and 85th Street. The McGraws paid the rent, the Mathewsons the grocery bill, and long into the evenings, McGraw would fill his obedient and attentive pupil full of baseball.

McGraw's relationship with New York's Fourth Estate was more mercurial. For the most part, McGraw was the apotheosis of the game to the New York writers. They hailed his skill at acquiring and handling players, his scrappiness and tenacity, his uncompromising insistence on excellence and his ability to exact it. He became known in the sports pages as "the little Napoleon," and one writer wrote in 1913 that he "combines in a rare degree the ability to . . . take advantage of an opportunity, to see into the future, and to plan for what may happen as well as what is happening."

But, having come to expect excellence from McGraw's Giants, the writers grew predictably impatient, then inevitably critical on those occasions when the excellence was not forthcoming. McGraw, in turn, grew short, then abusive, and, on a couple of occasions, downright irrational in his dealings with the writers. The most dramatic of these concerned McGraw's 1917 statements about National League president John Tener, after Tener had suspended and fined him for fighting with umpire Bill "Lord" Byron in Cincinnati. McGraw told Sid Mercer of the *New York Globe* that Tener sat at home in Harrisburg not attending to his job, that he was biased towards the Philadelphia Phillies, and, like everyone else in the National League, he was out to get the Giants. Mercer wrote up the story, even took the extraordinary step of sharing it with Sam Crane of the *Evening Journal;* but then, aware of the incendiary nature of the story and concerned for McGraw's welfare—the two men were friends—he submitted a copy of the story to McGraw for his inspection and approval. McGraw responded: "I don't give a damn what you say for me. I guess it's all right. Life isn't worth worrying about little things like this."

John Tener and the National League owners didn't see it as a little matter. They called McGraw into the office when the team returned to New York to explain his attack on the league president. McGraw, out of character, took the coward's way out and denied the remarks attributed to him in "certain scurrilous newspaper articles. . . . "

"I did not make these statements or give out by intimation any utterance that might be construed to in any way reflect upon the ability, honesty and integrity of the president of the league."

The National League, not at all out of character, also took the coward's way out. They accepted McGraw's repudiation of the story and proclaimed the matter closed.

The sportswriters were outraged. Sid Mercer was an able and popular reporter, and to call his credibility into question was to cast aspersions upon the entire sportswriting profession. The Baseball Writers Association of America formally requested that the National League reopen the investigation. Grantland Rice led the call. "Everyone connected with the signed statement repudiating recent interviews wherein McGraw gave his opinion of President Tener—this includes McGraw, Tener and every National League club owner—knows that the manger of the New York club was correctly quoted as to all the main issues," wrote Rice.

> The attempt to throw the scandal back upon the newspapers will hardly stand. . . .
> To say that these interviews were fakes and "scurrilous fabrications" is an insult to any degree of intelligence. . . .

There is still this side to consider: Before this repudiation of Mc-Graw's was accepted and given out, why couldn't National League club owners have secured testimony from the newspaper writers involved?

They knew the denial was a joke and that innocent parties were being slandered.

If they had desired to get at the real truth of what happened and what was said, the way was easy enough.

The sportswriters got the matter reopened, the league reversed itself; McGraw's suspension and fine was upheld and the sportswriters were vindicated.

The careers of Rice and McGraw were inextricably intertwined. McGraw managed the Giants in nine World Series; Grantland Rice bore witness to all of them. McGraw was sport's largest figure in the years before Babe Ruth came along; Rice was sport's most influential and widely read chronicler. The two men saw one another on a weekly, at times daily, basis for more than twenty years. They got along well, and Rice remembers McGraw on occasion being solicitous of a sportswriter's needs; he told the story of being stumped for an off-day story during his years on the *Mail* and having McGraw sit down beside him on the train and feed him a dandy. But they were never close; when he would travel with the Giants it was the gentler, more introspective Mathewson to whom Rice would be drawn—Mathewson whom he would invite along for a round of golf.

By the end of his first baseball season at the *Mail*—the Athletics beat the Giants in a rematch of the 1905 World Series—Rice has settled comfortably into the cycles of life in New York. There was much for a young couple making fifty dollars a week to do in New York. Fifty cents would buy a one-pound porterhouse-steak dinner, replete with rolls, potatoes and vegetables. The best seats at the best theaters to the most popular shows cost a dollar and a half a pair; and for three dollars a hansom cab could be hired for the entire evening—transportation to the show, to the restaurant afterwards, and home via the circuitous and romantic route through Central Park.

The after-hours scene on Park Row was just as rich in atmosphere and reasonable in price. Hitchcock's, in the Tribune building, served a beef 'n' beans dinner for ten cents, and Hesse & Loeb's, on the fourteenth floor of the World building, let reporters eat and drink on credit, and it was not unusual for a reporter to ruefully sign over his entire paycheck to Hesse & Loeb's to settle his account. On the ground floor of the World building was Perry's Drugstore, which was really a saloon and the most favored of the neighborhood establishments by the vast majority of Park Row reporters.

On workdays, Floncy would sometimes accompany her father to the office and frolic about the newsroom. Frank Adams reported once that "Miss Florence Davenport Rice, whose father, Old Grant Rice, uses a typewriter, approached our shrine yesterday and evinced surprise that we used the o.f. [old-fashioned] fountain pen. 'You don't work, like Daddy,' said Florence. 'You just write.' "

Freed from the miscellaneous duties he'd carried at the *Tennessean,* Rice did find some time to offer up a line or two to Frank Adams's column. As much as anything else it was a case of being seen with the right people. In the winter of 1913 Adams contrived to print a continuing play in his column, a new scene every day, contributed by the likes of Samuel Hopkins Adams, Will Irwin, Irvin S. Cobb, Rex Beach, James Montgomery Flagg and Rice. In this one case, at least, seldom has such an enormous assemblage of talent collaborated on such a forgettable work.

Rice's stories never failed to get good play in the *Mail,* especially during the baseball season, when not even major disasters could keep the late scores and stories from their customary spot atop the front page. FEDERAL FLIER WRECKED / GIANTS LOSE read one dual headline in the summer of 1911; and in the accompanying stories, *Mail* readers learned that "Sixteen people were killed when a Washington-Boston express jumped a viaduct and plunged down an embankment near Bridgeport. . . . " while in Manhattan, "Ineffectiveness and wildness on the part of 'Southpaw' George Wiltse lost the first game of a series for the Giants against the Pirates here this afternoon. Wiltse didn't have a thing and whatever he served up was pelted all over the lot for base hits." The train wreck story ran down the left-hand column, The Giants story down the right; *Mail* readers were left to decide for themselves which was the greater tragedy. (One year later the sinking of the *Titanic* did push the story of the Giants game off page one; but the score was still there.)

The cycle of the seasons was the same in New York as it had been in Nashville—baseball, college football, vacation, hot-stove league baseball, spring training, and then the whole thing all over again. The crush of news in the big city and the perceptible quickening of the times had heightened the intensity of his duties, though. "There was a time when this sportive life had its intermissions. Now there isn't a bench along the rail from January 1 to December 31," he wrote in December of 1913. The off-season stories during these years took Rice and the other sportswriters away from the playing fields and into the front offices and boardrooms. There were stories on a threatened strike by baseball players in 1917, that story stemming at least in part from the short life (1914–15) of the Federal League—the first incursion on organized baseball in more than a decade, and the last such insurrection to this day. Rice never took the Federal League very seriously—neither did anyone else for that matter. "The Feds are getting more than their allotted space," said Rice during the league's birthing days. "Just how much of this they will retain the day after Wagner faces Mathewson or Cobb faces Johnson is another canto. For in the winter a sport scribe's fancy lightly turns to anything he can peg a typewriter at or dig a fountain pen into."

In the winter of 1913 came the biggest of these off-season stories— the shocking revelation that Jim Thorpe had taken money for playing in a summer baseball league and would be stripped of his Olympic medals. The story broke in January of 1913, while Rice was on vacation, and Thorpe was making plans to compete in a Fordham College-sponsored track meet at the Twenty-second Regiment Armory in New York. It came five months after the Stockholm Olympics, three months after the referee who worked the Carlise-Army football game said that "the game has never seen [Thorpe's] like," and just three weeks after a reader had asked Rice who the greatest living athlete was. "The individual laurel spray goes to Jim Thorpe, with no one close," answered Rice in his column. "The Indian's combined record at Stockholm and over the gridiron puts him forward, not only as the leader of the year, but as the greatest all-around star within the memory of its contenders."

The Thorpe story first appeared in the Worcester, Massachusetts, *Telegram*. Charles Clancy, manager of the Winston-Salem club in the Carolina League told the paper that Thorpe had played for his team during the 1909 and 1910 seasons. Within four days Amateur Athletic Union potentate James E. Sullivan had convened a hearing, exacted a pitiful sackcloth-and-ashes confession from Thorpe, stripped him of his amateur standing, and ordered the Olympic medals returned. Virtually every quarter of the American press parroted Sullivan's smug assertion that justice had been served. "Laws are laws, and we must obey them," said Sullivan. "We cannot play favorites."

But of course they did. They played favorites by winking at the abuses that abounded amongst college athletes nationwide. They played favorites by publicly humiliating and swiftly executing Thorpe, while leaving uninvestigated and unchallenged assertions such as the following from Grantland Rice, in his first comments on the Thorpe case after returning to the office:

> We know of at least four star college athletes from past seasons who toiled through the summer in about the same brand of bush scenery as that which enticed Thorpe off the trail. None of them played for any salary or received any coin for their services. But at the end of the season, two sold their suspenders to the manager for $800 each, and the other two bet their manager $900 in the last game played that neither would make over nine errors in the lone battle left. Strangely enough both won their bets, as neither had over five chances to handle.

You have to read between the lines of Rice's comments to glean his feelings on the subject. These following remarks came at the bottom of a miscellaneous column. Nevertheless, they are among the strongest challenges to the hypocrisy of the Amateur Athletic Union—a hypocrisy that riddled the organization throughout its existence and riddles its descendants until this day. "[T]he best definition of amateur is one who can get away with it and not be nicked with the goods," wrote Rice.

> The difference between Thorpe and several dozen others who rank high in the amateur world is that the Redskin was caught with the merchandise. This doesn't so much excuse Thorpe as it makes him a minor offender where many have been more flagrant and are still as publicly spotless as the well-known snow en route to earth.

Scarcely had the AAU had the chance to file its decree banning Thorpe from amateur competition than Thorpe had signed a truly professional contract—to play ball for John McGraw and the New York Giants. Such was Thorpe's prowess as an athlete that no one—McGraw included—knew what position he'd play in major league baseball, still four different clubs bid for his services. The Pirates wanted him as a pitcher, but McGraw felt his strength was as a hitter and put him in the outfield. Big Jim hung around the majors for six years, mostly with the Giants, providing lots of copy but few thrills. He called it quits after the 1919 season, at the age of thirty-three, not because he was washed up or frustrated—indeed 1919 had been his best year, he hit .327 in sixty games with the Braves. It was just that a group of Midwestern businessmen were planning to spawn a new professional football league in 1920, and they wanted the old Carlisle hero as the centerpiece. He played until he was forty, enough of the speed, the deftness, the power and the all-around football sagacity lingered so that the National Football League, as those Midwestern businessman has christened their league, gained a measure of respectability and a foothold on the American consciousness.

Near the end of both of their lives, Rice did some lobbying—privately—with the International Olympic Committee on behalf of restoring Thorpe's medals and titles. Nothing came of it. Avery Brundage, the head of the IOC, a teammate of Thorpe's on the 1912 Stockholm team, and as much of an anachronism as the 1912 AAU, replied with a "so what . . . it's dead and forgotten" attitude, according to Rice. "The act that barred Thorpe could never be justified," wrote Rice in 1954. "What right did the AAU have to Thorpe's private gifts, fairly won in those 1912 Olympics. They merely robbed the Indian in cold-blooded fashion." You get the feeling, again by reading between the lines, that he wished he'd done more back in 1913 when his voice might have made a difference.

Rice's voice was a national voice by this time; it had been emerging gradually since his days in Atlanta, and was by now perhaps the most recognized sports by-line in the country. He had begun writing for a Philadelphia publication called *The Sporting Life* in 1906, and the informal newspaper exchange system in place during those years had brought snippets of his verse and writing to any number of American newspapers. Now he was a becoming a regular in the big national magazines as well. Soon after his arrival in New York he began work on a 5000-word article on the effect of nerves on winning and losing in sport. He argued that nerves were divided into three parts—the physical, the mental and the psychological, and used anecdotes from some of the more noted contests in recent years to demonstrate how these various elements of the nerve factor came into play. It was hardly a scientific study, and even from the distance of these years his assertions would seem susceptible to challenge. But the piece was written with conviction and authority and it is noteworthy on the strength of its subject alone. A look at the introspective, or intangible, aspect of sport was an angle that would continue to intrigue Rice throughout his lifetime, and it was to become a consistent theme in his magazine work.

The article on nerves—which carried the how-you-played-the-Game couplet as an epigraph—appeared in the September 1911 issue of *American Magazine,* among the largest, and, in terms of editorial content, perhaps the finest of contemporary American magazines in 1911. It was the home of the "muckrakers," writers like Ida Tarbell and Ray Stannard Baker and Lincoln Steffans who had pioneered investigative journalism near the turn of the century with exposés on Standard Oil, the insurance industry and municipal corruption. *American* also had the *Chicago Tribune*'s Hugh Fullerton writing regular articles on the finer points of baseball, and its circulation would peak at more than a million in the 1920s.

In 1912, Rice began his long, fruitful association with *Collier's,* for which he would write for more than thirty-five years, producing more than five hundred articles, including the pieces that accompanied his selection of the All-America football teams from 1925 to 1947. He began modestly, with a regular one-page collection of verse and aphorisms—"And then again, fools rush around and score the winning run where angels are among those 'left on bases.' "—that appeared variously under the headings "Pick of the Game" "Run of the Game," "Pickups and Putts," or "Pickups and Punts."

This growing renown, his growing value as a journalistic property, made his eventual departure from the *Evening Mail* inevitable. Despite the presence of Rice, Adams and Goldberg, the *Mail* simply could not make circulation inroads against the older newspapers. Then, as now, circulation figures run in direct proportion to advertising revenues, and advertising revenues determine the size of the salaries a newspaper can pay. The *Mail* simply didn't have the money in the cash drawer to keep their talent when the richer papers came bidding. On January 1, 1914, Frank Adams left, taking his column to the *Tribune*. One year later, on January 1, 1915, Rice followed him to the paper Horace Greeley had founded in 1841. In addition to the prestige the *Tribune* name offered, the paper offered something far more tangible—a far-flung and successful syndicate service, and the chance for Rice to have his column in newspapers all across the country. He left the *Mail* with a wish for his readers that he must have also seen as a hope for himself: "[O]ver the uncharted way that 1915 has to offer, here's hoping that you may not have enough bad luck to grow hardened nor enough good luck to grow soft."

The *Tribune* brought out all the trumpets and flourishes for Rice's debut, lest there be any doubt as to just who their new acquisition was. Adjacent to his first column on January 1, the *Tribune* printed a letter—apparently solicited—from American League president Ban Johnson to *Tribune* sporting editor G. Herbert Daley. "He is clever, keen and versatile," said Johnson of Rice. There was a similar tribute that first day from Howard Mann, the sporting editor of the *Chicago Post,* and another on the second day from Ed Bang of the *Cleveland News.* And on Sunday, January 3, the *Tribune* ran a full-page ad announcing Rice's arrival. "The *Tribune* has been speeding up its pace in all departments and the Sporting Pages must necessarily swing into the fast-travelling line," it said. "GRANTLAND RICE is just the man to give the desired momentum. He has done great work in the past. With the 'atmosphere' and the opportunities he will find here we feel certain this young genius will do the best work of his career."

The accompanying sketch by Arthur William Browne shows a pensive, statesmanlike Rice, with penetrating, deep-set eyes, and a sober, yet somehow benign, gentle countenance. Fleshing out the page were thirty-eight gushing paeans from the likes of athletes Francis Ouimet and Walter Johnson; coaches Walter Camp, Fielding Yost and Connie Mack; and fellow writers Will Irwin, Damon Runyon, Bozeman Bulger, Sid Mercer and Hugh Fullerton. Their comments were of the sort generally reserved for a man's funeral: ". . . without a peer . . . ," ". . . has no superior . . . ," ". . . he is a corker . . . ," ". . . different from the usual writer . . . ," ". . . one who has the knack of touching the pulse of the sport-loving public . . . ," ". . . well-versed in all the angles of sportdom . . . ," ". . . has more than any other sporting writer in the country and knows how to use it better . . . ," ". . . has always stood for what is best in sport . . . ," ". . . the best writer of sports in the United States . . . ," ". . . his ideals are high, his logic is wonderful and views are untrammeled."

Two of the respondents had the good sense to provide a bit a levity.

James Montgomery Flagg, who in just a few years would ensure his own immortality by drawing the "Uncle Sam Wants You" recruiting poster, wrote:

> While I dislike the game of baseball more than any one in America, and will promise not to read a word Grantland says about it either on weekdays or Sundays, I know seven or eight people who will, and I am convinced that he will tell these deluded fans what they want to know in the way they would like to read it. If at any time he should turn his talents to something really interesting he will find a constant reader in
>
> JAMES MONTGOMERY FLAGG

Ring Lardner, whose fame was just beginning to blossom on the heels of the Jack Keefe "Busher" stories currently appearing in the *Saturday Evening Post,* and soon to be Rice's dearest friend, wrote his tribute in the curious vernacular for which he was becoming noted:

> editor for the N. Y. city tribune. Gents. dear Sir. I herd where you got this here grandland Rice riteing for you on & after the 1 of jan. & I wisht you would stick my name in your maleing list if you got such a list like that & and male me your paper every day from the 1 of Jan. evry day accept Sunday excluseive. male the paper to what my name is & then stick the address on to the out side—I will get it o. k. & the address is box 163 riverside. Ills. & the box to which I reffer to is p o box & not no ice box. you can send me the bill how much it is to the same address like what you send the paper to & I will pay what ever is it only I work on a paper myself & may be you make reduckshuns & give me the best price you are able only weather you give me a bargun or don't give me no bargun but hold me up for the reglar price I got to read your paper & see what this here grandland Rice got to say. & oblige.
>
> Ring Lardner

Frank Adams, who greeted Rice warmly in his own column of the first, could not resist a poke of the excesses of the full-page ad. In his column of January 5 he had a script for a vaudeville sketch:

> —Did you see yesterday's *Tribune* where they had a whole page about Grantland, sayin' compliments about him and everything?
> —Yes, Frank, I saw that. . . .
> — . . . Well, didn't you think it would make him feel like a breakfast food, Orson?
> —Like a breakfast food, Frank? Why, no.
> —Ask me how, like a breakfast food?
> —How do you mean, Frank, "like a breakfast food"?
> —Why, puffed Rice.
>
> *[Exit]*

A switch from a local to a syndicated column carries with it some hazards. Often, what gains a columnist notoriety and wins him syndication is an ability to capture the reader's feelings and passions on those people and games that matter the most. Without exception, what matters the most to sports fans and newspaper readers are the comings and goings of the local team. And that is precisely the sort of thing that doesn't travel well in a syndicated column. A syndicated column has to have a certain antiseptic homogeneity; it must be free of a Southern accent or a New York elitism.

Rice had never written a parochial column at the *Mail*. Ty Cobb and Tris Speaker were featured as often as Mathewson or McGraw. So there were few adjustments to be made as he began writing for a string of papers that would quickly number more than eighty by the time he left to fight in the Great War, and would level off at between eighty and one hundred after the war and remain there for the next thirty-nine years. Still, there are subtle differences between Rice's work on the *Mail* or the *Tennessean,* and his columns during the early years on the *Tribune.* There was a greater sureness to his convictions now, a wider breath of issues in his columns, a greater preponderance of explanatory pieces. He was particularly given to more pieces on the off-field aspects of sport—rules changes in golf, and what they'd likely mean at both the club level and the championship level; the Organized Baseball-Federal League squabble; baseball players' disgruntlement with the reserve clause, as well as exactly just what the reserve clause was. There were also pieces on the role of courage in winning, and the difference between physical and mental courage; and just what everybody's individual roles and objectives were in the vernal minuet known as spring training.

He was unsparing in his opinions. He professed—again and again—boredom and annoyance with the courtroom engagements and mudslinging of Organized Baseball and Federal League moguls. He was particularly disdainful of Jess Willard, who had taken the heavyweight championship away from Jack Johnson in 1915, calling him "a drab outline against a dull gray sky."

By the end of the first year the column was truly national, attracting letters from around the country. Most of them took some sort of umbrage with matters like the seemingly obscene salaries athletes were then commanding and receiving, or the status quo attitude on the superiority of Eastern college football. On the latter point, Rice, with his roots in the South, was only too glad to extol the merits of the football teams from the provinces and agree with letter writers who pointed to comparative scores to show the merits of a Michigan or a Nebraska or an Oregon State. Autumn Saturdays, however, would find him not in East Lansing or Corvallis, but in Cambridge, Princeton or New Haven. Eastern college football was still a few years away from completing its heady run at the front of the pack.

On the former issue, the matter of baseball salaries, Rice was wholly on the side of the players, as he had been since his minor league days and would be through the end of his life, when the salaries for the DiMaggios and Williamses were pushing $100,000 a year. He was continually reminding readers that baseball careers were generally over by the time a player had reached his thirties, and that baseball had not left them prepared for any other line of work. (In 1915, the idea of a pension for major league ballplayers was as incomprehensible and as illogical as space travel.) Rice was firmly against multi-year contracts, however, feeling that they engendered a complacency and a lack of hustle and ultimately cheated the fan. In keeping with this attitude he was also supportive of owners who showed a propensity to spend large sums of money to acquire players in order to improve the product. In this, he was part of the chorus of voices excoriating Connie Mack of the Athletics for selling off the players that formed the soul of his championship teams like so many trinkets at a flea market. But Rice got the point across differently from many of his colleagues. Rather than damning Connie Mack for selling, he was more inclined to praise Charles Comiskey for buying. And therein lies the sort of distinction that separated the "Gee Whiz!" school of sportswriting from the "Aw Nuts!" school.

The division had always existed; it is a natural, inevitable, even atavistic distinction. As a society of fans, we invest our emotion in sport, and emotion leads us to irrational extremes—blind love and hero-worship when things are going well; bitterness, anger and affixing responsibility for deep and genuine hurt when things go poorly. The Gee Whiz! writer is the optimist, anticipating heroics and triumph, yet somehow able to be both heart-broken *and* exhilarated in defeat. The Aw Nuts! writer is the skeptic who writes with a how-can-you-let-these-guys-break-your-heart-they're-scarcely-worth-the-bother sort of defiance. The Aw Nuts! writer—like the Aw Nuts! reader and fan—doesn't love sport any less than the Gee Whiz! type; he just refuses to wear his heart on his sleeve. Put another way, Gee Whiz! people are the ones who cry at weddings, sad movies and when the national anthem plays at the Olympic Games. Aw Nuts! people, however moved, wouldn't dare.

Whenever sportswriting is discussed in terms of the Gee Whiz! and Aw Nuts! schools, Grantland Rice is named as the apotheosis of the Gee Whiz! And, as the years pass, and the Gee Whiz! falls increasingly out of favor, he virtually stands alone. In truth he had a lot of company during these years. Heywood Broun, writing baseball for the *Tribune* when Rice joined the paper, was a Gee Whiz! So was Runyon, at least at this stage of his career. And so were Joe Vila and William B. Hanna of the *Sun*.

Ring Lardner is generally cited as Rice's opposite, the apotheosis of the Aw Nuts! school. His "busher" stories in the *Saturday Evening Post* were revealing of a coarser side to baseball players and life in the big leagues. But there is an exuberance and naivete in Jack Keefe's boorishness; the stories were enormously popular. Such was not the case with Midge Kelly, the hero of "Champion," one of Lardner's most enduring stories, published in October 1916 to an uneasiness on the part of the reader. The seamy side of big-time sport was driven home most searingly in "Champion," right from the opening paragraph, which lands on the reader like a sucker punch to the jaw.

> Midge Kelly score his first knockout when he was seventeen. The knockee was his brother Connie, three years his junior and a cripple. The purse was a half dollar given to the younger Kelly by a lady whose electric had just missed bumping his soul from his frail little body.

What follows is a disquieting tale of a drunken lout who beats women, tanks fights, skips out on debts, and cuckolds his manager, all the while earning the approbation of newspaper writers, who eagerly swallow the offerings of Kelly's manager, and convey a portrait of a clean-living family-man. In a trenchant indictment of contemporary sportswriting in the story's peroration, Lardner suggests that the sportswriter could have written more accurately had he spoken to those whom Kelly had wronged.

> But a story built on their evidence would never have passed the sporting editor.
> "Suppose you can prove it," that gentleman would have said, "It wouldn't get us anything but abuse to print it. The people don't want to see him knocked. He's champion."

As graphic an example to the Aw Nuts! genre as "Champion" is, it was, after all, fiction. Nobody as thoroughly reprehensible as Midge Kelly ever escaped having his transgressions as a human being called to the attention of America's newspaper readers. And it is difficult to compare journalism and fiction. A better contrast between the Gee Whiz! and the Aw Nuts! comes from comparing Rice with his *Tribune* colleague, W. O. McGeehan.

Mentioned in virtually the same breath as Lardner as the embodiment of the Aw Nuts!, McGeehan had come to the *Tribune* staff from the *Journal* in February 1916. A lively and popular raconteur, McGeehan was called "at heart a sentimentalist" by Stanley Walker. Only those who knew him personally would see the aptness in that. For in his column, McGeehan was an unsentimental debunker. "No amount of build-up could convince him that a yellow or inept fighter was a potential champion," said Walker. "Nor could he see any resemblance between a fourth-rate sports event and the second siege of Troy." McGeehan was given to calling the people in the seats "customers," never "fans." His weapon was the stiletto, not the bludgeon; he drove home his point with a whimsy and a subtle yet unmistakable sarcasm.

The summer of 1916 marked the arrival on the national scene of golfer Bobby Jones; at the tender age of fourteen he advanced to the quarter-finals of the U. S. Amateur championship at Merion. And the sports pages of the *Tribune* printed during those weeks a striking example of the contrast between Gee Whiz! and Aw Nuts! at their best. Grantland Rice, who was at Merion, was moved to the following:

> Considering the personal equation and all other intimate details, a chunky, chubby broth of a lad, just fourteen years old, with a pink, round face and big, blue eyes, gave the greatest exhibition of golf today that we have every seen.
>
> His name is Bobby Jones, Jr., of Atlanta, Ga., and whatsoever he may do against Robert A. Gardner, amateur champion on Thursday, his name is already written on the sporting scroll, where only the select, those who combine surpassing skill with lion-hearted courage, have any right to belong.

In a column after returning to the office, Rice continued:

> At his age the game in this country has never developed anyone with such a combination of physical strength, bulldog determination, mechanical skill and coolness against the test. He is the most remarkable kid prodigy we have every seen—and here and there in sport we have looked upon one or two.

McGeehan didn't go to Merion. He stayed back in the office and read what Rice wrote. In his column on September 22, McGeehan wrote a screenplay called the "Adventures of Beatrice Buggs, Featuring Izzy the Photographer, in Lost at the Nineteenth Hole." In Reel One, Beatrice Buggs, "the beautiful girl reporter," comes into work to find a letter "written in a childish hand."

> Dere Beatrice.
> I am up at the Merion Kricket Club playing golf with a lot of old men. They are orfully soar at me and I think they meen to do me dirt. Plase help me, Beatrice. I am only a littul boy fourteen years old but I am a bare at golf if these old birds don't give me the works.
> > Your littul frend,
> > Bobby Jones

Beatrice Buggs goes to Merion, and finds Bobby Jones poisoned by
the elderly men who form the gallery. "Golf is not a child's game," says
one. Whereupon Beatrice assumed a Bobby Jones disguise, won the tournament and "exposed the childishness of the game of golf." Overall the
piece is strained, and it certainly doesn't stand up against the prescience of
Rice's assessment of Jones's probable impact upon the game. Still, it has a
certain appeal, and when taken together with Rice's hosannas, certainly
illustrates the difference in attitudes these two leading practitioners of contrasting schools brought to their work.

Rice was traveling more now. He was not yet the peripatetic creature
he would become in the 1920s, but spring training, the World Series, college football, and major golf championships like the U.S. Open and the
U.S. Amateur kept him away from the office a good part of the year. So
too did his own excursions to play the game of golf—less than a decade
after taking up the game as a relief from the tensions of his job at the
Tennessean, Rice was an avid and able golfer, shooting at or near par in
assorted second-rank tournaments and celebrity events. Kate would frequently travel with him, and they in turn would frequently run into Ring
and Ellis Lardner, sowing the seeds of a friendship that would flourish
after the Lardners moved east after the war.

Rice's second book of verse, called *Songs of the Stalwart,* was published, to kind reviews, by Appleton in the fall of 1917. The syndication
was a success, and his articles for *Collier's* brought his writing to readers
that didn't have access to one of the more than eighty newspapers that
subscribed to his column. By the fall of 1917 he was the most famous and
best-paid sportswriter in America.

But his tranquil and idealistic life was about to be turned upside down.
For while Rice was writing of Mathewson and McGraw and Bobby Jones,
and writing lyrical odes to the virtues of courage and fortitude and to the
melancholy of being separated from home, the headlines on the front page
of the *Tribune* had lost their innocence, and taken on an increasing darker
hue: Sarajevo, the *Lusitania,* Ypres, Jutland, the Somme. Finally, in April
1917, the headlines proclaimed that America, too, was at war. At the age
of thirty-seven, Grantland Rice felt duty-bound to become a part of it.

7

Of Clotted Gore and
Dreamless Sleep

The men of Grantland Rice's generation had not known the dead of the Civil War. They knew the survivors, men like Henry Grantland; and they were suckled on stories of valor and heroism and of war as a rite of passage. Their boyhood heroes were calvary officers. Sticks were burnished sabers and gleaming Springfield rifles. Every bush was a breastwork and every protected hollow a fort.

When they reached their maturity they were treated to the American incursion into Cuba, that "splendid little war" as it was once called. Though relatively few men fought in the Spanish-American War, those who did were appropriately lionized, for reporters outnumbered soldiers on the island that summer of 1898. Bloodshed was limited, and the newspapers had drummed into the American consciousness the conviction that this was the most noble of incursions, a crusade to liberate an oppressed people. That the reality was more complicated did not intrude upon Americans' innocent view of the rightness, the necessity of the war, and the valor and heroism of the men who fought it.

Turn-of-the-century Americans were not the only generation to hold a romantic view of war, of course. To a certain degree these foolish notions have been endemic to all of history's humans. (Why else would we keep fighting wars?) But the romance was particularly acute in the United

States in the years before the Great War. It was complemented—if that is the right word—by sense of destiny and invincibility Americans held. Our cause was always just, our means always honorable, our triumph always inevitable. Listen again to the words of "Over There," George M. Cohan's anthem to American swagger and naiveté:

> Johnny get your gun, get your gun, get your gun,
> Take it on the run, on the run, on the run,
> Hear them calling you and me,
> Every son of Liberty.
>
> Hurry right away, no delay, go today,
> Make your daddy glad, to have had such a lad,
> Tell your sweetheart not to pine,
> To be proud her boy's in line,
>
> Over there! Over there!
> Send the word, send the word, over there,
> That the Yanks are coming, the Yanks are coming,
> The drums rum-tumming everywhere.
>
> So prepare, say a prayer,
> Send the word, send the word, to beware,
> We'll be over, we're coming over,
> And we won't come back,
> 'Till it's over, over there.
>
> Johnny get your gun, get your gun, get your gun,
> Johnny show the Hun you're a son of a gun,
> Hoist the flag and let her fly,
> Like true heroes do or die.
>
> Pack your little kit, show your grit, do your bit,
> Soldiers to the ranks, from the towns and the tanks,
> Make your mother proud of you,
> And to liberty be true.
>
> Over there! Over there!
> Send the word, send the word, over there,
> That the boys are coming, the boys are coming,
> The drums rum-tumming everywhere.
>
> So prepare, say a prayer,
> Send the word, send the word, to beware,
> We'll be over, we're coming over,
> And we won't be back,
> Till it's over, over there.

This naiveté also extended to seeing war as some form of ultimate athletic contest. There grew an inevitability to American involvement during the European years of the war, and consistent with this American bravura, there was also much speculation on how the American "team" might fare—and how they might best prepare. Most of the country's sportswriters subscribed to the theory that sport was a good preparer for war, talk of the war filled the sports columns. Most of it was pretty silly, of course; and the silliest of all came at the very beginning, when the inevitability was still a bit distant and the speculation was thus theoretical. Grantland Rice's hypothesizing on sport as the great preparer for war began in January of 1915, and the first sport he offered as training was golf:

> [There are] at least 250,000 golfers under forty years of age. These golfers could march almost indefinitely as any one of them can play thirty-six holes up and down hill, covering at least ten or twelve miles without the slightest fatigue. Most of them are able to walk practically all day without becoming winded.
>
> Which is to say nothing of the young caddy army they are bringing up. There are at least 200,000 caddies in the USA under 21 years of age. These youngsters not only can walk all day, they can do the trick with golf equipment weighing twelve or fourteen pounds swung across their shoulders.

Some of it was intended to be silly. There was the time in 1916 when a French general reportedly commented on what a fine lot of grenade throwers American baseball players would make, and Rice reported Yankee infielder Paddy Baumann staring disconsolately out a hotel window, his "attitude . . . one of utter dejection."

"If we declare war I am going to have the sorest arm a ballplayer ever knew," said Baumann to Rice, "if I have to cut it off to get results." Rice went on to consider the possibilities of a spring training for grenade throwing, "where the players, standing in trenches, could peg back and forth until great speed and perfect control were attained." He went on to discuss on spit ball grenaders and lefthanders who might to used to confuse the enemy and outfielders who would be used for "long-range grenade throwing."

Most of Rice's dicta on the matter were absent any of this "levity and persiflage," however, and espoused that proud party line of the sports world—that athletic training did provide benefits for a country likely to be at war. The assertions were flippant and cocksure at the beginning, more considered as American involvement drew near. "There is no touch of sport in war," he wrote in 1917. "The two games are entirely different. One is a physical development. The purpose of the other is physical extinction for the enemy.

> But sport has its place in development for war, for, despite all the new inventions and new armaments, physical stamina and vitality are still qualities which have their place.

Other conditions being anywhere near the same, a nation in fine physical shape would have that much of a start.

Rice drew the line at suggestions that war would actually be helpful to sport. One of his correspondents suggested that the war *"among those who are left* [emphasis added], will develop a more hardy race through outdoor life and constant exercise, and that in this way it will help develop more athletes." People actually thought this way in 1915.

"This war will help sport as much as dynamite and destruction can help anything," wrote Rice in reply. "Any continent that loses five million of its best men is bound to lose many a present and future star, as the race is bound to sag." Rice did find a semblance of a silver lining in this dark cloud, though. He advised the reader that the rebound from the holocaust would not take long. "It was only shortly after the Civil War that baseball began its first march upward," he said, "and began to develop as the greatest game of them all."

It is only human nature to view acts of God and global cataclysms in personal terms: How will this affect me? What is this going to do to my world? What a person contemplates in this context, which aspects of the insular world concern him, can show a lot about attitudes, prejudices and character. Rice thought a lot of how the war would affect sport, and among the more remarkable of the speculations and proposals for postwar sport is a thoughtful and fair-minded essay that appeared in his column in January of 1917. It took note of the fact that before the war, most of the international sporting titles—polo, the Davis Cup, the British Amateur golf championship—were in British hands. The jingoistic notion had been put to Rice that as soon as the war was over, the Americans could swiftly "recover some of this lost ground." After some thought, Rice was inclined to believe that the "rush . . . to recover some of this lost ground . . . should not be too hastily made."

[The] reason is plain enough. Great Britain, if she cares for it, should at least be given time to get started again and to reorganize her playing form.

Most of her leading golfers, tennis players and polo players have been serving under the Union Jack. Many of these, including such stars as Arthur Wildin, Norman Hunter and scores of others, have been killed. The others undoubtedly have gone badly off their game.

All this while American entries have been engaged in the busiest sportive competition the game has even known beneath the Stars and Stripes. Our leading entries in golf and tennis, especially, have had the chance for great development, the opportunities to come forward at top speed. Obviously there could be no great glory to be obtained in rushing these trained forces at once into international competitive action, before our leading rival had a chance to adjust herself and get planted for the shock.

This should take no great while, but at least a fair chance should be given. If the war should end in the winter or spring, no international arrangements should be planned and carried through before the next year. For there would be very little credit in recovering trophies from an opponent too exhausted to make a fitting defence.

Three months later the prospect of taking advantage of an exhausted athletic rival became a moot point. Rice was in Georgia with the Yankees on April 2, when President Woodrow Wilson asked Congress to declare war on Germany. Rice had reported that in addition to the normal spring training regimen, the Yankees had undertaken some military drill under the direction of a Sergeant Gibson, who "was greatly surprised at the quickness shown by players in obeying commands or picking up instruction. He found them to be far ahead of the average recruit lot drawn from other realms of civilian existence."

Rice was making his way north with the Yankees when Congress complied with Wilson's war request four days later. Baseball season would open on schedule, and Rice would spend his summer writing about whether Ty Cobb could regain the batting title he had lost the previous season to Tris Speaker; or whether Babe Ruth, the young lefthander on his way to a 24-win season with the Red Sox, had supplanted Walter Johnson of the Senators as the best pitcher in the American League. But the column was tempered with talk about sport's role in the war, and offered up, as all of sport was offered up during these early weeks, with something of a sense of guilt, and a certain sense that these would be the last games for a while.

The curtain will be drawn across competitive sport in this country on the day that an American Army goes into battle and the casualty list comes in. The curtain of fire beyond an American charge will be the curtain of Good Night hung across all competitive sport.

This will not mean the end of all sport for physical upkeep in an individual way. Not even those along the firing line have gone this far. But sport as a spectacle—barring that for Red Cross purposes or other war funds—will fade out.

Sport, seen through the smoke of battle, isn't sport after all.

Colleges in the East cancelled their baseball seasons upon Congress's declaration of war; and among the first in line at the recruiting stations were a rash of beefy men who had in 1916 formed the heart of the football teams at Harvard, Yale and Princeton. Such patriotism immediately gave rise to discussion of cancelling the college football season in 1917. This was not what the War Department wanted. The leaders in the War Department were of the same mind as Grantland Rice on this one: that a recruit in good physical condition would likely make a better soldier, thus anything that enhanced physical conditioning should be encouraged. Secretary of War Newton D. Baker called the leaders of the Collegiate Athletic Association to Washington in July to encourage them to continue intramural and intercollegiate competition.

The college leaders' immediate concern was finding players. Every single football letterman at Harvard had answered the first call to arms; Yale and Princeton's numbers were less absolute but similarly impressive. In his column Rice proposed that the 1917 football teams be limited to athletes who had never before played a varsity game, and that any schools which did have returning athletes use them as assistant coaches. "All teams would start from the same spot," he pointed out, "there would be fine development of new talent, and those who need exercise most would get their opportunity without having to compete in vain against seasoned material." The colleges accepted Baker's suggestion and continued sports; they were lukewarm to Rice's suggestion that they do it wholly with inexperienced players—though the reality of competing with the Army for able-bodied men left them with little else. Rice refused to pick an all-star team that fall, saying that it would be unseemly, though he received a rash a letters from annoyed readers who felt differently.

Major League baseball made no concessions to the war at first; they played out the 1917 season normally, and American League president Ban Johnson went so far as to ask the War Department to exempt major league players from the draft. His request was met with harsh public criticism from players who saw it as an affront to their manhood, and was withdrawn before it could suffer the ignominy of being denied. In 1918, the major leagues scheduled a full season, but with rosters reduced by the draft and attendance decimated by the influenza epidemic, they truncated the season by thirty-odd games, playing the World Series in early September.

In late May of 1917, with the status of sport still very much in limbo, Rice joined Bill McGeehan, Hugh Fullerton, Max Foster and five others on a fishing trip on the Cains River, some fifty miles north of Fredericton, New Brunswick. It was a time of instruction and reflection. Canadian sportsmen, solicitous of the prominent Americans in their midst, were eager to share the Canadian experience on sport and war. As part of the British Commonwealth, Canada had been at war since 1914, and more than a million men had been killed or wounded. Canadian football, stocked principally with native Canadians, had all but dried up, with the first-liners off to war and the fans reluctant to turn out for exhibitions featuring the leftovers. Baseball, however, with imported American players providing the bulk of the manpower for the minor league franchises in Toronto and Montreal, had thrived. Rice was told that among those Canadians who remained at home there was "an unusual craving for what might be called recreative divergence." Interest was also sustained by the fact that the newspapers continued to devote generous space to the game. Rice's Canadian source theorized that "if the time comes when your newspapers begin to cut down heavily in sporting publicity," interest in baseball would wane.

This last point must have provided fodder for a good deal of spirited campfire conversation: To what degree did newspaper coverage of sport reflect interest, and to what degree did it create interest? And if it did create interest, to what degree should it be creating interest in so trivial a undertaking during a time when the country faced perhaps the gravest crisis in its comparatively short history? The evidence that Rice wrestled at length and in vain with this question is in the columns from the summer of 1917—a schizophrenic blend of happy speculation on how umpires might better cope with truculent managers like John McGraw, and anguished musings on other men, who

> . . . have come to the day
> When the big debt is due in the smash of the fray
> And if it be only death that they owe,
> Or if it be only to suffer and grow,
> They are ready to pay.

But if that fishing trip to New Brunswick yielded a measure of sober speculation on what the months ahead held, so too did it yield memorable camaraderie and memories of innocent good times. The campfire talk included conversation on spring trainings the men had known, and talk about "Gaffer Bill" McGeehan, who after ineptly trying to gaff a salmon in the time-honored way, gave up in frustration and scrambled down the bank of the Cains and tackled it instead. At journey's end the expedition stopped in Montreal for a couple of days, and Max Foster, then one of the prominent by-lines on the *Saturday Evening Post* and an internationally renowned fisherman, elicited the attention of a Montreal newspaper interested in his views on fishing. Foster was out when the call came, and Rice took the call and received the reporter. Posing as Foster, he spoke of the frustration of not being able to find enough worms ("Foster naturally would never have been so plebeian as to use worms for bait," noted Rice) and referred to rods as poles, and waders as rubber boots and made countless other déclassé references. Rice delivered the interview straight-faced; the reporter dutifully wrote it up, and the embarrassed Foster was livid when it appeared—though only for a time.

The changes that were sweeping American society swept through the sports department of the *New York Tribune* that summer too. Shortly after his return from the New Brunswick fishing trip, Bill McGeehan accepted a commission in the Army. Heywood Broun left the sports department to accompany the advance troops of the American Expeditionary Force to Europe.* Bill Wright and Frank O'Neil of the *Tribune*'s sports staff were also in uniform by year's end, as were assorted newspapermen from the city's other sports desks—Bozeman Bulger, Innis Brown, Donald Day and Jack Wheeler.

Rice was nearly thirty-seven years old, and not subject to the draft, which affected men aged twenty-one to thirty. He was married and the father of ten-year-old daughter. Still, a sense of duty called him to the war; a pride in his athleticism gave him the confidence that he could survive as a foot soldier. He was not alone in this sense of duty. It was particularly apparent in the scions of privilege—"gentlemen" such as the men of the Harvard football team; or Franklin D. Roosevelt, Harvard '04, Undersecretary of the Navy in 1917, who asked to be relieved of his bureaucratic duties and commissioned; or Scott Fitzgerald, ex-Princeton '17, who ordered his custom-made officer's tunic from Brooks Brothers and fretted that the war would be over before he could get over to Europe, as it was; or poet Joyce Kilmer, Columbia '08, who enlisted despite being father to four children and husband to a pregnant wife, and who one year later would be one of America's most mourned casualties; or actor John Barrymore, who grew so morose when his attempt to enlist at the age of thirty-five was rejected because of varicose veins that he went on a screaming drunk and threw dinnerware about the dining room of the Astor Hotel.

A number of athletes enlisted amidst much publicity—Red Sox manager Jack Barry and outfielder Duffy Lewis; Braves shortstop Rabbit Maranville; golfer Francis Ouimet.

Everything in Grantland Rice's nature and heritage pointed him towards enlistment, too. Not only was there the tug of duty, there was the tug of adventure, of romance. However illogical this may have been, it was decidedly real, and something that defied introspection. Cognizance of its absurdity did not lessen its pull.

> With all its misery and death,
> Its battered hills and shuttered plains,
> With all its drift of poisoned breath,
> Its lashing gales, its sodden rains—
> We know—with all its bleak despair,
> With all its phantom exodus,
> That somewhere—somewhere Over There
> The Great Adventure beckons us.

*Broun was dockside when the first troops disembarked in France, and recorded the first words of the first American soldier on French soil to be: "Do they allow enlisted men in the saloons in this town?"

We know in Flanders' bloody sod
    How deep they sleep in endless dreams;
We know how many crosses nod
    By silent hills and shadowed streams,
But thru the ghostly drifts that play
    We know thru bugle, fife and drum,
That somewhere Over There today
    The Great Adventure whispers "Come!"

We know that rare thanks we should feel
    So far from any foe's advance,
Safe from the shrapnel and the steel
    Which rains its fury down on France;
But in our hearts we also know
    The old content's forever gone,
Where on some far dream's undertow
    The Great Adventure calls us on.

By all the ghosts of No Man's Land,
    Thru all its fury, flame and flood,
On thru the anguish each must stand
    In wallow drifts of mud and blood,
On thru whatever hells may wait,
    With marching feet and rolling drum,
Beyond the final grip of Fate
    The Great Adventure whispers "Come!"

He chose the 115th Field Artillery, a Tennessee National Guard unit that was commanded by an old Nashville friend, Colonel Harry Berry, and federalized in the summer of 1917. He tidied-up his personal affairs, putting his by-now not-inconsiderable cash and securities in a trust to care for Kate and Floncy in the event he didn't return. At the office, he put twenty-odd columns "in the can"—enough to take the *Tribune* and the syndicate papers to the end of 1917—and left New York for Camp Sevier, near Greenville, South Carolina, on Monday, December 10, 1917.

Rice arrived in Greenville in the midst of a hellacious Southern storm, which swirled for thirty-six hours and left six inches of snow covering the ruts and stumps and softening the harsh landscape of recently cleared land at the camp. Camp Sevier, like most of the World War I training sites, was a city of pine shacks and white canvas tents hastily constructed during the summer and fall to accommodate the thousands of draftees and volunteers. While the snow masked the naked ugliness of the camp, it also brought accompanying sleet and a chill wind. The tents where the men were quartered flapped and whistled in the December wind and filled with smoke as the men—fighting colds and frostbite—struggled to stoke a fire in the stove with the green firewood cleared from the parade field short weeks before. Rice caught a cold upon arrival that he later insisted lasted until the Armistice.

On Sundays, during the week's only idle hours, the men would crowd into the Y.M.C.A. building at camp, which provided books and newspapers and magazines and soft chairs to read them in; paper and pens and tables to write home on—and all of it in a snug haven that was warm and dry and a refuge from everything except thoughts of what lay ahead.

> Of course each knows, deep in his soul,
>   That all who leave will not come back—
> That some must pay the closing toll
>   And "go West" on the twilight track;
> That Fate has marked from pawn to king,
>   The name of each who has been drawn
> To look upon his final spring
>   When April's sky rides out from dawn.

Rice made more productive use of the Y.M.C.A. building than most, for in addition to letters home he also continued to produce a column for the *Tribune* and her syndicate. Beginning January 1, the supply of canned columns exhausted, he cut back from six days a week to three. The "camp fire" flavor the *Tribune* had promised its readers and subscribers included verse on the angst of reveille and an essay on developing a philosophy for coping with the "annoyances and discomforts and routine." But the campfire column was short-lived. It fell victim to Rice's fourteen-hour days at Camp Sevier and was abandoned in mid-February.

Rice had enlisted in the 115th as a private, but as soon as he was sworn in he was promoted to sergeant and placed in charge of a training platoon. His age and maturity alone would have qualified him for this immediate promotion, but it was also impossible to ignore his celebrity. He was surely one of the few privates whose arrival warranted the lead story in the weekly camp newspaper.

He was also recommended for officer's training, and though he struggled a bit with the math proficiency expected from an artillery officer—it was a bit more complex than figuring batting averages and he was, as he pointed out to an officer overseeing his studies, more than fifteen years removed from his Vanderbilt math classes—he was commissioned a second lieutenant in February, just as the first traces of the South Carolina springtime were beginning to come into evidence. In addition to assorted other duties, he was made the unit's athletic officer, and told to clear a field for a baseball game against the 114th Field Artillery. He was directed toward a "patch of trees" that he described as a "solid green forest."

"It's got to be ready [in two weeks]," he was told. "[General Gus] Gately ordered it." Rice was empowered to commandeer as many men as he needed, and the next morning he had 280 men working with "picks, axes, saws and dynamite," felling the trees and blowing the stumps to "kingdom come." The field was ready in time and the game was played—Rice failed to record who won.

There were few games on Rice's field, for in late April, their stateside training complete, the 115th was assigned to the 30th Division and sailed from Patrick Henry, Virginia, aboard the transport *George Washington.* They zig-zagged across the Atlantic in a tortuous three-and-a-half-week passage, arriving in Cherbourg on May 24. Upon arrival in France, they were scheduled for further training at Besançon, near the Swiss border, south of the Argonne Forest. But Rice received orders cutting him away from his unit. He had been assigned to Paris, to the staff of the military newspaper, *The Stars and Stripes.*

It did not take any deep or profound thinking to recognize the potential utility of a newspaper in a community of 300,000, the size of the American Expeditionary Force in February 1918. But newspapers at the front had a poor track record; the British and French both tried. But the Americans were to succeed where their allies had failed because of the vision of a young second lieutenant named Guy Viskniskki, a censor at field press headquarters in Neufchateau, and before the war an assistant to Jack Wheeler in his fledgling syndicate service. Viskniskki's proposal to skeptical superiors was for a paper that was different from the house organs put out by the British and French. He argued that "a lively, slightly irreverent, plain-spoken newspaper, which did not smell to Heaven of propaganda and which was not choked up with deadly official utterances" was what was needed. He further convinced higher-ups that such a newspaper could not be put out by Army officers; it had to be put out by newspapermen. There was no shortage of erstwhile newspapermen serving with the A.E.F. Among the first people Viskniskki found was Pvt. Harold W. Ross of the 18th Engineers, who would become the paper's managing editor—and earn himself literary and journalistic immortality after the war as the founding editor of a quirky little magazine he called *The New Yorker.*

When he arrived at *Stars and Stripes,* Ross was a twenty-six-year-old high school dropout who had worked for a dozen different newspapers before the war, including the *Atlanta Journal* and the *San Francisco Call.* He was an unlikely soldier, unkempt and independent, and thus perfectly suited to resisting the pressure that would come his way from colonels and generals with their own notions of how a military newspaper should be run. The newspaper he produced was witty, eclectic, erudite without being pompous, as eagerly awaited at the front as mail call. It had frank news of the war, news of home (including a fairly comprehensive sports page); a generous selection of doughboy-produced poetry; and advertising for military tailors, custom bootmakers and the Paris branch of Brentano's bookstore. It cost ten centimes, which not only poured money into the coffers of various company funds but also served to convince its readers that this was a *newspaper* serving them and not a propaganda sheet serving their commanders. The doughboy in the trenches took further comfort in the fact that aside from officer-in-charge Viskniskki, the names on the masthead were those of humble enlisted men—privates and sergeants. Joining Ross in the office were privates Hudson Hawley of the New York *Sun* and the 101st Machine Gun Battalion and John Winterich of the *Springfield Republican* and the 96th Aero Squadron. The *Stars and Stripes* correspondent at the front was Sgt. Alexander Woollcott, formerly the drama critic for *the New York Times* and a clerk in Base Hospital No. 8 before he was requisitioned by Viskniskki for the newspaper.

Officers were welcome around the *Stars and Stripes* shop just as long as they had something to contribute and remained aware that, around the newspaper, rank had no privileges. Capt. Franklin P. Adams of the *New York Tribune* joined the staff for a time in the spring and summer. So did Lt. Steve Early of the *New York Times,* who some years after the war would be best known as President Franklin Roosevelt's press secretary.

Stars and Stripes had been in business for three-and-a-half months when Rice joined the staff in late May. Its circulation was approaching 500,000, and the editorial offices had moved from a single back room of the Field Press Headquarters in Neufchateau to a rather substantial suite of offices in the Credit Mobilier on the Rue Talbout in Paris.

Ross quite naturally put Rice in charge of the sporting page, and Rice moved quickly to leave his mark on it. He killed it off. In an unsigned piece announcing the sports page cancellation, Rice reiterated a point he had made back at the *Tribune*—that the time for spectator sports had passed when the first American entered combat. "There is but one Big League today for this paper to cover," he wrote, "and that league winds its way among the S.O.S. stations scattered throughout France and ends at the western front." Also carrying Rice's stamp, though again not his signature, is an August editorial condemning professional athletes who had chosen shipyard work as a means of eluding the draft. "With thousands of their countryman charging machine guns, working under shell-fire or grinding away back of the lines, it seems beyond belief that any well trained athlete, fit for service, should be guilty of such yellow-hearted cowardice, traitors to their country's good, and worse than traitors to their own souls."

With the sports page gone Rice wrote a bit of verse for the paper, and went off to report on the front, which in those weeks was centered on the village of Château-Thierry. But life in Paris was largely pleasant. In addition to running *Stars and Stripes*, Harold Ross ran a poker game at an out-of-the-way restaurant called Nini's, located on the the Place du

Terre on the top of Montmartre, cross town from the *Stars and Stripes* offices. The games would run all day Saturday, the day after the weekly paper was published and distributed. Rice, Ross, Woollcott, Winterich, Adams, Early and Ross were all regulars. They were joined from time to time by war correspondents such as Richard Oulahan of the *New York Times,* and Rice's old cronies, Heywood Broun of the *Tribune* and Ring Lardner of the *Chicago Tribune,* who had arrived in Paris in August—taking his *Tribune* colleague George Seldes aback when he showed up at the *Tribune* offices and said "the Colonel [*Tribune* publisher Robert Mc-Cormick] sent me to France to cover the comic side of the war."

However diverting card games with his friends may have been, and no matter how challenging and fulfilling and necessary was the work of *Stars and Stripes,* the fact remained that Rice had not come to Europe to be a correspondent. He had certainly had that option open to him at the beginning and had rejected it, just as he had rejected the pursuit of other options that could have landed him a commission and command of a desk or a tour group somewhere. No, he had enlisted to play a part in the shootin' war, and he intended to see it through. The 115th remained in cantonment through the summer and Rice was happy enough with his work on the paper. But when the 115th got its orders to move up in September, to take part in the Allied capture of Saint-Mihiel and in the Argonne campaign that would end the war, Rice was determined to go with them. He prevailed upon the powers that be to cut him orders liberating him from Paris and *Stars and Stripes* and reassigning him to the 115th, moving into the mud and misery and what-war-is-all-about reality of the Argonne forest.

Rice may have been one of the more famous men in the American Expeditionary Force, but he was nevertheless a humble lieutenant and when it came time to rejoin the 115th, he was on his own in negotiating the 150 miles between Paris and Flirey. He hitched rides where he could, and walked when he had to. According to Will Irwin, who saw him off, Rice was toting "enough equipment to quartermaster half the boys at the front." Included in the mobile inventory were blankets, fry pan, burner, shoes, socks, shelter half, rifle, ammunition, and "that infernal gas mask." It was raining as he left Paris, a rain that would persist throughout what became a three-day odyssey in search of his unit. He lightened his load quickly: "I shed stuff like a moulting turkey sheds its feathers," he said. After nearing the Argonne, he spent most of the three days walking. The roads were choked with ambulances and supply wagons engulfed in the mud. The ditches were lined with the dead and wounded. The Allies had been advancing steadily all summer and there had already been rumors of impending capitulations by some of Central Powers. But not everybody was optimistic. Rice met a Frenchman who told him that the war would never end, because the generals didn't want to go back to being waiters and taxi-drivers.

Rice slept a few hours in the shell of a farmhouse one night. The second night he crawled into an old dugout and lay down next to another soldier already encamped there. At dawn, he discovered his roommate was German, and had been dead for a couple of days. He reached Flirey on the morning of September 16, the day after the Americans and the French had secured the town of Saint-Mihiel. The 115th wasn't there but C. Leroy Baldridge, an artist for *Stars and Stripes,* was, and Baldridge drew a sketch of Rice in a regal and pensive pose, using his imagination to draw the face clean-shaven and the tunic clean and dry.

Rice next made his way through the mud up to Avocourt, where, on the third midnight of his journey, he finally hooked up with the 115th, just as they were about to set out for the high country of Montfauçon, at the edge of the Argonne. Rice was greeted warmly by the men of the 115th. These were mostly Tennessee farm boys, and the sophisticated Rice was somehow a bridge between their simple lives in Tennessee and the worldly horror in which they were now entangled. A member of the 115th called him "the flaming spirit of the company."

"He marched to war with a zeal and an ardor that quickened the heartbeats of his comrades," continued this one comrade. "To him it was a holy cause." Holy cause or not, he was a natural leader, and the men of the 115th were glad to have him back as they muscled their howitzers and caissons into position near Montfauçon.

As an artilleryman, Rice never knew life in the trenches of the front, never knew the wave-of-nausea, haven't-the-strength-to-stand, my-God-my-brain-isn't-going-to-let-my-body-do-this feeling of doubt that comes in the seconds before going over the top. He never suffered the trauma of seeing those around him fall like blades of wheat before the scythe. He himself would downplay his role in the war, saying that the artillery had it easy there back of the trenches, throwing shells and only occasionally having to sweat an incoming artillery round or a strafing by an enemy biplane. But the strafings were real and frequent, and the men of the artillery watched the dog-fights between the fabric-winged biplanes with the interest of those who have a keen and abiding stake in the result. They found the mud—the relentless, shin-deep, hard-to-tell-where-this-ends-and-I-begin mud—could be an ally. An incoming round landed five yards away from Rice and his men once as they were trying to pull their cannon out of a gully. The mud sucked in the shell and smothered its explosion; instead of being maimed or killed, Rice was merely splashed with more mud.

Stationed near the rear of the front as he was, Rice also bore there-but-for-the-grace-of-God witness to the cannon fodder who marched to the trenches with that curious mix of hubris and horror. He would see the same men a few days or weeks later, their numbers shrunken, their horror swollen. "I don't think many men come out of a war with their ideals and idols exactly the same," he said in consummate understatement of the grisly parade he saw at the front.

After the Armistice, Rice was moved to Third Corps Headquarters in Souilly and assigned as press officer. He was too much of a journalist to make a good flack. He alerted reporters to a group of fifty elderly German prisoners who had been promised cigarettes and American rations if they surrendered. After they surrendered there were no treats forthcoming and they felt bitter and betrayed. His commanding officer was upset that the story got out; it made the Americans out to be double-crossers, he said, and told his press officer he wanted no more stories like that called to the attention of the press.

The Third Army was to be the army of occupation in Germany, and two weeks after the Armistice, Rice found himself in a "Teutonic land of gray fogs and white shadows and mists and endless rains." Once final headquarters were established in Coblenz ten days before Christmas, the demands on him eased somewhat and his was able to resume his craft. He wrote some verse, and mailed it back to the *Tribune*. He was also particularly curious, as he made his way about the countryside, as to how sport fit into the lives of the erstwhile enemy. He was astounded to discover that sport had virtually no place in prewar and wartime Germany.

> It was a strange sight to move mile after mile through a country—on through village after village and city after city—without meeting any arrangement to handle a sporting event of any sort.
> Even France, almost wrecked and shattered by war, still had her golf courses open, many of her lawn tennis courts in full swing, her boxing pavilions under way—all apart from the hundreds of quickly improvised baseball diamonds American troops built up along their line of march.
> But Germany had nothing of this sort. With the German it was war, work, food and beer. He had no recreation on the side to improve his philosophy or erect in his soul a foundation of fair play.

Disregarding the smug and metaphysical tone of the final paragraph—the conqueror searching for the moral flaw in the vanquished—this was a revealing bit of reporting. This seeming ignorance and indifference to that which had dominated the nearly four decades of his life was a fascinating puzzlement to Rice. He quizzed captured German soldiers, the peddlers and housewifes at the village markets, and he watched the reactions of the townspeople when American soldiers would invariably—and soon after their arrival—spend their idle moments in pick-up baseball and football games:

> The expression upon the startled Teutonic faces was something worth inspection. Old and young—male and female—they began to gather in small groups as if watching some strange animal cavorting through a series of astonishing antics.
>
> They had expected this advance delegation to arrive with all the pomp and fanfare of conquering militarism—and yet here was a victorious vanguard engaged in the unseemly exhibition of a morning's frolic.
>
> It was all new stuff to the Germans—not only baseball and football—but the idea of sport itself.

Most of Rice's fretting after reaching Coblenz, however, was about matters far more mundane. The irritations of cantonment life, things he had in common with the other doughboys, like complaints about mail service—both incoming and outgoing—and about full-of-themselves officers who failed to display even a modicum of respsect or civility to the enlisted men. And the ubiquitous complaints about Army chow. He wrote Kate that everyone was "pretty well fed up" with the unceasing diet of beef and potatoes. "The commissary has no flour, butter, or bacon and only brown sugar that seems to be partly sand."

But what preoccupied him most was getting home and a reunion with Kate. The maddening part was that he had no idea when he would be mustered out, and nobody else in France seemed to be able to tell him. He wrote Kate about a possible reunion in Paris, and urged her to check into travel possibilities. But travel to Paris was quite impossible for Kate Rice in the weeks after the Armistice. At the moment that Rice was writing to Kate and asking her to come to Paris, she and Floncy were both bedridden with the flu, part of the epidemic that was raging across the world in 1918.

Kate had passed the months of Granny's absence as a volunteer in a Salvation Army canteen down on the Battery, but it had done little to blunt her loneliness. Her youngest sister Mildred, pregnant with her first child, and her husband Kenneth Luthy, moved into the Riverside Drive apartment, and that did help. Mildred nursed Kate and Floncy back to health during their bout with the flu; but she so exhausted herself that she and her infant daughter Jane left soon after for a stay with Mother Hollis in Americus. No sooner had Mildred left than her husband Kenneth, the advertising manager for *American Magazine,* took ill with the flu and died within a week. It was Kate's turn to be there for her sister.

So with a reunion on the Continent out of the question, Rice was left with no alternative but to await fitfully his demobilization. He was sprung from Third Army Headquarters in February and sent to Angers, near the

The Wallace University School football team, 1896. Rice is second from the left, back row. He was a 130-pound end, who insisted on playing four years at Vanderbilt, despite suffering cracked ribs and a separated shoulder. (Vanderbilt University)

Henry Grantland Rice of Nashville, age twenty. His Vanderbilt graduation picture. (Vanderbilt University)

Captain of the Vanderbilt varsity, 1901. A smooth-fielding shortstop who could hit for power, Rice elicited interest from the Nashville professional team in the Southern League following his graduation. His father and grandfather would not let him play. (Vanderbilt University)

Katherine Hollis of Americus, Georgia, as she looked when she met Grantland Rice in 1904. (Vanderbilt University)

First Lieutenant Grantland Rice, 115th Field Artillery, A.E.F. He was thirty-seven years old and already the most famous sportswriter in America at the time of his enlistment. (Vanderbilt University)

"Scribes and Pharisees," as Rice labeled them, covering the Yankees spring training in Savannah, circa 1912. Rice is on the far left. Next to him (l to r) are Ed
Curley, Unidentified, Bunk MacBeth, Bill Slocum, Fred Van Ness, Harry Schumacher and Fred Lieb. (Vanderbilt University)

Rice flat-out loved sport—playing, watching, writing, or talking about it. (UPI/Bettmann)

Over a half-century Rice hunted across the length and breadth of North America. The hills of north Georgia remained his favorite hunting grounds, and quail and wild turkey his favorite prey. (Vanderbilt University)

At the top of his game, circa 1916. A scratch golfer who played the game with the same elan and enthusiasm with which he wrote of it, Rice was named one of the "100 Heroes of American Golf" in a celebration of the sport's first one hundred years in 1988. (UPI/Bettman)

Rice made broadcasting history in 1923 when he did the play-by-play for the first World Series game to be carried live on the radio. The audience was estimated at a million and a half people. No man in the history of the world to that point had ever spoken to a larger audience. (UPI/Bettmann)

Dempsey's training camp in Great Falls, Montana, prior to his fight with Tommie Gibbons in Shelby. Rice stands along the ropes with four of his New York colleagues (l to r), Edgar "Scoop" Glesson of the *Tribune;* Rice; Herbert "Hype" Igoe of the *World;* Lawrence Perry of the *Sun,* and Harry Newman of the *Daily News.* (Vanderbilt University)

The Four Horsemen of Notre Dame, whose lives were changed forever by Grantland Rice's story in 1924—(l to r), right halfback Don Miller, fullback Elmer Layden, left halfback Jim Crowley, quarterback Harry Stuhldreher. (UPI/Bettmann)

There were tournaments in America with better golfers but nowhere was there a tournament with a field of better-known golfers than the annual Writers and Artists affair in Palm Beach. Here in 1928, are fight promoter Tex Richard, artist Rube Goldberg, heavyweight champion Gene Tunney, Rice, and Ring Lardner. (UPI/Bettmann)

Rice at the peak of his career in the early twenties. "Even jealous or disdainful colleagues had to be grateful for the positive image of the sportswriter he was projecting to the public." (Vanderbilt University)

Florence Rice, in a Hollywood publicity photo from the mid-thirties. Her success on screen and the stage was genuine and considerable, but she never quite escaped from the shadow of her father. (Vanderbilt University)

The friendship between Ring Lardner and Grantland Rice was one of laughter, gaity, and zaniness. Lardner was the wit, Rice the sunshine. Here they are on a Hollywood set in the early thirties, looking upon the peculiar with what might best be called "earnest bemusement." (Vanderbilt University)

Rice was more than twenty years older than Bobby Jones, yet the gap in years did not inhibit a friendship that flowered in the years after Jones retired from competitive golf and played the game solely for its own pleasure. Rice and Jones traveled together to all the major golf championships and generally found time to play the course before the tournament started. Here, Jones tees off Balustrol in New Jersey, a couple of days before the 1936 U.S. Open. Bridge columnist Hal Sims, novelist Clarence Buddington Kelland, and Rice wait their turns. Note the gallery Jones attracted every time he addressed a golf ball. (UPI/Bettmann)

Rice with Floncy and Kate, 1935. (UPI/Bettmann)

Rice in the 1940s, when over ten million newspaper readers had access to his column every day. (UPI/Bettmann)

In 1925 Rice succeeded Walter Camp as the man who chose the football All-Americans for *Collier's* magazine. By the time the team moved to *Look* magazine in the late forties, it was known as "Grantland Rice's All America Team." Rice's influence as a sportswriter ebbed in the final years of his career, but his status as a celebrity never did. He remained the most recognizable name in American sportswriting until the day he died. (Vanderbilt University)

They called this group the Village Green Reading Society, a group of New York sportswriters who gathered in rural Connecticut in 1949, ostensibly to discuss poetry. Seated in front is Joe Stevens. Second row (l to r), Tim Cohane, Rice, and Red Smith. Back row (l to r), Willard Mullin, Herman Hickman, Frank Graham, and Charlie Loftus. (Vanderbilt University)

Rice in the last weeks of his life. This photo was one of a series taken to be included in the photo collection in *The Tumult and the Shouting*. Rice chose an over-the-shoulder shot for the book, where the weariness in his face was not so evident. "I tire easily these days," he wrote in the caption. "Sometimes I think, perhaps, I've lugged too many typewriters to the top of too many stadiums." (Vanderbilt University)

coast. There were tens of thousands of men at Angers, waiting for berths on transports home—waits of up to six months in many cases. But Rice was no ordinary doughboy. Marching through the mud one day in the daily, pointless, keep-the-troops-busy close-order drill, Rice was hailed by a familiar voice. It was Jack Wheeler, an old friend and former *Herald* sportswriter, who, just before the war had gotten his newspaper syndication business off to a promising start. Wheeler had hooked up with Colonel Frank Knox, publisher of the *Manchester* [New Hampshire] *Union Leader,* and destined to become Alf Landon's running mate in the 1936 election and Franklin Roosevelt's Secretary of the Navy after that.

Knox was on a priority list for shipment home; he enlisted Rice and Wheeler as his adjutants, and the three newspapermen made their way down to the port of Saint-Nazaire. Their orders came to sail on the *Ryndam* in early February—Knox was to be commanding officer of the troops aboard, and Rice and Wheeler were to supervise the deportment of those troops. No mean task, given a shipful of ebullient soldiers heading home from a war, each man seemingly with his own supply of contraband cognac, each ogling the few nurses who were aboard. Knox was seasick for most of the voyage, so the ship virtually belonged to Rice and Wheeler.

The crossing began like a holiday cruise, what with the general sense of anticipation, the brandy, and the food prepared by the ship's cook, who in a past life had been the chef aboard J. P. Morgan's yacht. But it turned dark very quickly. The flu epidemic that had been raging through the camp in Angers followed the men aboard ship and exploded three days out. Rice and Wheeler spent much of the balance of the voyage preparing death certificates and overseeing burials at sea. When the *Ryndam* docked at Newport News in mid-month, on a drizzly day (if the sun ever shone during Rice's fourteen months in the Army, he did not acknowledge it), Rice remembered his emotions as thankful but subdued.

Rice choose Lake Placid as the place he would begin to forget, in the expectation that the chill Adirondack air would cleanse "the misery and stench of war" from his consciousness. But there, amidst the conifers and craggy summits and the pristine snow falling silently upon forests unscarred by shells and unwashed by the blood of lost youth, Rice began to understand that he could not forget—did not *want* to forget. "Dreams have a habit of jerking you back into your past," he wrote, and, for the rest of his life, his dreams were of France. He was fiercely emotional about Armistice Day, as only a veteran of the Great War could be. And each November 11 for the remainder of his career—regardless of what the sporting bill of fare held—his column would be built around a sober reflection of those who would remain forever in France. These were near epics, running sixty to eighty lines in many cases. And regardless of how many times he returned to the theme, the passion, the anguish, would not ebb. There is no sense of the mechanical in these verses, never the sense that these were merely lines dashed off to meet a deadline. These lines, from a fifty-six-line verse written in 1924, are fairly representative:

> One at a time they come,
> Old mates forgotten from the somber hunt,
> Where one can hear far off the rolling drum
> Of cannon thunder on the Argonne front;
> Old mates with shining eyes and welcome hand,
> Who've come again from distant fields and streams.
> Where they have lingered in an alien land
> As we came home and left them to their dreams. . . .
>
> Perhaps with wistful eyes
> They seek again remembered streets and streams,
> Or it may be that death has made them wise,
> To find contentment in their peace and dreams
> Which make the road's end of each pilgrimage,
> As some vast Inn at twilight sends its glow
> Across dark hills and trails where storms may rage
> For those who still grope blindly as they go.

When Rice returned to New York to begin work he discovered that the lawyer to whom he had entrusted his money before the war had invested it badly, lost it all, and committed suicide. The money amounted to more than $75,000, everything Rice had. But with the cash he had left in his pocket, Rice bought some flowers for the wretched barrister's funeral; then, he rolled up his sleeves and went back to work. The twenties beckoned.

Black Clouds on a Golden Dawn

*I**n the postwar flush*** there were dreamers who saw a future that would celebrate the similarities between nations and people, not fret about the differences. Sport had its place in this school of thought, though the future these dreamers saw was less utopian than jingoistic. Baseball would become an international game, it was said; it could not help but burgeon overseas after the doughboys had exposed their British and French brethren to its splendors. There was, however, never any talk about the Continental favorites blossoming in the New World as the result of this wartime exchange of cultures.

Rice espoused this possibility in some of his early writing after the Armistice, when he was still overseas. He was quick to put the notion to rest when he got back to work.

> The league of nations may get together on a peace programme, but getting together on a universal sport programme is another affair.
>
> After two years of intermingling the average Englishman still has no more use for baseball than the average American has for cricket.
>
> And the French have no use for either. . . .
>
> [C]hucking baseball down an Englishman's throat or ramming cricket down an American's windpipe will never be one of the results of an alliance.

Rice, who had returned to the office on March 1, spent three weeks working back into form with this sort of philosophical musing, prior to heading off to spring training. Another matter that caught his notice as he surveyed the sweep of American sport in the immediate weeks ahead was the reunion between those who had gone to war and those who had not. There is of course a nervousness to any reunion after a protracted separation, heightened when the cause of the separation is as wrenching as this one was; heightened even more when the prevailing attitude was that the man who had not gone—for whatever reason—was something of a shirker. A number of major league ballplayers had escaped the draft by finding work in war-essential industries, such as shipyards. Most of these furloughed major leaguers never got close to a welding torch or a rivet gun, though. They were recruited to play ball for the shipyard teams in local industrial leagues. Rice—as one who had gone—was clearly disdainful of those who has seized upon such a loophole to escape the fighting, and was plainly edgy about the prospect of a temperate reconciliation.

It may or may not interest those frenzied baseball patriots who ducked both service and the draft by jumping to some ship league that the forthcoming summer may not be entirely pleasant for them.

Both among the A.E.F. and the army at home the feeling toward these passionate patriots is not entirely one of abounding admiration.

And from over 4,000,000 men, quite a number of whom will drop in on a baseball game, more than a few outbursts can be turned out that will not be entirely in the way of loud cheering.

The ship league patriot figures, of course, that everyone will soon forget about it.

But this is one occasion where he may be wrong; a tip that he can take or leave, just as it suits his morbid fancy.

However long or short the doughboys' memories, or however keen their indignation may have been in the stands that summer, Rice's indignation had been spent. When the season began, Rice judged the athletes as he always had—by what they did between the lines.

Spring training was delayed three weeks and truncated a bit in deference to the hundreds of thousands of doughboys—including many ballplayers—still pouring through the ports along the Atlantic seaboard. But even the lightning demobilization that disgorged tens of thousands of soldiers into civilian life as swiftly as they could walk down the gangplank was not fast enough for some. By opening day the Brooklyn Dodgers still had five players stuck in the service. "The A.E.F. takes its own time about asking for waivers," noted Rice.

Rice set up spring-training headquarters that year in Jacksonville, where the Yankees and Dodgers were both training. His dispatches carry the relaxed air of a sepia-toned spring training, when Florida in March was still the virgin preserve of a few dozen men sweating out the lethargy of winter, under the benign gaze of a couple dozen sportswriters, grateful to be once again unloosed from the sweaty clatter of the city room and going about their business beneath God's soothing sunshine.

A narcoleptic alligator in the animal park across the street from the ballyard—billed as being more than 800 years old—amused Rice no end and he filed nonsensical daily updates on the reptile's naps. He was not the only writer who found material in the animal park. Bill McGeehan talked of arranging a spaghetti-eating contest between an ostrich and the stocky Yankee outfielder/third baseman Ping Bodie. Rice also noted that with Bodie looking like he'd win the Yankee third base job the Yankee infield would be comprised of Pipp, Pratt, Peckinpaugh and Ping; he asked his readers if that sounded more like an infield or a college yell.

Other preoccupations for Rice that spring were finding a drink in parched Florida—which had gone dry in advance of the July 1, Volstead Act deadline—and golf. Charles Ebbetts, the Dodger owner, was his part-ner in a regular morning game. "Why don't you throw it out?" he asked Ebbetts one day as Ebbetts thrashed futilely about a yawning bunker. "I am the owner of the Brooklyn team," replied Ebbetts regally, "not one of its pitchers." In between all of this, Rice even managed to watch and write about a little baseball. The result was a column with a pleasing me-lange that said for both writer and reader alike: It's good to be back.

In early April, Rice moved inland to Gainesville, to meet up with the Giants and Red Sox and barnstorm north with them. A tremendous aux-iliary source of cash for major league baseball during these years was the swing north from spring training with stops for exhibition games in the more populous Southern cities—cities such as Atlanta, Savannah, Charles-ton, Winston-Salem, Raleigh, Richmond—where fans would turn out in droves for their once-a-year glimpse of the major leagues. Teams would generally travel north in pairs—one American League team and one Na-tional—sharing the Pullman expenses and splitting the gate. In the postwar spring of 1919 the Giant-Red Sox match up was a promoter's dream. Ma-thewson was gone but the Giants still had McGraw and were still base-ball's signature team. The Red Sox, meanwhile, were simply the finest team in the game—World Series champions in 1912, 1915, 1916 and 1918. Moreover, in addition to their championship credentials, they boasted of the young Babe Ruth, a man whose skills and swagger had already earned him billing as perhaps the game's most appealing personality.

Rice's first considered writing on Ruth had come in 1916, and it was oh-so-considered. When others were beginning to suggest that Ruth—who had won eighteen and twenty-three games in his first two full seasons with the Red Sox—was the greatest left-handed pitcher of all time, Rice, the man who unabashedly called Bobby Jones "the most remarkable kid prodigy we have ever seen" after seeing him just once, was uncharacteristically circumspect on the matter of Babe Ruth. "[T]he time is not yet ripe to enlist him as the greatest . . . of the southpaw flock," he wrote. "[T]he test isn't one or two years. Ruth is still too young at the business to be classed with Rube Waddell or Eddie Plank or Nap Rucker. . . .

"He is young, powerful and always in good shape," Rice continued, and while the "always in good shape" part was more true in 1917 than it would be later, it was still an example of Rice's hyperbole and deference to the athlete's image. "He has the stuff. He should be the best pitcher in baseball this season, the most effective performer. But quite a stretch of time still lies between Babe and lasting greatness."

Now, in 1919, a portion of that stretch of time had passed. Ruth had won twenty-four games in 1917, thirteen in the war-shortened 1918 season, plus two more in Boston's World Series win over the Cubs. And it was apparent in this spring of 1919 that Ruth was destined to go down as one of the game's greatest. But it was becoming equally apparent that it would not be as a left-handed pitcher. In limited duty in left field and at first base, Ruth had whacked eleven home runs in 1918. That led the American League, and he had done it while playing in only slightly more than half the team's games. Ruth's prowess with the home run brought a new dimension to the simple act of watching the game. Heretofore home runs had been as rare as acts of God—something to be witnessed in wonder but never anticipated. Ruth brought the fervor of anticipation to his every at-bat. His delicate walk to the plate and his peculiar body—the barrel-like torso balanced precariously upon those spindly legs—belied the imposing, born-to-be-here batter's box presence. His swing, when he connected, was as fluid and as graceful as classical ballet; when he missed, it could be as violent, as tangled, and as disconcerting to witness as a train wreck. But it was never less than electric, even at this early stage. "Once committed . . . once my swing starts, I can't change it or pull up," said Ruth to Rice. "It's all or nothing at all." The swing was copied from Shoeless Joe Jackson of the White Sox, said Ruth. "His is the perfectest."

Ruth had by this time also complemented this prowess and presence with a penchant for headlines. Ever at odds with Red Sox manager Ed Barrow, Ruth had walked away from the team the previous July after being fined by Barrow and announced that he would not return and had signed a contract to play in the Delaware River shipbuilding league. Prior to the start of this spring training, he had held out—saying he wanted $15,000 for 1919, or a three-year contract at $10,000-a-year. He got the three-year, $30,000 deal, and scores of attendant headlines. The Babe never met a child he didn't like or didn't have a moment for. He never met a photographer he wouldn't pose for, or a reporter he wouldn't provide with a speak-his-mind-and-let-the-chips-fall-where-they-damn-well-may quote.

Rice had never met Ruth before joining the Red Sox–Giants road show in Gainesville. In Tampa, the day before, Ruth had slugged a Herculean home run—very soon such blasts would come to be called Ruthian—that the aghast witnesses estimated to have traveled somewhere between 500 and 600 feet. In later years, Barrow would fix the distance at 579 feet. Whatever the precise distance, it was nearly half again as long as the longest home run any of the gathered sportswriters had seen hit to that point. In batting practice in Gainesville, Ruth was reinforcing the Tampa shot with an impressive show, rattling about ten shots out amongst the automobiles that ringed the outfield at the Gainesville park.

His show over, he ambled over to the Red Sox dugout and manager Ed Barrow introduced him to Rice.

"You sound like you got a cold," said Ruth.

"I have, sort of," answered Rice.

Whereupon Ruth produced what Rice described as an enormous red onion and gave it to Rice. "Here, gnaw on this," he said. "Raw onions are cold-killers." And so went the first interview between America's best-known sportswriter and her best-known sports hero. "While Ruth talked I gnawed," said Rice, "with tears streaming from my eyes."

Ed Barrow was still insisting that Ruth would pitch and pinch hit during the 1919 season, though he played him exclusively in left field during the series with the Giants. He was the center of attention on the swing north. Officials in Winston-Salem declared a half holiday so that everyone in town might have the opportunity to see the World Series champions and their tempestuous star. The New York writers, in their first concentrated exposure to Ruth "were pop-eyed," according to Rice, wildly speculating on how many home runs Ruth would hit were he to play the full season in left field. Rice was enthusiastic in his praise, but not unreserved; Ruth hadn't yet convinced him he could hit a lefthander. "Babe Ruth has a mighty wallop," he wrote, "but the sharp breaking curve of a lefthander tied his full, free swing into knots."

John McGraw was a doubter too. "If he plays every day, the bum will hit into a hundred double plays before the season is over," heckled McGraw, the greatest of all hecklers.

The name-calling, the press attention, the inordinate crowds, all served to crank up the intensity of these Giant–Red Sox games well beyond the level at which springtime exhibitions were generally played. There were beanings, spikings and general ill-will. "How's that for a double-play ball, Mac?," shouted the Babe to McGraw after every impressive hit. As the tour approached Baltimore and tempers approached the point of explosion, McGraw and Barrow got together and agreed to calm things down, and the series proceeded without further incident, and the only one who went spoiling for a fight on the train was Jim Thorpe, then in his last days with the Giants and forever looking for a good-natured wrestling match.

It also became clear on that trip north that Barrow would not be able to keep Ruth out of the lineup on a day-to-day basis. He would pitch in seventeen games that summer, compiling a record of 9–5 and an earned run average of 2.97. But he would play 111 games in the outfield, and four more at first base, compiling a .322 batting average and blasting twenty-nine home runs, very nearly doubling the previous American League record of sixteen, set by Ralph "Socks" Seybold of the Athletics in 1902. In the early days of the season, Rice, finally a believer, would take notice:

> I've seen a few I thought could hit,
>   Who fed the crowd on four-base rations;
> But you, Babe, are the Only It—
>   The rest are merely imitations.

On that 1919 train ride north—aboard the "palatial rattler" where "ever and anon the resounding snore of the weary athlete mingles with the grinding creak of the train" and "the State of Florida blew in through the open windows and settled upon the clothes and in the eyes and ears of the tourists"—the acquaintance that began over an onion in Gainesville would flower into friendship. Rice and the Babe found a common interest in bridge and hearts, and would be partners in hundreds of games aboard countless "rattlers" crisscrossing America over the next fifteen years. Rice, as did all the other writers, found a bottomless fount of colorful copy in the irrepressible Ruth. But he found more. "Ruth, the man-boy, was the complete embodiment of everything uninhibited," said Rice. "He couldn't possibly fail—that was Ruth's credo." For a guest appearance on his radio show in 1928, Rice had scripted out some remarks for Ruth that called for him to cite the Duke of Wellington's line about the Battle of Waterloo having been won on the playing fields of Eton—preparing the remarks "in the interest of spontaneity" as Red Smith liked to tell the story. On cue, the Babe, never one for scripts, said with aplomb and authority: "Well Granny, as Duke Ellington once said, the Battle of Waterloo was won on the playing fields of Elkton."

Nonplussed, Rice told Ruth after the show that he could understand confusing Duke Ellington with the Duke of Wellington, but where in the world did Elkton come from?

"Well," said Ruth, "I married my first wife in Elkton [Maryland], and I always hated the gaddamn place," not at all bothered by fumbling a line on network radio.

It was this unbridled exuberance that Rice found most irresistible, a *joie de vivre,* and an innate decency that made the Babe a most pleasant companion, irrespective of his accomplishments on the field. On the eve of a World Series game in Chicago in 1932, Ruth sought out Rice to ride some sixty miles to visit a child in a hospital. "And if you write anything about it I'll knock your brains out," Ruth threatened him. Rice didn't.

Not surprisingly, they were closest following the Babe's retirement from the game, when the games they played changed from hearts and bridge to hunting, fishing and golf. On the day he retired from the game in 1935, Ruth gave Rice a call and told him, "Get out your golf clubs, kid. I'm ready for you now." And the two were frequent golfing companions in their later years, generally together in some form of fund-raiser or pro-am. Rice would become toastmaster to the world during the twenties, thirties and forties, emceeing events from New York to Augusta to Pasadena. He was constitutionally incapable of saying no to anyone who requested his presence on the dais. And be it an audience of businessmen with deep pockets or orphans with nothing in theirs, Rice found that Ruth was another man incapable of saying no.

When Ruth died in 1948, Rice's tribute in verse saluted that charismatic blend of personality and talent that defined the outer limits of twentieth-century celebrity.

> Game called by darkness—let the curtain fall.
> No more remembered thunder sweeps the field.
> No more the ancient echoes hear the call
> To one who wore so well both sword and shield.
> The Big Guy's left us with the night to face,
> And there is no one who can take his place.
>
> Game called—and silence settles on the plain.
> Where is the crash of ash against the sphere?
> Where is the mighty music, the refrain
> That once brought joy to every waiting ear?
> The Big Guy's left us, lonely in the dark,
> Forever waiting for the flaming spark.

A second friendship with an athlete who would become synonymous with the twenties was born that summer of 1919. Though Rice's friendship with heavyweight champion Jack Dempsey was slower to flower than his Ruth relationship: It was the mid-twenties before Rice and Dempsey grew truly fond of one another; it was later still—after Dempsey's fight career was over and he had embarked on his equally lucrative and prominent career as former champion—before they spent much time together socially. In fact, some words Rice wrote after Dempsey won the title from Jess Willard opened wounds in Dempsey that were a long time in healing.

Rice must have found it hard to root for anyone when Dempsey met Willard in Toledo on the Fourth of July. He had never much cared for

Jess Willard, the paunchy Kansan who had wrested the championship from Jack Johnson in 1915. Willard had been dutifully hailed in the press following his victory—here finally was a "White Hope" who had delivered boxing from the "darkness" of a Negro champion. But Willard proved a less-than-charismatic champion who also proved less-than-eager to defend his crown. This didn't sit well with Rice, whose notion of a champion was a man who was eager to stand on the mountaintop and challenge all comers. Throughout the prewar years Rice commented frequently about the number of "minutes" Willard had worked since winning his crown. Willard eschewed defending his title in favor of touring with a circus, all the while finding it difficult to push himself away from the table at mealtimes— habits that earned Rice's scorn. "Why doesn't he challenge one of his own elephants," said Rice. "In the matter of weight, at least, the match should be all to the fifty-fifty." When Willard did fight, Rice was scornful of his performance.

"If boxing, as now conducted in this ten-round, strictly business affairs, is brutal," wrote Rice after watching Willard successfully defend his title against Frank Moran in March of 1916, "then dancing should be stopped on account of its innate cruelty and savagery. There are times when even an expert can't tell which of the two sports is under way."

When Willard failed to show up for a New York tribute to French sailors in 1917, Rice's disgust was absolute. It was here that he referred to Willard as "a drab outline against a dull gray sky," saying that the champion's arrogance and irresponsibility were "an excellent tip-off on the fight game today"—words that carried an extra edge in light of a push to ban boxing in the state of New York, a push that was gathering momentum during those months.

But if Rice would secretly or not-so-secretly delight in the champion's tumble, were the scenario his to script he probably would not have written in the sturdy young Dempsey as the man in the white hat. For in those months when the ex-doughboy Rice was still separating the world into two groups—those-who-went and those-who-didn't—the twenty-four-year-old Dempsey was standing rather conspicuously in the line with those who didn't.

But if, as a fan, Rice was torn, it didn't matter; for as a reporter and writer he could revel in the story and the spectacle. It was a dandy, a benchmark really, an *étalage* that foreshadowed the giddiness, the gaiety, the pomp and the excess of the decade ahead. When Rice arrived in Toledo a week before the fight, he noticed and told his readers that whereas, in the Argonne the previous autumn there had been only twenty-two correspondents to record the history-moving heroics of 700,000 American soldiers, here in Toledo there were four hundred to write of two men whose struggles were of decidedly lesser historic import.

Included among the four hundred were Rice's *Tribune* colleague Bill

McGeehan, Bat Masterson, Damon Runyon, Otto Floto, Tad Jones, Rube Goldberg and Ring Lardner. They were necessarily tripping all over one another as they sought to service their newspapers. Under an unremitting sun, and in temperatures that topped out above one hundred degrees, the writers moved from camp to camp—pronouncing Willard as more fit than they had expected to find him, though not as fit as Dempsey. They found Dempsey the harder worker, but reminded readers that the more casual champion had six inches and more than fifty pounds on his harder-working challenger. Toledo was becoming the improbable crossroads of the nation as June neared July and the Independence Day fight grew closer. Rice and his companions in the press soon found attendance at the daily workouts something of a drudgery, "with nothing in either workout to cause the human pulse to hop a beat faster or to prod one into hoarse huzzas." Yet each day, swarms of spectators would pay out twenty-five cents each to stand in the sun and watch Willard and Dempsey go through their training regimens. Rice found it a peculiar human menagerie:

> There are sedate old gentlemen with chilly side whiskers, young girls with round and wondering eyes, ponderous looking matrons in mauve or pale heliotrope, prim-looking old maids, any number of kids around ten or twelve years old; bums, yeggmen, old sports—those who know nothing and those who know all of the seamy side of life—a cross section cut out of humanity and lured daily to the spot through the immense publicity imparted to the carnival.

The publicity that lured these crowds—and lured the reporters too—was the product of promoter Tex Rickard and Dempsey manager Doc Kearns.

Tex Rickard and Doc Kearns were made for each other. They also greatly distrusted and very nearly loathed one another; they were too much alike to ever get along. "Here was one con trying to out-con the other," said Dempsey of the two men who made him rich. Doc Kearns called Rickard a "crafty son-of-a-bitch" and "chiseling gambler." To Rickard, Kearns was a "sneaky no-good" or a "cheap conniver." But let the two men work together for a common purpose—making enough money to keep them both happy—and they would complement each other like the instruments in a sonata. They brought a Broadway-style press agentry to boxing. Like a courtesan working the room with subtlety and sophistication, they would tempt the wizened sportswriters with obvious promises. And the sportswriters, who knew exactly what they were buying, bought into it anyway. And with each line in the newspaper, interest grew, and as the interest grew, so did the take.

Rickard had lured Willard with a $100,000 guarantee. Kearns had lured Rickard into luring Willard with a promise that Dempsey would fight for nothing if it came to that. Rickard built a 90,000–seat wooden stadium for the fight—built it out of the best, freshest wood available. That was a mistake. As Rickard proudly gave tours of the stadium to the visiting writers, the late June sun beat down on the new wood, and the sap oozed out and congealed on the seats. On fight day, when fans went to leap to their feet in excitement, many of them discovered they couldn't; they were stuck.

Another Rickard mistake was building it to hold 90,000. Despite the whirlwind of publicity all roads did not lead to Toledo on July 4, 1919; and those roads and rails that did lead to the Ohio city on the shores of Maunee Bay were clogged with traffic, and several thousand people never made it to the fight. There would have been plenty of room for them though; on fight day the stadium was scarcely half full. So Rickard fell short of the million-dollar gate he was hoping for and had the sportswriters predicting. But with tickets priced at $60 for ringside seats, and scaled down $10 for the general admission high seats, he easily made enough to cover the $100,000 guarantee to Willard, the $27,500 to Dempsey (Kearns may have promised to fight for nothing but he never intended that to be the case) and the substantial overhead, with plenty left over for Tex Rickard.

In the pantheon of 1920s sports heroes—the truly larger-than-life firmament of Ruth, Rockne, Jones, Tilden, Man o' War and such—Dempsey stands at the fore, the man "who captured the public fancy in the most basic way," according to Joseph Durso. But on July 4, 1919, he was just a contender grateful for a shot. He had come by his shot honestly; he had fought and won nearly fifty times in the two years leading up to Willard. And his was an appealing story. He came by his fighting blood naturally; he was descended from the West Virginia Hatfield clan—of the Hatfield-McCoy fighters. He grew up poor and worked in the mines in Colorado and Utah and rode the freight trains to distant mining camps in search of his first professional fights. Believing that New York beckoned after some initial success, he went east in 1916 and presented himself to the city's sportswriters, looking for a plug and a lead on a fight. They all threw him out—all except Damon Runyon, who told him to "face the city slickers head on and bounce back every time I was knocked down." Dempsey left New York after a few months and drifted, his career directionless and foundering on the rocks of indifference until he chanced into a relationship with Jack Kearns.

Jack Kearns—he was "Doc" to Dempsey and almost everyone else in the fight game—was the consummate big-city hustler and dandy, the sort of fellow who would people Damon Runyon's stories. The first time Dempsey saw him he was dressed in a vested suit, a diamond stickpin in his tie, and wearing a gaudy pinkie ring. He was also stinking drunk and contentiously precipitating a bar-room donnybrook. Dempsey jumped into the free-for-all on Kearns's side—he was the smaller of the two principal protagonists. For his gallantry, Dempsey received an offer from the fast-talking Kearns to serve as his manager. "All the newspapermen will sit up and take notice by the time I'm through," said Kearns to Dempsey. "The name Jack Dempsey will be up in lights!"

Kearns sweet-talked tailors and hotel managers and victualers into providing their goods and services on credit. He would then use this bought-on-time, class-act image to secure good fights at good rates for his young fighter. And Dempsey was enough of a fighter to make good on his chances. "Most folks are willing to believe anything you tell 'em, no matter what," said Kearns. "It's all in how it's presented. You gotta sell them good, kid, make 'em buy anything and let them think they got the best end of the deal."

Kearns was perhaps a bit too brassy for Rice's taste and he kept the manager's name out of his column. But Kearns couldn't have minded, for Rice was saying all the right things about Dempsey. He heralded his fitness, his work-ethic, his zest for the fray, and his leaden punches that had laid waste to four dozen men on his travels to Toledo. Rice, as did just about everybody else, rated the fight as even, yet further predicted that with such aggressive fighters possessed of such hurtful punches, the fight was unlikely to go the distance. The experts didn't know who the winner would be, but they expected the win to come by knockout.

Were they ever right. In the thick 100+-degree air, amidst a sea of straw boaters giving the tony ringside crowd a few square centimeters of surcease from the high, scorching sun, Dempsey exploded at the two-minute mark of the first round, crashing a left into Willard's jaw. "The champion sat down heavily," wrote Rice, "with a dazed and foolish look, a simple half-smile crowning a mouth that was twitching with pain and bewilderment."

Six more times over the next fifty seconds, Dempsey sent the stupefied Willard tumbling to the canvas. On the seventh occasion, coming near the two-minute, fifty-second mark of the round, "nothing in the world but the bell could have saved him," according to Rice. "He sat there, dazed, bewildered and helpless—his big, bleeding mouth wide open, his glassy bloodshot eyes staring wearily and witlessly out into space, as the 114-degree sun beat down upon his head that was rank with perspiration and blood."

At the end of the round, referee Ollie Pecord raised Dempsey's hand in the air, and the new champion began to make his way out of the ring, as "Willard was dragged to his corner, as one might drag a sack of oats." But it was just a bit premature. In the din, nobody had heard the bell. Or perhaps it had malfunctioned. Whatever, the timekeeper ruled that the champion had been saved by the bell—"saved" in this case being a relative term; it had really forced Willard to stand once again, helpless before the maniacal ravages of the "tornadic" Dempsey.

He lasted an improbable two more rounds. "It was unbelievable," wrote Rice. "From less than ten feet away we looked on and refused to credit the vision of our eyes. It looked as if every punch must tear away his head, but in place of this the fountain [of blood] continued to gush, the features continued to swell, the raw meat continued to pop open in deep slits as the red surf rolled from his shaking pulp-smashed frontis-piece."

When the bell rang to start the fourth round, Willard's attendants threw a towel into the center of the ring to signify the battered champion's capitulation; it was mottled crimson with the blood from Willard's many wounds. The move was greeted with a stunned incredulity and some random boos from the crowd in Toledo, and immediately the belittling of the fallen champion began. The fans, perhaps disappointed at having borne witness to only nine minutes of fighting, saw the surrender as unseemly behavior for a heavyweight champion. "Did you ever hear of any other champion who refused to get up and take it?," Rice heard after the fight. "Why didn't he stand up like a man and take his medicine? If he had had any fighting instinct in him . . . he would have groped his way forward until he was knocked cold. He's nothing but a big quitter."

Rice and Bill McGeehan, who had sat next to him at ringside and wrote the page one story on the fight for the *Tribune,* were aghast at what they heard in the crowd as they moved around in the minutes after the fight. Both men had been moved by the pluck and gallantry that Willard had shown in accepting Dempsey's punishment, and both made it a point to say so in their stories. "While Dempsey gave one of the greatest exhibitions of mighty hitting as anyone here has ever seen, Willard, in a different way, gave one of the greatest exhibitions of raw and unadulterated gameness," said Rice. "Never in the history of the ring, dating back to days beyond all memory, has any champion ever received the murderous punishment which 245-pound Jess Willard soaked up in that first round and the two rounds that followed."

McGeehan said: "No one can say of Willard that he lacked courage. He held a big and stout heart to have carried him through the second and third rounds."

Part of the blood-hungry fans discontent stemmed from the sheer shock of so brief a heavyweight title fight; history had prepared them to expect a protracted struggle—a contest of endurance as much as strength. It had taken Jim Corbett twenty-one rounds to knock out John L. Sullivan. It took Bob Fitzsimmons fourteen rounds to take the title from Corbett. Jim Jeffries went eleven rounds before knocking out Fitzsimmons; and then Jeffries, coming out of retirement as the greatest of the "Great White Hopes" forced Jack Johnson to go fifteen rounds in defense of his title in 1910. Willard, in winning the title from Johnson in 1915, needed twenty-six rounds, or more than ninety minutes of actual fighting time. For a title fight to be over in just nine minutes of fighting was something completely outside the fans' experience.

But once they shook off the initial shock, the fans realized what the writers had realized immediately. They had not been cheated out of a afternoon's boxing, they had instead been witness to the dawn of a new age—the age of a champion who hit so hard that he had changed the very nature of his sport. "And how this Dempsey can hit," noted Rice. "When he hit Willard it was exactly the same as if some strong man had swung upon the ex-champion with a heavy hammer. He felt as if raw steel had broken through his skull. He fell before a man who must able to hit harder than any man that ever lived."

If, on the morning after he battered Willard stupid, Dempsey was not yet a champion for the ages, he was at the very least a champion for his own age. Here was a champion for a quickening time, a champion who left no room for question as to his superiority. He was like America itself—the warrior prince of a gallant warrior nation, was he not?

Ultimately, of course, that is exactly what the myth would become. But not immediately. The road to canonization contained some early detours, and the most protracted and perilous of these detours for the the new champion was mapped out by Grantland Rice. Of all the hundreds of thousands of words written that July Fourth, it was the peroration of Rice's fight story that made the biggest morning-after impression on the new champion. "I was the World Heavyweight Boxing Champion and didn't have a mark on me," said Dempsey. "But Grantland Rice dealt me a blow that hurt more than Willard's punches." The comments to which Dempsey refers were perhaps the greatest reflection of the sensitivity Rice had on the matter of those who had not served in the Great War. "[Dempsey is] the champion boxer—" wrote Rice,

> not the champion fighter. For it would be an insult to every doughboy that took his heavy pack through the mules' train to front line trenches to go over the top at dawn to refer to Dempsey as a fighting man. If he had been a fighting man he would have been in khaki when at twenty-two he had no other responsibilities in the world except to protect his own hide.
>
> So let us have no illusions about our new heavyweight champion. He is a marvel in the ring, the greatest boxing or the greatest hitting machine even the old timers have ever seen.
>
> But he isn't the world's greatest fighting man. Not by a margin of fifty million men who either stood or were ready to stand the test of

cold steel and exploding shell for anything from six cents to a dollar a day.

It would be an insult to every young American who sleeps today from Flanders to Lorraine, from the Somme to the Argonne, to crown Dempsey with any laurels built of fighting courage.

He missed the big chance of his life to prove his own manhood before his own soul—but beyond that he stands today as the ring marvel of the century, a puncher who will be unbeatable as long as he desires to stay off the primrose way and maintain the wonderful vitality of a wonderful human system.

Dempsey called it hitting him "below the belt." "He'd called me a slacker and it hurt like hell because I knew I wasn't." Dempsey had come by his deferment honestly; he had done some shipyard work in San Francisco, and had provided support for his family from his fight earnings. Moreover, he had raised money for the Red Cross and other war relief organizations with assorted exhibitions and public appearances. Still, the fact remained that he did not go and he could have, and if that fact rankled the likes of Grantland Rice, it rankled Dempsey just a little, too. He never confronted Rice over what he had written, on Jack Kearns's advice. "I'll set Rice straight on a few things and the whole thing will blow over," the manager told his fighter.

Whether Kearns ever did speak to Rice and however Rice may have received the entreaty is not known; but the whole thing definitely did not blow over. A couple of months after the fight, Dempsey's estranged wife Maxine wrote to a San Francisco newspaper claiming that Dempsey had falsified his draft papers and obtained his deferment fraudulently. She would later recant her story and admit that it had all been a ploy to extort money out of Dempsey. But her admission had not come until after newspapers across the land had taken up the cry of "Slacker" and a federal grand jury had indicted Dempsey on charges of feloniously evading the draft. On the witness stand, Maxine Dempsey broke down and admitted the hoax and Dempsey was exonerated. But it had all made for some anxious hours in the first months of his championship, and for a chilled relationship between Dempsey and Rice, until their work over the next few years continued to draw them together, and they found themselves liking one another despite their earlier differences.

No pain is so great as the pain of betrayal. More than seventy years after the 1919 World Series, the conspiracy by eight members of the heavily favored Chicago White Sox to throw the Series remains not only good box office but also a sort of timeless American morality tale. The scandal so rocked the American consciousness because the perpetrators were the pro-ideal American heroes, the defiled game the very symbol of American vitality. First, a world torn asunder by Middle Ages-like savagery and carnage, and now baseball—the sanctuary from the madness and complexity of the larger world, the arena where good and evil, winners and losers are so neatly defined, and where defeat is always tempered by the solace of the grand effort and the promise of tomorrow—now baseball is crooked? Say it ain't so Joe, indeed. Say it ain't so Joe, please; for what we need most is a deniability here—as long as we can *believe* in its purity, than it shall be so; and it can be as before, serving our needs.

At every turn Americans had been assured that the game was clean. "The game of base ball . . . is a clean and straight game," said President William Howard Taft, "and it summons to its presence everybody who enjoys clean, straight athletics." An article in *The Nation* said: "We do not trust cashiers half as much, or diplomats, or policemen, or physicians, as we trust an outfielder or a shortstop."

Fix the World Series? An incredible thought. One of the reasons the job was so bungled was the fact that the very men who were executing the fix—players and gamblers both—persisted in a disbelief that such a far-fetched scheme was actually possible.

In looking back at the Black Sox scandal, though, the wonder of it is not that it happened, but that it took so long to happen. The further wonder is that the people of the game and the sportswriters who wrote of the game could have have been so brazen to express such incredulity and outrage that it did. By 1919, the game and gambling were once again virtually synonymous—as they had been back in the rough-and-tumble early days just after the Civil War. There were sections in nearly every big-league ball park where betting was in full view, with the gamblers "loud and obstreperous," with wagers being shouted out like the cries of the hot dog vendors, money changing hands openly.

It grew particularly acute during the Great War, when the government shut down horse racing and the bookies and gamblers simply moved their operations to the ball parks. Club owners and members of the National Commission running baseball in those pre-commissioner days were disquieted by this activity, but they tolerated it, for they knew that some of these gamblers who filled the stands and crowded the lobbies of the hotels had reached the players on the field. To what extent they weren't sure; but they *were* sure that any suggestion of impropriety by the players would explode upon the public in a way that would be bad for business. Wanting nothing to openly tarnish the integrity of the game and compromise the gate receipts, the baseball moguls had sworn a tacit allegiance to the see-no-evil, hear-no-evil, say-no-evil school of management. "For the good of the game," went the refrain.

Newspapers were witting accomplices in the promotion of the gambling. Most printed weekly "run totals" as an aid to those who played the ubiquitous pools that flourished across the land. The pre-event stories from any big series, especially the World Series, would include an account of the wagering activity and odds. And newspapers also eagerly presented to their readers the whitewashed image the owners favored. When Hal Chase, called "the archetype of all crooked ballplayers" by historian Harold Seymour, was suspended by Giants manager John McGraw late in the 1919 season, McGraw simply told reporters when asked for a reason: "I cannot talk about the matter," and the reporters accepted the brush-off and let the matter drop. It would have had to have been a mighty naive reporter not to know that the the jig was finally up for Chase. The talented yet incorrigible first baseball had been fixing games since 1907, every manager he had played with had publicly accused him of throwing games, and one—Christy Mathewson of the Reds—had even brought him before National League president John Heydler, who dismissed the charges for "lack of sufficient evidence." None of this ever made the newspapers.

If reporters had been looking for the patterns of abuse and the potential for scandal, instead of looking to avoid it, they would have found in the 1919 Chicago White Sox a team vulnerable to temptation. This was a fractious team, split into two bitterly antagonistic camps. The eight men who gained notoriety as the "Black Sox" comprised one clique. Rough-hewn, fun-loving, of humble rural roots, pitchers Ed Cicotte and Lefty Williams, infielders Chick Gandil, Swede Risberg and Buck Weaver; outfielders Joe Jackson and Happy Felsch; and utility infielder Fred McMullen had all hung together even before Gandil had recruited them to be a part of the fix. At the head of the second clique was second baseman Eddie Collins—sophisticated, Columbia-educated, and the object of some jealousy because he made more than twice as much as most of the rest of the White Sox regulars. He had not pried his princely $14,500 salary from the niggardly Charles Comiskey; it was the salary Connie Mack had paid him when Collins was the heart of the Mack's stellar infield during the Athletics' championship years, and Collins had had the good sense to have it made a part of his contract when Mack sold him to the White Sox in 1915.

If there was a unifying element on the 1919 White Sox it was their hatred of Charles Comiskey, the cheapest owner in baseball. Comiskey paid Joe Jackson, arguably the greatest hitter in the game in 1919, just $6000, scarcely a third of Ty Cobb's salary. Ed Cicotte likewise received just $6000, barely a half of what pitchers with a fraction of Cicotte's statistics earned on other teams. Up and down the line-up it was the same story. "Many second-rate ballplayers on second-division clubs made more than the White Sox," said Black Sox historian Eliot Asinof. "It had been that way for years."

Comiskey scrimped on meal money and dry cleaning bills and when Cicotte was approaching thirty wins in 1917, a figure that would have earned him a $10,000 bonus, Comiskey ordered him benched, ostensibly to save him for the World Series. He finished with twenty-nine wins and no bonus.

The White Sox's frustrations with the parsimony of their owner reached a head in July of 1919 when they very nearly went on strike. When attendance dipped in 1918, owners agreed to pare back salaries for 1919, in anticipation of an off-year at the gate. But by mid-season it was apparent that baseball was on its way to a record season at the gate. And with the White Sox on their way to another pennant, and with their salaries out of line as usual, the players called a team meeting and asked manager Kid Gleason to present their case to Comiskey.

Comiskey refused to so much as discuss it, and his truculence set the players to seething. They received the news as they were dressing for a game and immediately threatened to strike. Gleason persuaded them to play, and they did. But the resentment lingered and festered as summer turned into autumn.

None of this made for much of a news story, because most of the reporters covering the White Sox saw things Charles Comiskey's way. As good as he was at making his players feel trivial and subordinate, Comiskey was at least that good at making the writers feel exalted and essential. The press room at Comiskey Park featured a daily spread of roasts, salads and liquor—all of it free—that would have ordinarily been beyond the imagination of a $25-a-week newspaperman. When the White Sox traveled, the reporters with them traveled first class, as guests of the owner. Few who had partaken of Comiskey's largesse remained immune from speaking kindly about him in their newspapers. Grantland Rice generally saw the White Sox only when they came to New York, and he had missed their 1917 World Series triumph, still he remained a Comiskey fan in his column; he was particularly supportive when the White Sox bought up Collins and a couple of lesser men from the Athletics in 1915.

It would have had to have been a mighty naive reporter not to have heard the rumors of the fix on the eve to the first Series game in Cincinnati. And Grantland Rice was no naive reporter. The Hotel Sinton, home to the White Sox and most of the out-of-town writers, Rice included, was abuzz with talk that the White Sox were in the tank. Even absent the talk, the gambling action in the hotel and throughout the city would have suggested to anyone that something was afoul. The White Sox, prohibitive favorites, commanding 7–10 odds at outset of the money changing, dropped quickly to even money as large sums of the smart money came in on Cincinnati. A check by any of the writers with their home offices would have told them that in New York and elsewhere the story was the same: the late money and the big money was coming in for the Reds.

But none of the rumors of impending impropriety was shared with the nation's newspaper readers. The reasons were simple; the reporters held a skepticism born of habit—seldom was there a big sporting event without at least a couple of accompanying fix rumors—and they had a

abiding fear of libel laws. To a reporter, rumors are rather like a fog—
blinding and all-encompassing, yet vaporous and ephemeral, and all but
impossible to package and share with someone who hasn't experienced it
first-hand. Then too, they really didn't know. Rice himself had printed a
story on September 27, before leaving for Cincinnati, that argued that the
Reds were going to be tougher than most of America believed.

> [T]here will be no slaughter. [Cincinnati manager Pat] Moran is
> fairly certain to get good pitching at every start. He has a club that will
> fight back with all it has to give. It has fought hard all the year, and
> when it met the Giants in two vital series there was nothing to it—the
> Reds tore in and swept them off their well-spiked feet.
>
> The Reds outclass the the remainder of the National League by a
> much wider margin than the White Sox outclass the remainder of the
> American League.
>
> These impressions are offered to remove the idea which so many
> seem to have that the Sox will have a romp. For they won't. The Sox
> should win through greater power, but only after a good, hard scrap.

So the pre-Series stories concerned no dissection of why the smart
money was going to Cincinnati, and were filled instead of stories about
how the White Sox were unhappy with the layout and condition of Redland Field—rightfielder John Collins got a lot of ink when he announced
after a workout that right field in Cincinnati was the worst sun field in
he'd ever seen and it had him worried. Rumors that Eddie Cicotte had a
sore arm received the widest play. He had been roughed up in his last
regular season start, and that fact gave the rumor some legs. It had been
started, of course, by the gamblers, to help deflect talk of a fix.

Still, for all of the innocence of their dispatches back to their papers,
amongst themselves the fix rumors were all the reporters talked about in
the last hours before the first game. Champ Pickens, a friend of Rice's
from Alabama, came into Rice's room on the eve of the first game, threw
his ticket on Rice's bed and told him he was going to the race track. He
had found himself thoroughly frustrated trying to figure out the Series
action and wanted no part of it. "This series is fixed," he said to Rice.

Hugh Fullerton, Ring Lardner, and Rice all sat down to question
Christy Mathewson—working as a journalist for the *New York Times* during this series (and already dying a slow death from the poison gas fumes
he had inhaled during the war)—on the different ways a player might
work to throw a game. Point by point, Mathewson, the man who had
tried to bring Hal Chase to justice two years before, went over the scant
inches that separated an out pitch from a gopher ball, over how an infielder bent on deception could hesitate for a mere fraction of a second and

make a routine play look difficult and thus allow a hitter to reach base safely. And as to the hitters, said Mathewson, it was simply impossible to tell how hard they were trying. Fullerton, who was particularly agitated by the rumors and his growing belief that they might be true, decided to circle in his scorebook any play he thought suspicious. He asked Mathewson to provide counsel in this matter.

The next afternoon, shortly past three o'clock, in the bottom of the first inning, Chicago's Eddie Cicotte threw his second pitch squarely into the back of Cincinnati lead-off hitter Maurice Rath. In New York, gambler Arnold Rothstein, the man who had bankrolled the fix, was "listening" to the game via a telegraph feed in the noisy, crowded Green Room of the Ansonia Hotel. Through his subalterns he had instructed Cicotte to hit the first batter to signal the fact that the fix was on. Satisfied, Rothstein left the Ansonia to put down another $100,000 on the Reds. Over the years this pitch has become the resonant symbol of the crookedness of the 1919 World Series. But in the Redland Field press box, Hugh Fullerton put no circle around the play on his scorecard. He and his colleagues made no mention of this ominous beginning when they wrote their game stories a couple of hours later.

The focal point instead came in the fourth inning. With one out and a runner on first, Cincinnati shortstop Larry Kopf bounced the ball back to the mound, an easy 1–6–3 double play ball. But Cicotte hesitated, then threw high to shortstop Swede Risberg covering second. The runner at second was forced, but there was no chance for the double play, and the Cincinnati fourth remained alive. In the press box the writers looked uneasily at one another, Mathewson nodded at Fullerton and the Chicago writer circled the play on his scorecard. But when the number seven and eight hitters followed with singles and then Cincinnati pitcher Dutch Reuther crashed a triple to the gap between Jackson and Felsch, Mathewson's suspicions vanished.

"When the bottom of the order starts hitting you, it just isn't your day," said Mathewson. How can we be sure?, the writers must have wondered. What *is it* that we're seeing here?

Cicotte didn't last the inning. He left trailing 6–1 (the final would be 9–1), leaving Rice's pre-Series assertion—"Cicotte will only be beaten in one of those 1–0 or 2–1 affairs. He isn't going to yield more than one or two runs"—to stand as one of the silliest predictions of White Sox prowess.

Ring Lardner, who had just left the *Chicago Tribune* to start a syndicated column out of New York for the Bell Syndicate, composed two pieces on the game. For his readers, a sample of his wry and sarcastic humor in a mild, moderate dose:

. . . The White Sox only chance at this point was to keep the Reds hitting [in the fourth] until darkness fell and make it an unfinished legal game. But Henie Groh finally hit a ball that Felsch could not help from catching and gummed up another piece of stratagem.

For his colleagues in the press box and his friends and former friends on the White Sox, Lardner showed a much more sharply honed sarcasm and bitterness. He composed a new lyric to the tune of "I'm Forever Blowing Bubbles." The song had been a part of the pre-game repertoire of the band at Redland Field; the melody was fresh in Lardner's mind. His lyric went "I'm Forever Blowing Ballgames" and he spent the balance of the Series warbling it in the press box, hotel lobbies and railroad dining cars:

> I'm forever blowing ball games,
> Pretty ball games in the air.
> I come from Chi
> I hardly try
> Just go to bat and fade and die;
> Fortune's coming my way,
> That's why I don't care.
> I'm forever blowing ball games
> And the gamblers treat us fair . . .

Rice built his story for the *New York Tribune,* and the eighty-odd papers then subscribing to his work, around the fourth inning rally and the smelly Cicotte play that triggered it. "GRANTLAND RICE TELLS HOW EDDIE CICOTTE TOSSED OWN GAME AWAY BY FAILURE TO FIELD BALL QUICKLY," said the sub-head on the page one story in the *Tribune.* "Eddie, instead of jumping swiftly for the ball, took his time with all the leisure of a steel striker," wrote Rice. "He made no attempt to hurry this ball along to Risberg for a sure double play. . . ."

That was all. No explicit assertion that he should have made the play, certainly no insinuation that he deliberately botched the play. And it all came in the thirteenth paragraph of the story. The preceding twelve inches of the story, as well as the following twelve inches, were filled with effusive praise for the Reds, particularly pitcher Dutch Reuther, who not only held the White Sox hitters to "three clean blows and three filmy, pale blue scratches," but contributed two triples and a single to the Reds' offense as well. It was a circumspect yet, in its own way, a telling article. It set the tone for Rice coverage throughout the Series. The counterfeit errors and deliberate lapses of the on-the-take ballplayers—the plays that Hugh Fullerton had circled on his scorecard—all came in for considered attention in Rice's writing. On virtually every tainted play or episode Rice played up the out-of-character aspect of it. On Lefty Williams's fourth-inning wildness that cost the Sox game two:

> Williams was wilder today than Tarzen of the Apes, roaming the African jungle from limb to limb. . . .
> Claude issued six passes to the Reds, and when he presented three of these in the fourth inning it was all over but the clouting. . . .
> The Sox struggled hard in spots, but could not overcome the combination of Williams's wildness, [Cincinnati starter Slim] Sallee's great steadiness and the handicap of an unfriendly Fate.

On Eddie Cicotte's two fifth-inning errors in the fourth game, Rice wrote:

> Strangely enough, the ancient wing held up and it was the ancient bean that went awry. If you happen to recall the details, it was Cicotte's slow fielding and bad throw which cost him five runs and complete annihilation in his first start.
> Today three mistakes in one inning, two misplays and one error of judgment, all lumped into a game-losing mass, cost him his second start.

But Rice's articles also pointed out that the Reds pitchers were stifling the White Sox hitters (with the singular exception of Joe Jackson, who even though he accepted $5000 from the gamblers, was incapable of pulling up when it counted and hit .375 for the series), and unlike the Sox, were taking advantage of the opportunities presented to them and were playing fine baseball. Or, looking at it in retrospect and putting it another way, as good as the Sox were, they were incapable of losing this Series on their own. At least not as subtly as they did it. They needed help from Cincinnati in the form of a good Series from the Reds players.

This aspect of the Series has been lost on history. But it wasn't lost on the seasoned observers who watched and recorded these games, and it only added to their confusion and frustration. Cincinnati was playing fine baseball; the White Sox looked more lackadaisical or overconfident than they did crooked or inept. Every man in that press box—Rice particularly—espoused and fervently believed that the beauty of sport was that David slew Goliath all the time. Were it not for the rumors there would have been nothing they saw that would have prompted rumors or doubts. Cicotte's fielding *maybe*. But in game seven of the best-of-nine series, with a chance to lose it all, Cicotte pitched brilliantly, winning 4–1. How to figure? The writers had seen no evidence to overwhelm the implausibility of the rumors. When confronted directly—and this was before the era in which reporters routinely visited locker rooms after the game, so such questioning was always away from the field and generally one-on-one, and thus more conducive to catharsis—the players laughed and assured the writers that: no, they were just off their game. So the writers—Rice, Runyon, Lardner, Wheeler, Fullerton, Mathewson, all of them—wrote of what they had seen, a Cincinnati victory. An implausible victory, perhaps, but not nearly so implausible as the rumors that the Sox had thrown the Series.

But this time the rumors would not go away, as they generally did after a few days. Rice came home and addressed them somewhat obliquely later in October, suggesting overconfidence may have caused the White Sox to play at less than their best.

The only past performance chart that counts is where both sides were trying. And the only effective way it can be used in completing the future dope is for both sides to continue trying.

The plastic dope can only show what they out to do if they give 100 percent of what they have. It isn't supposed to take into consideration any such psychological upheavals as the late series supplied for the edification of the astounded multitude.

On October 28, Rice finally mentioned the word "fix" in his column. It was in a generic discussion of the process of fixing, and made no reference to the recent World Series. In fact, he made it clear, without mentioning the rumors, that he still at this point found it difficult to give them any credence.

The easiest sport to fix or frame is a wrestling match. You can make a fixed wrestling match look better than a star bout strictly on the level.

The next easiest sport to fix is a boxing match. Here you have only to fix one entry.

The next sport in line is a horse race. This isn't always so easy in a big field or in a well matched field. But by getting to one lone jockey, considerable damage can be wrought at times.

Probably the hardest sport to fix is baseball. A gambler might buy off a pitcher and a shortstop. But the pitcher might be yanked out in a jiffy and the shortstop might draw only one or two chances during the game, with none of these at vital moments.

In baseball there are too many men to be reached to make it sure. And a baseball crowd isn't very easy to fool.

There has been cheating in every game—just as there has been in every business and in every profession. But as long as a game is closely watched and closely guarded and the penalty for cheating is lifted to the limit, corruption can be held well in hand.

The first newspaper mention of the rumors that had swirled around the Series came not from any of the journalists themselves but from White Sox owner Charles Comiskey, who released a statement to the press on October 15. "There is always a scandal of some kind following a big sporting event like the World Series," it read. "These yarns are manufactured out of whole cloth and grow out of bitterness due to losing wagers. I believe my boys fought the battles of the recent World Series on the level, as they have always done. And I would be the first to want information to the contrary—if there be any. I would give $20,000 to anyone unearthing any information to that effect. . . . " The story sort of languished at that, as indeed, Comiskey had hoped it would. Hugh Fullerton published a broadside in December that named no names and stopped short of directly charging that the 1919 World Series was fixed (his first draft did so but was deemed libelous by editors in both Chicago and New York and ordered toned down). But Fullerton was able to suggest that the Series had been tampered with to the benefit of gamblers and accused organized baseball of sticking its head in the sand instead of attending to the stink that surrounded it. "Professional baseball has reached a crisis," said Fullerton.

> The Major Leagues, both owners and players are on trial. Charges of crookedness amongst owners, accusations of cheating, of tampering with each other's teams, of attempting to syndicate and control ballplayers are bandied about openly. Charges that ballplayers are bribed and games are sold out are made without attempts at refutations by men who have made their fortunes in baseball.

Such heretical utterances earned Fullerton the scorn of the *Sporting News* and *Baseball Magazine* and the cold shoulder of the baseball establishment; it did nothing to force the hand of baseball men to investigate the rumors about the recent World Series; likewise it did nothing to provoke his brethren in the press into reporting what they knew or suspected, or into learning any more. Rice had been on vacation when the Fullerton piece broke and thus had no public comment on it.

On the day after Christmas, Harry Frazee sold Babe Ruth to the Yankees and that story commanded the headlines through the winter; and by the time spring training was drawing nigh, nobody was interested in recycling a fantastic set of rumors from the previous fall. The 1919 World Series faded from everyone's consciousness. From time to time Rice and the other writers would hear more stories, but it was nothing they could print, or wanted to print. So the story went untold by all. Even Hugh Fullerton had forsaken his crusade. Any one of a dozen reporters—Rice included—could have earned journalistic immortality by printing what he knew. But in the end they couldn't bring themselves to believe it had really happened. They mulled over in their minds what they had told their readers time and time again—that baseball was "above all suspicion and no game could be thrown." For they knew that the vast majority of the players were honest and they believed these players could detect any dishonesty and protect their own game. Had you asked any of this elite corps of American sportswriters what they honestly thought about the whole mess during the eleven months that rumors swirled but the lid stayed on the fix, most would have echoed Bill McGeehan. "In all sincerity, I believe that the world's series between the White Sox and the Cincinnati Reds was absolutely honest," he wrote.

> I do not see the justice of spreading the scandalous gossip of a few cheap gamblers who are liars and thieves at heart and whose mutterings do not merit the attention of clean sportsmen. I prefer to believe in professional baseball until it is demonstrated that it has gone the way of all professional sport.

It was so demonstrated in September of 1920, when Eddie Cicotte and Joe Jackson confessed to their part in the fix and a Chicago grand jury indicted the eight players involved. Rice's reaction was that of the betrayed loyalist. He was hurt, angry and vengeful. He accused the National Commission of gutlessly having ignored the rumors in the hopes that it would go away—though he oddly expressed no chagrin at having done the same thing himself.

He reserved his most unequivocal bile for the players, calling for "prison sentences for every ballplayer and every gambler involved." It is likely that these are the most angry words Grantland Rice ever wrote or spoke in his life:

> Those mixed up in this crookedness are worse than thieves and burglars. They are the ultimate scum of the universe and even the spotted civilization of the present time has no place for them outside of a penitentiary.
>
> The only complaint in this case would come from the thieves and murderers, who might have a just kick in being forced to associate with the crooked gamblers and the crooked ballplayers convicted of the dirtiest crime in sport.

There were no shades of gray in this issue for Rice, no vacillation in his pronouncements and opinions. When he heard that Shoeless Joe Jackson and Lefty Williams had received a warm ovation upon their return home to Greenville, South Carolina, following the indictments, he snapped,

"Anyone who would extend a welcome to crooked ball players . . . would endorse burglary and child murder." There was never any doubt in Rice's mind that the confessions of Cicotte and Jackson and their implication of the other six White Sox players had closed the matter of their ever playing baseball again. "The crooks, of course, are gone forever. If one of them was ever brought back the game would be finished unless the baseball colony at large is the most feeble-minded collection of humans that ever lived."

But Rice also correctly realized that organized baseball's biggest problem was that it had no captain on the bridge. The National Commission, a tribunal comprised of the American and National League presidents and Cincinnati Reds president Gary Herrmann, never really had any real power to establish and enforce rules or standards. It had been particularly toothless since the resignation of Herrmann prior to the start of the 1920 season. What was needed was a major reorganization, said Rice—"a clean sweep at the top and a vital change all along the line." He was speaking sentiments shared by many in baseball.

> Those handling the destinies of baseball have failed utterly. If they have done their best, their best is not enough to protect or save a game that is much closer to the brink than most of the club owners believe it is. Outside of the proven crooks there have been others who have been willing enough to sign up any shady characters and who in other ways have done nothing but bring discredit on the game.
>
> Possibly nothing at all will be done except to throw out the crooks and call it a day's work. If this happens we wouldn't give twenty cents for any franchise in America. For we can't believe under such conditions that newspapers will continue printing professional baseball news and that fans will continue to pay out regular money to support magnates who have no vision beyond the box office.
>
> If nothing is done except to throw out the crooks—if no further move is made to give the game a thorough cleaning and to bring it back upon a new foundation—and both the press and the fans still continue their support, there can be no further complaint, no matter what happens.

Rice proposed Pacific Coast League president William H. McCarthy as a possible head of a new and independent commission to rule baseball. "In every instance he was the one who took the jump without wading through a snarl of red tape and official fear," said Rice of McCarthy's role in handling gambling in the Pacific Coast League. But the baseball people had somebody else in mind, Federal District Judge Kenesaw Mountain Landis. A man of stern visage and demeanor, Landis had gained national prominence in 1907 by having the temerity to fine John D. Rockefeller $29 million in an anti-trust case. But he had earned the affections of major league baseball owners some years later by avoiding making a decision in the Federal League's anti-trust suit against organized baseball.

Rice could live with Landis. He had called for a national figure, and Landis was certainly that. He had called for somebody outside the game, as much for appearances as anything else. Landis was that. And he had called for a strong, uncompromising leader. Landis would certainly prove himself to be that. If baseball was not immediately cleansed of its troubles when the no-nonsense jurist took office, the *appearance* was that things were different, that Landis was not only in control but capable of dealing surely with any crisis, however large. This was enough to let Rice and the rest of the writers, and all of the fans, get back to the *game*.

But after the reconciliation any betrayed lover retains the smallest ache that is now forever a part of him. For Ring Lardner, the sense of abandonment was very nearly absolute. He continued to watch the game, often seated at Rice's side. And he continued to write of the game, for very nearly a decade more. But he watched now with vacant eyes and absent any zest for what he saw. "I had forgotten what terrible things worlds series were so I consented to cover this year's," said Lardner in a letter to F. Scott Fitzgerald in 1925. "I got drunk three days before it started in the hope and belief I would be remorseful and sober by the time I had to go to it."

Rice's was a more resilient psyche; the greatest of his hero-making and wide-eyed celebrations of games still remained in front of him. But he too carried an ache from 1919. "I felt as though I'd been kicked in the stomach," was all he could bring himself to say about it in his memoir. A complementary insight into just how much it hurt, perhaps, comes from the following thought, expressed a month after the scandal broke in 1920, in a reflective interlude after he had spent much of his anger, and before he tried to forget and move on:

"Sometimes we wonder if, after all, is there very much sport that is real sport beyond the amateurs. The rest may be called competition, or amusement, but just how much of it is sport?"

A Golden Age for Friends

*W**hen Warren G. Harding was* sworn in as President of the United States on March 4, 1921, Rice marked the occasion with a few lines of verse wishing him well, as he had done when William Howard Taft was inaugurated in 1909, and on the two occasions Woodrow Wilson took the oath of office. Unlike his predecessors, however, Harding was a reader of Rice's column, and after settling into the White House, he wrote Rice a note to thank him for his words. "I have seen your verses in the *Tribune,*" he said, "and wanted to drop you a line so that you would know of my grateful appreciation. Probably it is not important, but it will make me feel better if I have you know about it."

Rice wrote back, asking if he might have the opportunity to write about the President's golf game. Harding responded eagerly. "I am not very much of a player myself," he insisted, which was more true than modest, but a little bit of both, "though I get quite as much enjoyment out of it as do those who play vastly better. If you will run down to Washington almost any time you find it convenient and will let me know a day or two in advance I will take you on for a game and show you how an official can forget the problems which are his."

Rice arranged a date for Thursday, April 7, and took Ring Lardner along. They took an early train out of New York and arrived in Washington early enough to catch breakfast at the Willard. They were seated at a table next to Vice President and Mrs. Calvin Coolidge, "who, however, remained in blissful ignorance of same," according to Lardner's account. Harding's invitation included lunch at the White House, and Lardner's deadpan humor is what survives from the meeting.

"Rice is here to get a story," said Harding to Lardner. "Why did you come?"

"I had a good reason," replied Lardner. "I want to be appointed ambassador to Greece."

"Why?" asked Harding, obviously a good sport in all of this.

"My wife doesn't like Great Neck," explained Lardner.

"That's a better reason than most people have," allowed Harding.

That account came from Rice. But Lardner was there to get a story as well, and the story he fashioned is a monument to his irreverent, hilarious style. As Rice and the President talked golf, Lardner fell into conversation with Mrs. Harding. This is from his account:

> I and the 1st lady indulged in the light chit-chat which makes the present writer so sought after by fashionable N. Y. hostesses. For inst. when she said:
> "I expected you would be a man 20 years older" I just smiled and said banteringly: "Did you?"

Following lunch, the two writers were assigned a manservant to assist them in getting into their golfing togs. For all of their worldliness, valets were something outside the experience of both Rice and Lardner and both were mildly disquieted by the experience. "I and Grant realized that people that lives in the White House has got a servant problem like nobody else," said Lardner in his story. "The problem there is how to get the valet out of the room while you dress and neither of us were able to solve it." Lardner's pants provided as many laughs as his conversation. They had just come back from the cleaners and had shrunk to the point where they stopped some four or five inches from the tops of his shoes. Of the picture that the White House photographer took of the foursome prior to the start of their round, Lardner quipped, "I am afraid mine will be very risqué," and even Rice couldn't resist pointing out the pants when he published the picture in his memoir more than thirty years later.

They played at Burning Tree out in Chevy Chase, Maryland, Lardner and Rice against Harding and Under Secretary of State Henry Fletcher. This was a working round of golf for Rice, he had come to write a story on the President's golf game. The story he delivered to his readers a week later was a workmanlike study of the physical and psychological aspects of Harding's game—he might have been telling his readers about Chick

Evans or Bobby Jones instead of a twenty-five handicapper, for all his studious attention to technical detail. It was a column written for golf freaks, and there were few golf freaks who did not read Rice religiously. Taking note of the less-than-perfect weather—gray threatening skies, a chill blustery wind and a sodden course—Rice told his readers that the President "used a square stance for every shot—which is to say that his toes were on a line, his feet fairly well apart, with the ball on a line midway between the two heels. . . .

"Placing direction and control above distance, the President is content to use only a three-quarter swing, making no effort at any time to kill the ball."

Rice told his readers that the President's game, under normal conditions, ranged between 95 and 100; he did not tell them that on this day it was closer to 110. He told them of the type of putter Harding used ("a center-shafted affair with two steel prongs") and about his putting style ("a good style and a firm definite touch that gives the ball a chance. His main fault was a tendency to start his body ahead of his hands and arms, one of the most prevalent of all golfing faults.") Mostly, however, Rice spoke of Harding's character, as revealed through the game of golf. It was a most flattering portrait of a stalwart leader, a man who had matters both in control and in perspective:

> Another noticeable feature of his game was an utter lack of alibis.*
> At the second hole, after a good drive, the ball was lying in a cluster of thick, wet grass.
> From here he put his second shot into the bunker.
> "Hard luck," someone remarked, "but the lie was very bad."
> "No," he said. "It was just a miserable effort. I had no excuse."
> No matter into what trouble the high wind would occasionally take his tee shot or his approach, he continued to battle on without any idea of surrender. Even when it was apparent that the odds on some holes were all against him, he played with as much care and as much determination as if the chances were still even. . . .
> You can get a good line upon most men though a round of golf. As a soul exposer it has few equals. More important than his golf swing, it was interesting to find that the President was a regular human being—a hard fighter and a good loser—that he had a sound, sane philosophy and a sense of humor—that he had dignity without pretense or aloofness and a friendliness of manner that was in no sense cultivated or forced. He is a sportsman who can take the break of the game as it comes. A firm jaw and an unwavering eye are hardly the component parts of weakness. He doesn't expect to become Open golf champion of America, but he knows the world has discovered that a certain amount of

---

*Golfers with ever ready alibis were a particular and long-standing *bete noire* of Rice's.

recreation and exercise is as necessary to a sound body and a sound mind as food and sleep.

The sense of humor that Rice referred to was brought to the fore by the wry irreverence of Lardner. At the turn, the secret service man keeping score reported that Harding stood at 55, Rice at 44.

"Well," said Lardner, "I'd rather be Rice than President."

Harding also had the habit of driving and then walking on ahead, without waiting for the others to hit their drives. Harding was well down the fairway on one hole by the time Lardner hit his tee shot. He hit what Rice described as "a lusty ball with a slight slice." The ball caught the dead branch of an apple tree and the branch dropped down and hit the President in the shoulder. Harding waited, anticipating an apology. Lardner approached with studied nonchalance and said with aplomb: "I did all I could to make Coolidge President," whereupon Harding dropped his club and roared with laughter.

Rice allowed sometime later that "Harding was a poor President, but he was quite a fellow and, I think, an honest man." It is plain from his writing both after the match and in his memoir that he cherished the experience of playing golf with the chief executive of the United States, and plainly regarded it as one of the highlights of his career. But the stories he delighted in telling of that day were the stories of his friend Ring Lardner. Lardner plainly enriched the experience, as he enriched Rice's life for most of the twenty-odd years the two men knew one another.

Lardner and Rice were as close as brothers throughout most of the twenties and up until the time of Lardner's death in 1934, and the world had seldom seen a more unlikely combination. Or, at least those who have contemplated the friendship—most of them Lardner biographers and scholars—have been perplexed by the close bond between the two men. "Rice was as ebullient as Ring was reticent," said Jonathan Yardley in his elegant biography of Lardner, "as sentimental as Ring was skeptical, as corny and obvious as Ring was subdued and subtle. Ring must have gagged when he read Rice at his most florid and hero-worshipful, as in his notorious [Four Horsemen] lead."

Ring Lardner, Jr., described the differences similarly: "Ring was a debunker of sports heroes, Granny a glorifier of them. Ring was reserved and taciturn, Granny outgoing and loquacious. Ring's views tended to be pessimistic with a sardonic flavor, Granny's optimistic and tinged with sentiment. There was a whole side to Ring most people never saw because he couldn't be relaxed except with his family and a few close friends. Granny was at his ease everywhere and with everybody; he knew an astonishing number of people by name, but that was only a fraction of the number who knew and greeted him."

Lardner's son also pointed out, however, that both men shared a common foundation in sports, both men enjoyed games like golf, poker and bridge, both liked to gamble and both liked to drink. Friendships are more often forged of shared interests than of shared personalities.

Perhaps the puzzlement over the affection between the two men is that Lardner was not a man who suffered fools gladly, and so many of the fools in the world are possessed of the hail-fellow-well-met personality so akin to Rice's unaffected ebullience. Rice's exuberance and gregariousness were genuine; it was phonies that Ring Lardner did not suffer gladly, and Rice was no phony.

Rice knew the Lardner that his son said few people got to know, though Rice allowed that he "never quite knew that I knew him," despite their deep-rooted, abiding relationship. "To contain Ring would be like wrapping up a wraith," he said. Still, he knew that there was a compassion in Lardner that the more casual of his acquaintances never saw and that the readers of his work never detected. "How can you write if you can't cry," said Lardner once to his family after someone had contemptuously dismissed Charles Dickens as sentimental and sloppy. While Lardner would have never fashioned any of Rice's material himself, it is entirely likely that he did not gag when he read it.

Both men shared equal, considerable fame, and, given that both were pigeon-holed—as society is wont to do—into the category "sportswriters," from whom little was expected in the way of thoughtful, compelling writing, both enjoyed considerable reputations as artists as well. Ironically, given the divergent paths of their reputations in the years since their deaths—Lardner moving to the fore of American letters while Rice has become merely a bauble amidst the dusty memorabilia of the twenties and thirties sports worlds—in their own times, Rice enjoyed slightly the better reputation as the artist. Rice's books of verse and Lardner's books of stories sold about the same. Critics were generally kind to both men. The critics' praise of Rice, however, was unequivocal—one praised his "realism" and his "viewpoint and philosophy of life." Another said: "No sports writer has the literary firmament that Rice possesses." Still another said he "must . . . be regarded as one of the first poets of his native land. Indeed, a ranking in future years may set his place definitely higher. There is at the present time no American poet of action that is his peer."

Meanwhile, much of what we treasure in Lardner today—his remarkable ear for the argot of the uncouth, characters stripped of shallow, popular-fiction contrivances—escaped the early critics of Lardner's work. It was not until the distinguished Edmund Wilson reviewed *How to Write Short Stories* in 1924 that critics began to realize that there was something of substance in these stories of ballplayers, boxers, waitresses and salesmen. Prior to that, and to some degree long after that, Lardner's choice of characters, and his humorous treatment of them, befuddled the critics, who according to one critic writing in the mid-twenties, "doubt whether Mr. Lardner will survive as a master writer because of the phraseology employed in most of his works. It is probably true that some of his writings that are today considered among his best will not survive the future changes in American idioms. . . ."

None of this played any role in the Rice-Lardner relationship. Theirs was a friendship free from any sense of competition, jealousy or insecurity about the other's writing. In fact there is no evidence that they ever discussed their respective careers. That seems incredible, given the volume of things they had in common and the extraordinary opportunities they had for conversation. Both men wrote a syndicated column, both contributed to the same major magazines, both frequently wrote of the same events. And they did a large portion of this writing while spending time traveling together during the 1920s. Writers in general are seemingly incapable of refraining from discussing their scribblings with other writers. But if Rice and Lardner ever discussed their feelings and thoughts about writing in general and their own in particular—as surely they must have—those thoughts and feelings remained private. When Rice wrote his memoirs he remembered not the man who, during the years that Rice knew him, was honing what was perhaps the most distinctive and one of the most original voices in American fiction. He remembered the man who one-upped the President, the man who deflated a pompous New Orleans buffoon who was impressing upon Lardner his impeccable, aristocratic pedigree, by simply waiting until the New Orleans gentleman was finished and them explaining to him: "I was born in Niles, Michigan, of colored parents."

Rice had first met Lardner shortly after coming to New York from Tennessee in 1911, when Lardner was a baseball writer for the *Chicago Tribune,* traveling—during different seasons—with the Cubs and White Sox. In 1913, he took over the long-running *Tribune* column, "In the Wake of the News" and, freed from the relentless travel and deadlines of the beat writer, he began to contribute the Jack Keefe "busher" stories to the *Saturday Evening Post.* His national voice thus began to emerge at precisely the same time Rice's was emerging; their paths began to cross frequently at baseball games, football games and prizefights, and the professional relationship had flowered into a solid friendship by the start of the war. After the war, when Lardner moved east to Great Neck, Long Island, and began writing a syndicated column with Jack Wheeler's Bell Syndicate, the two men became boon companions and inseparable traveling partners. When Kate Rice and Ellis Lardner came to know one another, the bond between Lardner and Rice was strengthened even further. "[I]t was Kate and Ellis who really forged the bond between the two couples," said Ring Lardner, Jr. "Like Granny, [Kate] had the knack of getting along with all kinds of people, but she had a special regard for Ellis's integrity and judgment, looking up to her as, relatively speaking, an intellectual."

The Rices visited the Lardners in Great Neck, the party world that served as the inspiration for Fitzgerald's *The Great Gatsby*. And the two men were frequent golf companions in the greater New York area. But most of their time together was spent traveling—to Ivy League football games, where they would sometimes meet up with Jack Wheeler and Damon Runyon and their wives; to the out-of-town openings of plays in New Haven and Atlantic City, to sporting events and stories of any kind. Virtually each winter throughout the twenties, the Rices and Lardners vacationed together. They would go to California, or Florida, or the Caribbean, and sometimes all three. In 1926 the trip ran more than three months, stretching from Florida to New Orleans for Mardi Gras, to California for some golf, and finally back to Florida for spring training. "The Rices have been with us right along and we are all still speaking," said Lardner in a letter to Scott Fitzgerald.

Floncy Rice was along on the 1926 trip, and had occasionally accompanied her parents and the Lardners on others. As she had moved through adolescence, Floncy flowered into a radiant young woman, with her father's blue eyes and blonde hair, and, as one gossip column of the times put it: "perfect features, a fresh complexion and an alluring figure [which] combine to form genuine pulchritude." Beset by suitors and self-indulgent to a fault, Floncy had a social life that fascinated Ring Lardner. He would listen with amusement to her dramatic babbling on the perils of juggling a half-dozen boy friends, each of them the unquestioned, unchallenged, absolute love of her life—if only for a few hours. "I can't breathe," she would exclaim as she tried to give voice to her infatuation. Finally, sometime during or just after that 1926 trip, Lardner could resist no longer; he built a short story around Floncy Rice's impetuous love life. The story took not only its inspiration from Floncy's conversations, it took its title as well; it was called "I Can't Breathe." It appeared in *Cosmopolitan* in September of 1926, and it is held today to be one of the finest of all Lardner's stories. It takes the form of a diary kept by the unnamed heroine while she's staying at an inn in the country with her aunt and uncle, "both at least thirty-five years old and maybe older." She is eighteen years old and has been engaged to be married, by her own count, six times. She is currently engaged to Walter, and apparently to Gordon too. Then she meets Frank at the resort and he too asks her to marry him and she says yes. She can't understand how she got into all this, of course. It is a combination, she insists in her diary, of misunderstandings on the part of her suitors, and her innate compassion that prohibits her from hurting anyone's feelings. In truth, of course, it is all a result of her conviction that she can keep them *all* on the string, and there is no reason she shouldn't:

I played golf with Frank in the afternoon and we took a ride last night and I wanted to get in early because I had promised both Walter and Gordon that I would write them long letters, but Frank wouldn't bring me back to the Inn till I had named a definite date in December [for us to get married]. I finally told him the 10th and he said all right if I was sure that wasn't a Sunday. I said I would have to look it up, but as a matter of fact I know the 10th falls on a Friday because the date Walter and I have agreed on for our wedding is Saturday the 11th. . . .

Life is so hopeless and it could be so wonderful. For instance how heavenly it would be if I could marry Frank first and stay married to him five years and he would be the one who would take me to Hollywood and maybe we could go on parties with Norman Kerry and Jack Barrymore and Buster Collier and Marion Davis and Lois Moran.

And at the end of five years Frank could go into journalism and write novels and I would only be 23 and I could marry Gordon and he would be ready for another trip around the world and he could show me things better than someone who had never seen them before.

Gordon and I would separate at the end of five years and I would be 28 and I know of lots of women that never even got married the first time till they were 28 though I don't suppose that was their fault, but I would marry Walter then, for after all he is the one I really love and want to spend most of my life with and I wouldn't care whether he could dance or not when I was that old. Before long we would be as old as Uncle Nat and Aunt Jule and I certainly wouldn't want to dance at their age when all you can do is just hobble around the floor. But Walter is so wonderful as a companion and we could enjoy the same things and be pals and maybe we would begin to have children.

But that is all impossible though it wouldn't be if older people just had sense and would look at things the right way.

Eventually a fourth beau calls and she sees him as her way out, and, of course, as her real true love. "A whole year and he still cares and I still care." According to Jonathan Yardley, "I Can't Breathe" succeeds because Lardner "exactly caught the self-indulgence and self-dramatization of a late adolescent who is getting too much attention for her own good and is too shallow to know how to handle it. He sensed precisely how much to put into the diary and how much to leave out, so that the story moves at a clip appropriate to the speed of its narrator's misadventures." It succeeds also on a darker level that Lardner could not have anticipated when he crafted the story during the spring of 1926, when Floncy had just turned nineteen. It was premonitory; Floncy Rice, an indifferent student at Smith College for a year, Queen of the Dartmouth Winter Carnival in 1927, was married and divorced twice before she was twenty-three years old. Her life was grist for the gossip columns as she proved herself a blithe, reckless spirit on the New York night club scene—Grantland Rice's long friendship with restauranteur Toots Shor began during these years when Shor was a bouncer in a Greenwich Village club and took an avuncular interest in protecting Floncy from herself and the more unseemly of the men who fluttered around her.

She also became quite an able actress in time, after owing her start to Ring Lardner. Floncy had been stagestruck since early adolescence; Granny and Kate discouraged her ambitions at first, but soon realized that resisting their headstrong daughter was futile and gave their blessing. Lardner, who devoted great energy to writing for the stage during the twenties, wrote in a role for Floncy in his 1928 collaboration with George S. Kaufman, *June Moon*. In fact, after resisting Kaufman's early entreaties to collaborate on a play based on Lardner's story "Some Like Them Cold," Lardner finally capitulated on the condition that the play contain a part for Floncy Rice.

The play was a parody of the Tin-Pan Alley songwriting business and Floncy played a cynical song publisher's secretary, a substantial supporting role. She received a great deal of pre-opening ink, all of it owing to her kinship with her famous father. "She has taken to acting like a duck to golf," said Lardner. Floncy also received—as did Kate and Granny—the unique Lardner needle in the cast notes he wrote for the Playbill:

> FLORENCE D. RICE is the daughter of Grantland Rice, the taxidermist. Miss Rice's parents have no idea she is on stage and every time she leaves the house to go to the theater, she tells them she has run down to the draper's to buy a stamp. On matinee days she writes two letters (that's what they think). She is very proud of her wire-haired fox terrier, Peter, because on the night the play opened in Atlantic City he sent her a wire.

The proud parents accompanied the troupe to tryouts in Atlantic City and Washington, and when the play opened on Broadway on October 9, 1929, they anxiously awaited the first-night reviews for their twenty-two-year-old daughter. The notices came in unanimously kind. "Florence D. Rice is neat as the disillusioned secretary," said Brooks Atkinson in the *Times*. "Florence Rice could handle a larger role," said Max D. Davidson. And Arthur Pollack of the *Brooklyn Eagle* wrote: "You will find . . . no better bit than that provided by Florence D. Rice."

June Moon enjoyed a run of 273 performances on Broadway. But, despite her initial success, Floncy was still far from a polished actress. Granny encouraged her to study her craft, and this, finally, was one parental suggestion she embraced. She embarked on a string of acting, voice, and music lessons. She worked in summer stock in East Hampton and elsewhere and soon began landing replacement roles on Broadway, including one in the George S. Kaufman-Moss Hart smash, *Once in a Lifetime*. Finally, in 1933 she landed a starring role in the Howard Lindsay comedy,

She Loves Me Not. Percy Hammond called her "outstanding," in this role in his review in the *Herald Tribune,* "lovely to look upon and four-square in her every syllable and movement." She used this play as a springboard to Hollywood, where she enjoyed a steady success playing the ingenue and assorted supporting roles in a string of "B" pictures throughout the thirties and forties. But for all of Floncy's genuine success, she never fully escaped being her father's daughter. Her clippings—a formidable collection, rivaling, in its heft, the collection of clippings on her father in the family papers—seldom failed to mention some manner of the fact that she was "the daughter of Grantland Rice, the noted sportswriter." That rankled. While father and daughter remained close, and while Floncy's public statements always bespoke a gushing pride in her father and a gratefulness for having grown up as his daughter, there was the occasional manifestation of her yearning for her own professional identity. "There's one more thing I'd like to say," she said in an ad-lib close to a carefully scripted radio interview on WOR in 1934. "I hope that one day I won't be known as the daughter of Grantland Rice, but rather that he'll be known as the father of Florence Rice."

That never happened.

The bond between the Lardners and Rices had grown so strong during the twenties that in 1928 the two men jointly purchased a four-acre parcel of grass and dunes and bluff at the Atlantic's edge in East Hampton. There they built side-by-side summer cottages, "separated by an unbroken stretch of lawn with no indication of a property line." The Rices' house was a towering four-bedroom, four-bath Cape, with servants' quarters, an expansive living room and a sprawling deck. "Perched on a porch on our dune, we could stare straight out and into the bull rings of Lisbon," said Rice, "or perhaps it was the clearness of the gin cocktails. At any rate, nothing but gulls, whales and water separated us from Portugal and Spain."

On this modest little estate, Rice constructed what he called "an all-around sporting club." There was a nine-hole, chip-and-putt golf course, as well as croquet, archery and horseshoe pitching set-ups. The Lardners and Rices moved easily and frequently between the two houses, and on summer Sundays Rice was often host to a veritable who's who of the New York sports and journalism worlds. Some thirty to fifty people were likely to make the trek out to East Hampton to avail themselves of Rice's humble little sporting club and his expansive hospitality. At various times through the years, Gene Tunney, Jess Sweetser, Damon Runyon, Harold Ross, Frank Crowninshield, Percy Hammond, Irvin S. Cobb, Rube Goldberg, Clarence "Bud" Kelland, Jack Wheeler, Hal Sims, Bruce Barton, Dan Topping, among scores of others, were drawn to Rice's home. Most were drawn again and again. The organization of these affairs fell to Kate,

as did most of the hosting. Granny would greet his guests, get them started
on their games, make certain their glasses were full, and then, as often as
not, he would repair to the Maidstone Golf Links nearby for a few quick
holes of the real game.

When the Rices and Lardners were by themselves, they would play
bridge of an evening, or just sit on one of the porches and gaze out towards
Portugal as they talked. At night, before retiring, Granny would indulge
in a full, round, two-inch-thick slice of watermelon, washed down with a
few fingers of bourbon or gin. There was a pleasing order to his life in
East Hampton. He was surrounded by friends—or acquaintances at least—
wherever he went, but in East Hampton the circle of friends—although
still quite large—was select. "There have been many other friends," said
Rice of his guests in East Hampton and of the Lardners in particular, "but
these have contributed more to the happiness of my wife and me over past
years than perhaps any others. For they came to us when the sun was just
above the meridian, before sunset was due."

In the first days of March 1931, the splendor that was the Rice and
Lardner compound in East Hampton came within a few fragile feet of
being washed out into the Atlantic. A tremendous winter storm struck the
eastern end of Long Island, and as the East Hampton newspaper reported
it, "the Rice and Lardner homes are taking the brunt of it." When the
storm began there was a hundred feet of sand and another hundred feet of
lawn between the Rice and Lardner homes and the high water mark. By
the time the wind and ocean had abated several days later both houses
teetered precariously over the edge of a twelve-foot cliff carved out by the
water's wrath. The porches and portions of the foundations had been washed
away.

Nobody was home at the time of the storm. The Rices were both in
Florida for spring training; the Lardners were in a New York hotel. Work-
men placed pilings under those parts of the houses hanging over the cliff
to prevent them from toppling into the sand, but there were several anxious days for the owners, as another storm hit on March 8, destroying a
pile driver that was aiding in the rescue. It took two weeks before both
homes were up on blocks and pulled back a hundred or so feet, behind a
dune that afforded some protection from future storms. The storm cost
Rice $26,000 depression dollars—money he nonetheless could afford, for
at this stage of his career, his income from all sources—the column, magazine work, radio appearances and the movie shorts he produced and narrated—exceeded $100,000 a year.

Ring Lardner battled tuberculosis, alcoholism and insomnia throughout the early thirties. The great vitality slowly ebbed and by the summer
of 1933 he was confined to home, spending much of his time in his bed-

room and on the adjacent sleeping porch. The only people he would see outside his family were Kate and Granny, and the two couples would play bridge on the sleeping porch. Granny would shuttle in and out of East Hampton, traveling to events, and spending a day of two in New York each week writing. He was in New York, at home in his Fifth Avenue apartment on the evening of September 25, preparing to leave the next morning for the Jack Sharkey-Tommy Loughran fight in Philadelphia, when Kate called to tell him that Ring had died a couple of hours before. "The news didn't come as a shock," said Rice, "rather as a heavy wrench."

Since 1908, when Rice's old Atlanta friend Joel Chandler Harris had died, he had generally marked the passing of a friend with a verse in the column. In the years ahead, as more and more of his friends and colleagues passed away, these poetic eulogies became more frequent and among the most widely acclaimed of his work. They generally played off some theme of Charon, the boatman from ancient mythology who ferried the dead across the River Styx. Nothing from this very tender oeuvre carries greater feeling than his lines on Ring Lardner:

> Charon—God guide your boat—
> On to the journey's end.
> Keep it safe and afloat—
> For it carries a friend.
> One who has given the world
> Drama and wit and mirth—
> One who has kept unfurled
> The flag of a cleaner earth.
>
> Charon—the night is dark—
> Watch for a port ahead—
> Stick to the wheel of your barque—
> Charon a friend is dead;
> The friend of a shattered age,
> Standing upon Time's brink—
> The friend of the printed page,
> For those who could read—and think. . . .
>
> I wish you could know his worth,
> Out in the realm of ghosts—
> What he has meant to the earth,
> What he has meant to hosts
> Of those who can understand
> The message of brain and heart,
> Flashed upon sea or land
> With only the master's art. . . .

10

Fame, and Other
Earthly Rewards

t some point during the twenties—it is impossible to pinpoint exactly when—Grantland Rice passed from being merely a sportswriter to a combination media conglomerate and public figure the equal of Dempsey or Ruth or Grange or any of the men he covered. He became synonymous with sports information. He threatened to monopolize the field.

Not only did his daily column appear in between eighty and one hundred papers with a combined circulation in excess of ten million, his newspaper writing also included game coverage of the big events and a regular Sunday golf column called "Tales of a Wayside Tee." He appeared weekly in *Collier's* and regularly in a dozen other publications (and advertisers would frequently use his likeness, endorsement or some of his copy in their ads). He edited the Spalding baseball guides and various golf guides. He was all over the radio and it was hard to go the movies without catching a Grantland Rice-produced short at the start of the bill.

He failed to recognize the extent of his own fame. Once during the twenties, at the National Tennis Championships at Forest Hills, he made his way to the locker room with a posse of writers. As they flashed their press cards on the guard at the door, Rice discovered that he was without his. "I'll wait out here for you," he told his friends and made ready to sit down and wait when the guard told him he knew who he was and he certainly didn't need any press card to prove it. It was a scene that repeated itself throughout Rice's career. He was constantly losing his press pass, and constantly surprised when his face alone would allow him entry through the press gate. At a World Series game in St. Louis in the 1940s, Rice was at the press gate, rummaging in vain for his credential, which he had left back at the hotel. Frank Graham of the New York *Sun* told the guard what had happened. "Please, Mr. Rice. Come in. You don't need a press card," said the guard. Rice was convinced that Graham had coerced some special favor from the guard and remained slightly incredulous. "How'd you do that?" he demanded of Graham as they made their way up the stairs to the press box.

When he traveled, he generated profiles in the local papers like a Hollywood starlet on a promotional tour. Syndicated features like "People You Read About" and "Who's Behind the Name?" profiled Rice, and in all the words written about Rice in his lifetime it is impossible to find a single one that is harsh. No exaggeration. No hyperbole. No one ever apparently uttered or published an unkind word about Grantland Rice in the more than thirty years he spent as a major public figure. Today, this alone would be enough to earn him some scornful, or at least sarcastic, profiles. But we were a gentler people in Rice's day, less willing to devour our public figures simply because they were public figures. We were more naive perhaps, but without question gentler.

"Don't ever lose a chance to meet Mr. Rice," said syndicated columnist Robert H. Davis in 1924, and those who had the good fortune savored the encounter. On a golf course during the early twenties an elderly gentleman playing in a foursome ahead of Rice, out of sight in a deep valley in the fairway ahead, got conked by a drive hit by sportswriter Dan Parker. The feisty codger came scrambling out of the ravine swearing and demanding an explanation. "Who the hell hit me?" he wanted to know.

"Probably Grantland Rice," said Parker. "He loves to come in contact with strangers."

"What!" said the man, his demeanor changing swiftly from anger to anticipation. "You don't mean Grantland Rice the sport writer!"

"That's the one," said Parker. "The only one."

"Well, I've always wanted to meet him. Introduce me."

The man was so excited, honored even at having been hit by the famous Grantland Rice, that Rice didn't have the heart or the energy to put the facts in order.

Rice helped to make history in 1922 when he provided the play-by-play for the first World Series game ever to be carried live on the radio. From its infancy, broadcasting recognized the value of name-recognition in its announcers; Rice was the natural choice, and probably would have been even if the *Tribune* had not been involved in putting the broadcast package together.

At the time, the broadcast was a greater radio landmark than it was a sports landmark; nobody foresaw the dramatic impact that broadcasting would have on sport. A special three-stage amplifier, built specifically for the event, was installed at the Polo Grounds, to carry Rice's voice the twenty miles back to the WJZ studios in Newark with sufficient current to activate the transmitter there. There had been successful remote broadcasts prior to the World Series—two Tommie Leonard fights and a New York Philharmonic concert from City College had been successes within the past year—but nothing of this scope. The transmitter at WJZ was rebuilt, quadrupling the normal range of its signal. The other New York area radio stations all agreed to suspend their programming during the game so that WJZ might operate without interference. Optimistic engineers suggested there was "every prospect that Mr. Rice's voice will be heard as far as the Mississippi, while there is no doubt that he will be heard throughout the entire eastern section of the country." There was hope that the signal would reach Pittsburgh with sufficient strength to trigger the transmitter at WJZ's sister station, KDKA, further extending the WJZ signal. Reality fell somewhat short of that; still, Rice's call of the game was carried with stunning clarity over a three-hundred-mile radius, reaching an estimated one-and-a-half-million people. In the history of the world to that point, no man had ever spoken to a larger audience.

The broadcast was a *tour de force* by the modest standards of the nascent industry, not only for its technical landmarks, but for the nuances in the content of the broadcast as well. Rice described the pre-game activities—the entry of Commissioner Landis and his honored guests, Christy Mathewson, Jack Dempsey and General Jack Pershing. The call of the crier announcing the batteries to the stadium crowd was carried "out over the ether" to the radio audience as well. Listeners heard the ebb and flow of the crowd's roar, punctuated in quiet moments by the sound of the bat hitting ball and cries of the Polo Grounds peanut vendors hawking their wares near Rice's microphone. And above it all they heard the docile drawl of Rice, describing the Giants' 3–2 win over the Yankees.

Wearing earphones under his fedora, gripping the microphone—a tall slender item, resembling a speakeasy-era telephone—in his hand, and broadcasting from a field box next to the Yankee dugout, Rice would wait until the play had concluded and then deliver a succinct, poetic report to his listeners. The radio audience would hear the crack of the bat and then the shouts of the crowd and wait breathlessly the few seconds for Rice's account of the play. As the Polo Grounds crowd was split about evenly between Giant and Yankee fans there was no suggestion of the tenor of the play from the noise. The radio people had a notion that this wasn't the best way of approaching this new art of play-by-play. "The broadcasting officials wanted me to keep talking, but I didn't know what to say," said Rice. But from the listener's perspective, remember, Rice was painting on a clean canvas; far from being an annoyance, his primitive style proved a delight to the rapt audience, the frenzied seconds of the delay only serving to heighten the listener's interest. "I would hear the crowd let out a terrific roar," said one listener, "and it would seem ages before I knew whether it was a single or a three-bagger that had been made or whether the side had been retired. Of course it was only a matter of seconds before we got the announcement, but the interest was so intense that it seemed longer. . . . It was in a way too realistic. . . . It was a wonderfully successful broadcast."

Every radio store in Manhattan had a set on the sidewalk and crowds pushed and jostled to get close enough to hear a bit of the novelty. In Newark, the crowd outside the WJZ studios grew so large that it clogged traffic, and police had to be called in. In Hackensack more than two thousand people crowded around a single set.

Across the street from the Polo Grounds, employees of the Ninth Avenue Elevated Railroad gathered on the roof of the employee clubhouse, and not only watched the action on the field below but listened to the noise that came to them simultaneously from the stadium at their feet and the radio set at their ear. Writing of the broadcast on the front page of the *Tribune* the next day, radio editor Jack Binns called it "the most remarkable project ever undertaken in the annals of communication. Not a single detail of the game was missed." The Sunday rotogravure section of the *Tribune* carried a picture of Rice with his microphone.

Rice's voice would become among the most familiar in America over the next three decades—the full span of network radio as the American hearth. He would, for the most part, leave the play-by-play to others after the Polo Grounds debut. But he had a weekly interview show on NBC throughout the twenties and thirties; he would do special preview reports on big sporting events like the World Series, the Olympics and major prizefights; and he was frequently a guest on the variety and interview shows of others, reading his verse or proffering his insight. It was his celebrity that radio wanted, not his broadcasting presence or charisma. His was a voice ill-suited to radio success, a little flat, atonal, sometimes awkwardly modulated and unmistakably Southern. But his name had marquee value of the first order.

Movies, as much as newspapers, magazines and radio, had helped to spread the Rice name throughout the land. In 1920 he and a filmmaker named Jack Eaton entered into a partnership to create a series of one-reel films, a staple of the movie business in those years as a preliminary to the feature attraction. They christened the enterprise Grantland Rice Sportlight Films and produced films at the rate of one a week, "a pace that had us all mumbling," said Rice. On top of the pace of making the films, they had to hustle to find distributors for the product. Exhausted, they gave it up after a year, and Eaton went off to run a theater in Denver. In 1925, however, John Hawkinson, the cameraman for the original Sportlight series, approached Rice about resurrecting the concept. Rice worked out a distribution deal with Pathé pictures, and secured financing for series of three films that would serve as an audition of sorts for Pathé. The films were accepted and Rice and Hawkinson signed a contract to produce one ten-minute film a month for Pathe. Sportlight Films quickly became one of the most successful one-reel production companies in America, appearing in more than seven-thousand theaters in the United States and another thousand abroad at the height of their popularity in the thirties and forties. Their success was based in part on Rice's contacts and celebrity, to be sure. But they also succeeded on the strength of Hawkinson's skill and innovation with the camera. An interior decorator, musician, and painter, Hawkinson married the sensibility of the artist to the élan of the adventurer to produce pictures that were original in their use of light, perspective and motion, and stunning and powerful in their effect.

Rice sometimes narrated and hosted the films, but as the years passed his appearances in the films and his hands-on involvement in the production grew more limited. Throughout, however, he served as a combination studio mogul, executive producer, talent agent and press agent. He'd arrange for prominent athletes to appear in the films. In the early years this was enough. To get a look at Dempsey's face—or Ty Cobb's or Red Grange's—was enough to bring people into the theater. Rice and his partners soon learned that newsreel-type looks at big-name athletes were not going to be enough. So they began to delve a little deeper. Mirroring the sort of theme articles he would sometimes write for *Collier's,* Rice and Hawkinson produced a series of theme-based films. One was called "The Knack," breaking down and analyzing—in slow motion—the mechanics of certain athletic skills, such as throwing a football, serving a tennis ball, or putting a golf ball. Another series was called "Stamina" and featured athletes like marathoner Clarence DeMar. Profiles of headline athletes such as Ruth, Jones, Tunney, tennis star Suzanne Lenglen and polo stalwart Tommy Hitchcock appeared under the rubric "Top Notchers."

But Rice and Hawkinson soon discovered that participant sports made for even better films. "We needed novelties," said Rice. "Women, we were quick to learn, are heavy movie goers and the novel side of a sport seems to have the greater appeal for them." So Sportlight Films became a sort of ancestor to "The American Sportsman," the long-running ABC-TV show that showed various American sports celebrities in assorted hunting and fishing environs. Sportlight featured salmon fishing in Canada; quail shooting in Georgia and the Carolinas; puma and crocodile hunting in Florida (including then-rare underwater footage); salt-water game fishing; grouse shooting in Scotland; lion hunting in Africa; and, in one memorable early show, moose hunting in New Brunswick.

Bill McGeehan was the featured hunter in that 1926 film that Sportlight entitled "The Call of Wild." The script called for McGeehan to track a moose through the "tundra, muskeg and blueberry bushes" of New Brunswick. But McGeehan had a bad heart and wasn't up to any extended trekking through the tundra, or even the blueberry bushes. To complicate matters, the troupe saw nary a moose. So they improvised. "They 'rented' a particularly noble, if dead, moose from a native guide," said Rice, "and propped him up in the densest bush with stilts, baling wire and ropes. Then Bill banged away with his muzzle loader at the defunct critter."

Twice during his lifetime, Rice and Sportlight films were honored with an Academy Award for best short subject, further enhancing Rice's reputation as more than simply a sportswriter. There were other suggestions of art. The year 1924 brought the publication of two books, a collection of columns from 1923, published by Putnam and called *Sportlights of 1923,* and his third collection of verse, entitled *Songs of the Open,* brought out by Century.

Reviewers and readers continued to treat Rice's verse kindly, though in many of the reviews there was more than a hint of: who'd've believed a sportswriter could have written anything beyond the punch-drunk prattle of the sports page, much less something as thoughtful and tender as this.

"Grantland Rice's new book, *Songs of the Open,* is a sturdy clarion of fellowship, humor and work," said the reviewer in the *New York Post.* In the *Boston Evening Transcript,* another unsigned review bestowed praise of the more-than-just-a-sportswriter variety. "[H]ere as elsewhere he makes poetry of the diamond and gridiron, the prize-ring and the field of war. But it is more than the poetry of struggle and battle, it is literally the songs of out-of-doors, of April and October, of woodland fields and lakes—'the songs of the unfettered.' "

But, like much of the poetry he chose to collect, the verse in *Songs of the Open* has an underlying current of the melancholy, the sense of the tenuousness and the fragility of achievement, of fame, of life. Here are some lines from "Ex-Champions":

> "It is better to have won and lost
>   Than never to have won at all—"
> This may be true enough, and yet,
>   The far heights yield the greater fall:
>
> I've watched when the fickle crowd
>   Arose to give the victor cheers.
> The haunted look within their eyes

That turned back through the vanished years,
The fame that leaped to sudden glow
To fade within their sullen state,
As if they, too, had come to know
How soon the laurel withers there.

When acclaim meets celebrity in the book world, the result is generally a pretty successful sales run. While neither *Songs of the Stalwart* in 1917
nor *Songs of the Open* in 1924 made the best-seller lists, each enjoyed mul-
tiple printings and a sustained life, sufficient to enhance Rice's bank account as well as his reputation.

The year 1923 had been a fertile one in sport, and a natural year to
collect Rice's history of it in book form. Bobby Jones won his first U.S.
Open title in a playoff at Inwood; Bill Tilden won his fifth straight U.S.
Tennis Open at Germantown; and the Yankees edged the Giants four games
to two in another Polo Grounds World Series—Casey Stengel winning
game one for the Giants with an inside-the-park home run, winning game
three with another home run, but proving no match for the heroics of
Babe Ruth, who hit .368 and tied Frank "Home Run" Baker's Series record with three home runs in leading the Yanks to the championship. All
of these stories gave Rice ample opportunity to stretch his imagination,
and all are included in *Sportlights of 1923*.

Of Jones, who had lost the chance to win the title in regulation when
he took a double-bogey six on the final hole, Rice wrote: "The red badge
of courage always belongs upon the breast of the fighter who can break
and then come back with a stouter heart than he ever had before." With
Casey Stengel's heroics in the World Series, Rice could not resist a comparison with "Casey at the Bat." "Mudville is avenged at last!" he wrote,

Oh, somewhere in this blighted land the sun is dark and gray;
Somewhere the winds are sighing and the children no more play;
Somewhere the bands are silent and the scene is like a pall,
But Mudville's raising hell again since Casey socked the ball.

Babe Ruth too, moved him to derivative flights of melodrama, with
a verse called "The Return from Elba." But above all 1923 was Dempsey's
year, and of all the events that Rice covered and included in his book,
Dempsey was the central figure in both the most peculiar and the most
vivid. The most peculiar, at least in terms of venue, was the Dempsey-
Tom Gibbons fight in Shelby, Montana, "a one-street town of straggling
frame houses and fewer than a thousand souls." In a misguided fit of inspiration, the town fathers of Shelby promised Dempsey $300,000 if he
would defend his title in Shelby, in the anticipation that the resulting pub-

licity would fuel a great western land boom. The less-than-visionary Bab-
bits of Shelby had trouble raising the three hundred grand, but not the
publicity. Where Dempsey went, the scribes followed. Rice shared a Pullman out of Chicago with Broun, Runyon, Hugh Fullerton and assorted
other writers. When they reached the dusty plains of Shelby—a surreal
scene where "Cowpokes, their spike heels kicking up the alkali dust, bought
drinks for millionaires and the millionaires mingled with Blackfoot Indians, many of them in full tribal gear"—the writers set up living and working headquarters in another Pullman pulled off onto a siding, and began
their customary Pavlovian reaction to the Doc Kearns publicity machine.
Not since Compiègne had so much news come from a humble railroad
car. Unfortunately for the Shelby folks hoping for a land boom, much of
the news concerned Kearns's consternation over the final $100,000 of
Dempsey's guarantee, still missing as fight time approached. Kearns "of-
ficially" called off the fight seven times, according to Rice. "Endeavoring
to keep fresh, up-to-the-minute bulletins pumping over the wires, Broun,
Runyon and the rest of us had long since gone nuts," he said. As a result
of this uncertainty, special trains scheduled to bring fight fans in from San
Francisco and Chicago were cancelled, and when Dempsey and Gibbons
climbed into the ring beneath a high hot sun on the Fourth of July, there
were fewer than 10,000 souls in 50,000 seats in the freshly built wooden
coliseum. (Though another 15,000 were waiting outside, in the hope that
the gates would be thrown open as the fight moved along, as they were
at the start of the fifteenth round.)

The Gibbons fight in Shelby was Dempsey's first title defense in two
years, and his first of two that year. In the less exotic environs of the Polo
Grounds in September, Dempsey took on Luis Firpo, a fighter as unlike
Gibbons as Manhattan was unlike Shelby. Whereas Gibbons was a skilled,

savvy fighter whose punches packed little punch, Firpo was "an untrained giant who knew nothing of boxing, nothing whatsoever, an untrained giant who knew nothing except to fight on and on in his raw, crude way." But this crude, untrained South American possessed a gameness that allowed him to rise from the mat seven times in the first round, whereupon he landed a right to Dempsey's jaw, sufficient to lift the champion off his feet and send him hurtling ingloriously threw the ropes. Dempsey landed in a heap on the press table, four seats down from Rice, his feet high above his head, "a dazed look upon his bloody face."

He returned to the ring a man enraged, and through the haze of his own blood and stupor, knocked Firpo out one minute into the second round. It was four minutes of fighting that Rice measured as perhaps the most thrilling exhibition of drama, courage and athletic wherewithal as he would ever witness in his half-century behind a typewriter. Reflecting on it some years after, he said, "It had everything. A champion knocked out of the ring, his title gone, his almost instantaneous return, followed by his knockout of Firpo and his title saved. It was the high water mark of drama in combat."

Writing on it in the next morning's *Tribune* Rice called it "four minutes of the most sensational fighting ever seen in any ring back through all the ages of the ancient game." Given time to elaborate upon that thought as he rewrote his morning-after piece for the *Sportlights* book, Rice called Dempsey-Firpo,

> A throw-back to the Neolithic or Pleistocene Ages—a cross-section lifted from the first dim dawn of mankind—a primeval battle to survive—all bounded within four minutes through the most sensational fighting ever seen in any ring back through all the ages of the ancient game—. . . .
> There has never been another fight quite like it. For sheer melodrama, thrills, heart throbs and hair-lifting there may never be another as closely packed with vital action.

Sportlights of 1923 has the feeling of the first volume in a planned series of books, an annual review of the sports world akin to Burns Mantle's *Best Plays* series. But it never came to be. *Sportlights of 1923* is the sole volume of collected prose published during Rice's lifetime. The book did not command much review attention, and it it not difficult to see why. There is a perishability to daily journalism, even inspired daily journalism on historic events. There is the sense in the reader that he is chewing on stale goods; and collections of journalism have historically foundered on the shoals of indifference. A book must bring a reader more than a newspaper. Though he did refine and polish the prose in his stories and eliminated the datedness in the pieces, Rice did not deliver any appreciable perspective in the book, or work it into any sort of seamless narrative. Thus it was less of a summary of 1923 and more of a scrapbook. It also may have been the only one of its kind because Rice was quite simply too busy keeping up with the present to spend any time looking back at yesterday.

Rice added another credit to his sportswriter-poet-radio announcer-film producer résumé in 1925 when, in collaboration with actor Frank Craven, he wrote a play entitled *The Kick Off*. Craven was an actor who enjoyed sustained, if limited, success until his death in 1945; he was also a Rice crony and traveling partner for much of this time. "No one could take his place," said Rice of Craven, "on the golf course, on a trip, or standing bravely in front of a dry martini." As near as can be determined, Craven provided the story line and general stage presence of *The Kick Off* while Rice provided the anecdotes, dialogue and characterizations. It made its debut on November 16, 1925, at Pittsburgh's Nixon Theater, with a Rice rooting section comprised of Ring Lardner and W. O. McGeehan and an audience of University of Pittsburgh and Carnegie Tech undergraduates and the visiting delegates to the Amateur Athletic Union convention then under way in Pittsburgh.

The story line of the play followed the passage of a season in the life of a college football team, a season that began in crisis and ended in triumph. It was a comedy, with a second act emphasis on how the lessons and principles of football can be applied to life beyond the campus—in this case the circus. There was some live on-stage scrimmaging and an eve-of-the-big-game College Inn scene which provided an excuse for a medley of college fight songs. Craven played the lead in the play and McGeehan and Lardner were pressed into service as old grad extras in the College Inn scene. McGeehan also doubled as the *Herald Tribune* theater critic and delivered a ringing endorsement of the play in the next day's paper. "Judging by the cheers, it's due for a touchdown on Broadway," said a part of the headline over McGeehan's review.

> *The Kick Off* is an appeal to the undergraduates and the old grads, those collegiate Peter Pans who are wise enough never to want to grow up. . . .
> It is a play for youth, which seems to be the majority party in the United States at the present time, whether it is the fresh-face youth of the undergraduate or the bald and frivolous second youth of the old grad. This clientele ought to furnish a large and permanent cheering section for *The Kick Off*.

McGeehan's optimistic assessment of the play's Broadway future was apparently not shared by the men who financed Broadway productions in 1925; the play died aborning in Pittsburgh, and with it Rice's career as a playwright.

It is not likely that Rice pined for Broadway what-might-have-beens; it was his teenage daughter who was stagestruck, not he. He found his fulfillment in the acclaim and the opportunities that stretched before him for his newspaper work. In the summer of 1924 this took him back to France, this time for a happier world spectacle—the games of the VIII Olympiad in Paris. These were the "Chariots of Fire" Olympics of British sprinters Harold Abrahams and Eric Liddell; the Olympics of American swimmer Johnny Weissmuller; and, in particular, the Olympics of Paavo Nurmi, the lithe, wraith-like Finn who won gold medals in the 1500-, 5000- and 10,000-meter races, shaming the competition in all.

If Rice's writing over this nine-day period in Paris may be taken as a measure, the Olympic Games in these more innocent days can be said to be an embodiment of the Olympic ideal—to shrink the world and make us all more aware and more appreciative of other cultures. Coverage of the Games has always had a jingoistic air, and Rice's coverage of his first Olympics began with the promise that an unabashed nationalism would flow from his pen too. In sizing up the Olympic 100 meters, Rice spoke exclusively of the American quartet of Jackson Scholz, Charles Paddock, Chester Bowman and Loren Murchison. "[T]he final contest to name the world's best sprinter will be one of the great competitions of the year," he wrote. "Paddock and Scholz finished stride for stride [at the Olympic qualifying meet] with Bowman, of Syracuse, an inch or two ahead. When these men meet the decision will usually be a matter of inches, and the nation that can present a sprinter capable of outracing them to the wire will deserve more than the usual applause."

Well, Great Britain did present a sprinter capable of outracing the Yanks to the finish line. And Rice, in rendering victor Harold Abrahams "more than the usual applause" also rendered more than the usual incredulity. "No one would have believed that Jackson Scholz, Chester Bowman, Charles Paddock and Loren Murchison, bunched in one final test, could be run into complete submission, but Harold Abrahams, Cambridge greyhound, put the miracle across," he wrote, framing the race in the context of an upset before acknowledging fully the Abrahams accomplishment on its own merits.

"Abrahams ran in Olympic record time on a track not exceptionally fast, proving the full merit of his sensational performance, one of the greatest track achievements ever known. This is the second time in modern Olympic history that an American had failed to win at this distance, and the defeat of so many American stars has added much to Abrahams' fame."

Four days later, however, when Abrahams's Cambridge classmate Eric Liddell won the 400 meters, Rice was sufficiently imbued with an Olympian sense of the family of man to celebrate Liddell's gold medal, not as the triumph of the Brit over the Yank, but as the triumph of the glorious manifestation of enormous athletic gifts:

He was away from the line as a startled deer leaps for woodland shelter, and once in front he began to extend his lead foot by foot with every fiber of his taut system extended to the limit as the flying [American Horatio] Fitch made a brave but vain effort to come on to victory.

Fitch and the others . . . were pursuing the will o' the wisp through deep shadows that no human hand may ever touch. Liddell was smashing an ancient record and record breakers are rarely caught. So the bagpipes sent their stirring notes through the cloudless skies above and the fluttering kilts took an extra whirl in honor of one of the greatest runners any nation ever sent to glory.

Rice's clearest espousal of this sort of we-are-one-world philosophy that was evolving during these games comes in his writing on Paavo Nurmi. The remarkable Finn demonstrated to Rice and the throng of 60,000 who crowded into Colombes Stadium that it was possible to root as hard for the son of another land as it was to root for a son of your own. On July 10th, Nurmi routed the field in the 1500, winning by fifty yards in Olympic record time. Fifty minutes later he was back on the track in the 5000, running fluidly, effortlessly, beautifully. "Out to the front he glided to the accompaniment of a thundering tumult," wrote Rice of Nurmi's move to the front, five laps into the race. Checking his watch at every lap, Nurmi ran two strides ahead of countryman Willie Ritola in a "clocklike, effortless spin." When Ritola challenged at the end, Nurmi checked him with a quick glance over his shoulder, and then simply accelerated, showing no weariness from his efforts an hour earlier. At the wire he was a stride ahead of his teammate, and he had broken his own world record by more than four seconds. At the tape he simply kept running; "he ran for his sweater on the turf nearby, jogging swiftly back to his quarters with a roar of applause following like the thundering of surf in a storm.

"To smash two Olympic and one world's record under such conditions stamps Nurmi as the marvel of all track marvels whose like will never again be seen within this generation," wrote Rice, and for once he was understating what he had witnessed. Nearly a score of Olympic generations has come and gone and we have not seen Nurmi's like. Over the next four days, as Nurmi won a third and then a fourth gold medal in the 10,000-meter cross country race and the 3000-meter distance relay, Rice was sketching a picture of Nurmi that threatened to make him as popular with American readers as Babe Ruth:

Although the Olympic's greatest star he is in no way temperamental. He is greatly liked in his own country and by all who know him among the other nations. If a car is not ready to take him from Finland's training quarters to the stadium, Nurmi, never complaining, sets out to walk this extended distance, even though a long race be just ahead.

He has deep religious beliefs, a strong spirit of complete self-sacrifice and discipline. He is twenty-eight years old and had given up the last ten years to rigorous training for the honors he has brought to his native land after the manner of the ancient Greeks for glory of the commonwealth.

That the Olympics' greatest hero to the American public came from a land distant and peculiar to most Americans was a lesson not lost on Rice. The man who had begun his Olympic correspondence with the suggestion that this would be an American rout—and the tacit implication that an American rout would be an affirmation of the proper athletic social order—concluded that the athletic world was a richer place after all if there was greater parity and keener competition. "Sport needs the revival and athletic growth of many nations," he said. "No one nation should have anything like a monopoly, since such a monopoly is also bad stuff for the victors. Too much success at times may be worse than many failures. It's the battle to get there, not the arrival, that builds up the breed."

The Rices sailed home from France—Kate and Floncy had accompanied Granny to the Games—to a busy fall. Floncy left for Smith College in September and Granny and Kate moved from their Riverside Drive home of the past thirteen years to a spacious new flat at 1158 Fifth Avenue, overlooking Central Park at 97th Street. There, in the company of their two devoted servants—Kate's Swedish maid Helga Christenson and Granny's chauffeur Charles Goering—they would remain for the rest of their lives.

Granny's contract called for a winter vacation, so there was no time for any extended relaxation in the wake of their European sojourn. He resumed his tour of American sporting events. In September, Bobby Jones won the first of his five U. S. Amateur championships, with Rice as witness. In October it was college football and the tour de force on The Four Horseman. As famous as that piece was to become, however, it was matched in its quality and contemporary impact by another story Rice wrote that same month, just eight days earlier.

In the seventh game of the 1924 World Series, Walter Johnson—thirty-seven years old and beyond his peerless prime, the loser of two games to the Giants already in the Series—came into the game on just one day's rest and went the last four innings of a twelve-inning victory that gave the Senators their only World Championship in their ignoble history. Rice's story on that game is testament to his own unfailing capacity to share in the celebration of a tempting athletic moment—no matter how many such moments he might witness in the course of a year. Each was as special to him as it was to the athletes and fans to whom it was unique, as he demonstrated in his story on Walter Johnson's long-awaited and richly deserved moment of triumph. "For it was Johnson, the old Johnson, brought back from other years with his blazing fast ball singing across the plate . . . that stopped the Giants' attack," wrote Rice.

"Here once more was the mighty moment," he wrote of Johnson's ninth-inning entry into the game, "and as 38,000 stood and cheered, roared and raved, Johnson began to set the old-time fast one singing on its way." This was a Johnson who "turned back to something lost from his vanished youth," gushed Rice, "calling back stuff from a dozen years ago. . . . [Y]oung blood was coursing through his veins again."

As special as Johnson's performance was, however, it was the crowd's appreciation of the moment that Rice found most electric:

> Washington won just at the edge of darkness. . . . As Earl Mc-Neely singled and Muddy Ruel galloped over the plate with the winning run in the last of the twelfth, 38,000 people rushed on the field with a roar of triumph never heard before, and for more than thirty minutes, packed in one vast serried mass around the bench, they paid Johnson and his mates a tribute that no one present will ever forget. It was something beyond all belief, beyond all imagining. Its crashing echoes are still singing out across the stands, across the city, on into the gathering twilight of early autumn shadows. There was never a ball game like this before, never a game with as many thrills and heart throbs strung together in the making of drama that come near tearing away the soul, to leave it limp and sagging, drawn and twisted out of shape.

During these years the Baseball Writers' Association awarded a jeweled gold watch annually to its "Most Valuable Writer." The award generally recognized a body of work and went to a beat writer—a writer assigned to team throughout the season, writing the daily game story. Syndicated columnists weren't commonly eligible. In 1924 the Baseball Writers broke with custom and instead recognized what they regarded as a single exceptional story—Rice's story on Walter Johnson and the seventh game of the 1924 World Series. It was a surprisingly rare recognition for Rice. Throughout his career he received any number of trumped-up awards given by civic organizations and alumni associations to coax him onto the dais for some manner of banquet or fund-raiser. But unlike journalism today, when writing prizes are as commonplace and as inevitable as bowling trophies, peer recognition throughout the first half of the century was exceedingly rare, and worthy of notice.

In that same fall of 1924 the management of the *Tribune* bought out the assets of the *Herald,* its great nineteenth-century rival that had withered and atrophied since the death of James Gordon Bennett, Jr., in 1918. It began publishing the paper in October as the *Herald Tribune,* and they began a serious but ultimately futile run at supplanting the *Times* as the most comprehensive and influential paper in the city. Publisher Ogden Reid recognized the value of Rice's column to the paper, and though he already had Rice under contract through March of 1926, in early 1925 he began looking to extend that contract and locking Rice into the *Tribune* through the end of the decade. The contract Rice signed in July of 1925 guaranteed a minimum salary of $1000 a week ($52,000 a year) with provisions for even greater earnings should the syndication of Rice's column grow further. This was a bank president's salary, not a newspaperman's; $50 a week would still be considered handsome wages by most denizens of the city room in 1925; $100 a week made a man darn-near wealthy. Here was Rice making ten times that. And the *Tribune* salary represented just a part of his earnings. He still had *Collier's, American Golfer,* Sportlight Films and myriad other miscellaneous checks coming in.

Rice's contract and salary got no attention, but to put it into perspective, $52,000 a year was exactly what Babe Ruth was making. And Ruth's salary, of course, came in for considerable attention—precisely because it was so out of line with everybody else's; it was more than three times the second highest salary in the game. In all of American sport during the decade of the 1920s, Jack Dempsey is perhaps the only man to have out-earned Grantland Rice.

The contract recognized Rice's value to the *Tribune* in another way as well. It gave him overall responsibility for the direction and tone of the sports section. It made him an associate editor of the paper, "having . . . general authority and supervision of the Sporting Department and pages of [the *New York Herald Tribune*] and of the material on sporting topics therein published." This did not put him in the office, with responsibility for getting out the sports section each day. In fact, Rice (or his lawyer) was shrewd enough to have incorporated into the contract language that called for his supervision of the sports department "provided that [he] shall not be required to render such services during fixed hours or at any particularly designated time." Should the situation at the *Herald Tribune* change, Rice was not about to let some loophole in the contract chain him to a desk. His responsibilities took on more of a consulting nature. And while it was not in his personality to throw his weight around, he nonetheless had a power to match his presence and visibility at the paper.

When Walter Camp died in 1925 there was much speculation as to who would replace him as the man who searched American gridirons each fall to find the eleven finest players—the All Americans. Camp had helped write the rules of the game. Coach, author, administrator, he carried the title "father of American football" next to his name in most public mentions. He had published his list since 1889, and while alternative All America lists cropped up from time to time, none ever carried the impact of

Camp's. It had been a December staple of *Collier's* since the early 1900s, and when Camp died the magazine promised to "look far and wide" and to give "much study to find his successor." While *Collier's* may have given some thought to how to handle the succession, the promise of a protracted and considered search was so much public relations, for the choice was obvious and he was already on the *Collier's* payroll.

Camp's All America team was an anachronism by the 1920s. When he began the team in 1889, and throughout much of its life, Camp could be reasonably assured that he had seen the finest football players in the land by simply looking out his office window in New Haven and studying the players who played for and against his dear old Yale. But college football had been a national phenomenon for two full decades by the time of Camp's death. Moreover, in the years since the war, everybody was aware of it. Picking an All America team was no longer a one-man job. *Collier's* needed someone to dust the Ivy League mustiness from the All America list. While there may have been people in America who knew the game of football better than Grantland Rice, there was no one who knew as many people who knew the game of football. Nor was their anyone with a broader national reputation or a greater reputation for fairness and impartiality. *Collier's* was getting not only his industry and contacts. They were getting his fourteen-carat reputation as well, a reputation that was at least the equal of Camp's in its reach and its luster. In short, not only would Rice be effective in the job, he would also be accepted.

"Grantland Rice is . . . not only the most scholarly sport scribe we know of, but he is a thoro student of sportdom from any angle you care to challenge him," said Otto Floto of the *Denver Post* when *Collier's* announced Rice as Camp's successor.

> Possessing a superbly trained brain, [he is] a wonderful judge of man and the task he has undertaken, impartial to a degree, fearless in every way, a man not swayed by friendships or dislikes in making his calculations, a keen observer of all he sees, and with the kindest disposition we know of. So how are you going to beat him? How are you even going to compare him with any other expert on his new job? . . .
>
> Because of this we believe a real All-America team will be selected, with which nine out of every ten impartial experts will agree.

Rice claims to have balked when approached about the job. He didn't need the money; he didn't need the challenge; he didn't need the aggravation. Picking All America teams was a no-win task. Those who were chosen would not feel especially beholden; they felt they belonged. Those who were left off felt unfairly slighted. "I squawked loudly," said Rice of his response to *Collier's* president Lee Maxwell's entreaty to succeed Camp. "I didn't want any part of the job."

But of course he did. More than once during his life, when faced with an interviewer's question as to what sport was really his favorite, Rice—who generally tried diplomatically to avoid answering that question—allowed as to how football really tugged the hardest at the heartstrings. Perhaps it was because, unlike golf and baseball, football had not been easily mastered and it thus retained the grudging respect that any unconquered foe commands in the heart of a good athlete. "Due to the ingredients . . . courage, mental and physical condition, spirit and its terrific body contact which tends to sort the men from the boys . . . football remains one of the great games of all time," he wrote. He saw the game—particularly during these innocent years—as inseparable from the college experience and the passage of a young man.

> I'm the soul of the college spirit
> And the maker of a man.

When charged with his task, Rice began by sending close to a hundred telegrams to the nation's football community. To Rockne at Notre Dame, Zuppke at Illinois, Yost at Michigan, Heisman at Rice, McGugin at Vanderbilt, Warner at Stanford, columnists and editors at papers that subscribed to his column and at papers that did not. He asked each of his correspondents to provide him with confidential information on who, among those they knew and had seen, they felt the best players at each position were. Most were quite eager to help, and their detailed letters and telegrams began arriving at the *Collier's* office in mid-November. The mail also brought unsolicited letters, hundreds of them, from well-meaning fans and alums who provided detailed information and arguments on behalf of this player or that. Rice read them all. He married this intelligence to what he had seen during his travels throughout the fall, and made his selections alone. His first team was headlined by Red Grange of Illinois, who earlier that fall had received his immortal nickname, "The Galloping Ghost," courtesy of Rice's nimble typewriter.

> A streak of fire, a breath of flame,
> Eluding all who reach the clutch;
> A gray ghost thrown into the game
> That rivals' hands may never touch,

wrote Rice of Grange, whom he had traveled to Champaign to see that fall.

Increasingly now, Rice's travel to football games reached beyond the predictable Princeton, New Haven, Cambridge corridor. Not only were the more interesting games now found in the hinterlands more often than not, Rice now had an obligation to see some of the players his mail was alerting him to. But it was too big a country, too large a task for any one man, even a man who was beginning to take on larger-than-life qualities in the sportswriting world. He very quickly added what he called the *Collier's* board. This was comprised of writers H. G. Salsinger of the *Detroit News,* Clyde McBride of the *Kansas City Star,* Braven Dyer of the *Los Angeles Times* and O. B. Keeler of the *Atlanta Journal,* among others. Together this board would thrash out the nominations and arrive at an All America team every year. The members of the board changed from time to time but the system stayed intact for twenty-one years.

"If it was the 'eleven best' the fans wanted, we tried to give them just that without any particular favoring of 'geographical distribution' of players to fit the national scope of readership," said Rice. "In 1928 Dink Templeton, my West Coast operative, told me, 'There's nobody out here worth a damn!' That year the team went no farther west than Minnesota. . . . My West Coast mail was both terrific and terrible. However, I respected Templeton's judgment. . . ."

If the members of the board could not arrive at a consensus, Rice generally had the final say, and if he wasn't sure he had some tricks that he developed through the years. He knew for instance that his friend Jock Sutherland of Pitt built his offense around his center. He wanted the best athlete on the team at center, for football depended on the running game, said Sutherland, and the running game began with a reliable snap of the ball. Rice, in turn, knew "that if I was in doubt in picking All American talent, I was dead sure to get a good center from Pittsburgh when Jock ruled the Panthers."

In 1947, Tim Cohane of *Look* magazine, in conjunction with the Football Writers Association, persuaded Rice to bring his All America team over to *Look.* The team would be known as the Grantland Rice All America team, but the legwork and selections would be done by the 500-member Football Writers Association. The team lives till this day. It no longer bears Grantland Rice's name, and *Look* of course is long since gone, but the Football Writers Association still chooses an annual All America team, the only one among the dozens of annual All America lists that can trace its roots back more than a century, the direct lineal descendant of Walter Camp's original.

And there are other vestiges of Rice's once-enormous presence in the college football world that have survived the ravages of time and America's short memory. From 1964 to 1972, the NCAA honored Rice's contribution to the development of the game's popularity by naming one of its regional college-division bowl games the Grantland Rice Bowl★; and his gray fedora and typewriter are a part of the College Football Hall of Fame in Ohio.

★In 1973, the Grantland Rice Bowl and the other college division bowl games were replaced by the playoff system for NCAA divisions 1-AA, II and III.

11

A Game for the Age—
A Champion for the Ages

G**olf is twenty percent mechanics** and technique," wrote Rice in 1920. "The other eighty percent is philosophy, humor, tragedy, romance, melodrama, companionship, camaraderie, cussedness and conversation."

He might have forgotten that golf was also part escape, for that was what lured him to the game. In the beginning, golf for him was primarily an escape from the nuttiness of his sixteen-hour days on the Nashville *Tennessean,* like tending to the garden behind his home on Broad Street, or stealing some time of a morning to play with Floncy before heading off to his desk in the city room. But even as recreation and escape, golf provided something that hearth and family did not; it provided a release for the athletic fires that had always burned in Rice. He was twenty-nine years old the first time he swung a hickory-shafted cleek, and eight years removed from the semi-pro baseball tour that marked the close of his career as an athlete in 1901. While refereeing football games and umpiring baseball had always given him a connection with sport beyond the typewriter, like the typewriter these were peripheral, vicarious connections. Beneath Rice's gentlemanly Southern demeanor he was still the dogged competitor, the skinny kid determined to make his college football team.

In golf he found a sport that brought together the disparate parts of his life and personality.

To the athlete in Rice the sport provided the challenge of mastering the "mechanics and technique." There was a grace in the lanky, still-lean body, a natural coordination that conquered the many maddening parts of the most unusual motion of hitting a golf ball. Rice's swing—like the swing of a Ouimet, a Hagen, a Jones, and unlike the swing of the typical week-end country-clubber—was a fluid seamless whole. There was also the work ethic that left him capable of hitting the thousands of balls that were the prelude to any "natural" swing. In addition, Rice possessed the introspec-tion—especially in the heat of a close match—to know when the circum-stances called for going for the flag with a mid-iron and when prudence dictated that he lay up with a mashie. He was also possessed with the temperament to realize that a skulled niblick, or a four-foot putt that rims the cup and spins out were now and again inevitable, the byproduct of human imperfection, and he did not carry his mistakes with him from one hole to the next. As a result, Rice soon came to play the game with a tournament golfer's skill, flair and consistency. On championship courses, in celebrity tournaments in front of sizable galleries, Rice shot par or near-par golf on literally hundreds of occasions between 1915 and 1935. It is enticing to speculate on what he might have achieved had he devoted 100 percent of his energies to mastering the game. Might his legacy be that of Francis Ouimet, Walter Travis, Chick Evans, Jess Sweetser, Bob Gardner or any of the other great champions in the era before Hagen and Jones? Perhaps. Perhaps not. It is legitimate speculation, however, and that in itself is sufficient testament to the very impressive quality of his game.

Ultimately, however, Rice worked so much on his game and cared so much about the quality of his game not because of the 20 percent that fulfilled his needs as an athlete and competitor. It was the other 80 per-cent—the philosophy, the romance, the melodrama, the camaraderie and all—that so filled his needs as a writer and a person that fired his devotion to the game, and inspired him to write of it with such passion and energy. "Golf gives you an insight into human nature, your own as well as your opponent's," said Rice. "I've got a host of columns from the locker room—not only about and with name golfers but about and with headliners of every sport and business. Peeled down to his shorts, a highball in one hand and an attested score card in the other, it's hard for a man to be anything but himself."

Rice made readers aware that there was philosophy in a hedonist like Walter Hagen no less than in a gentleman like Bobby Jones. He made his readers understand that the act of hitting a stationary golf ball while a hushed gallery bore polite witness was no less a piece of athletic where-withal than was the act of hitting a baseball thrown at ninety miles a hour while 40,000 fans screamed and jeered. To the increasing number of

Americans who were trying the game, his writing held a mirror to the humor and frustration inherent in playing the game infrequently yet expecting to play it well. And it is here—as a popularizer of the game, the man who took golf off the society page and put it onto the sports page, the man who helped readers and fans recognize Bobby Jones as a hero of the same magnitude as Babe Ruth—that Rice's golfing legacy lies. And in its own way, it is a legacy that is every bit as important to the history of the game in America as the legacies of Ouimet and Jones.

Rice had been playing the game for just about a year when he moved from Nashville to New York. He joined the Englewood Golf Club in New Jersey, site of the 1906 U.S. Amateur Championship and the 1909 U.S. Open. Englewood lay just beyond the Palisades, a short ferry ride across the Hudson from his home on Riverside Drive. Rice had gotten into the habit of writing in the morning during these years, then hustling himself over to Jersey. By the time he left the *Mail* for the *Tribune* in 1915 his handicap had vanished. The newspapermen—business office people and writers alike—had a Tuesday press league during those years. Rice was invariably the medalist in those low-key events, played on a variety of Metropolitan-area courses. Only his travels out of town on assignments gave anyone else a real chance.

In January of 1916 and again in January of 1917 Rice played in the annual advertising industry golf tournament in Pinehurst, North Carolina—a tournament that resulted in a lot of free advertising for the then-budding Pinehurst resort. Rice won the tournament in 1916, when it was a seventy-two-hole medal play event. He was runner-up in 1917 when it was a match-play contest. He doubled as the *Tribune* correspondent frequently when he played in this kind of tournament. This meant that when he played well, there was invariably a hole in the story, as modesty prevented him from pointing to his own triumphs. In writing of the quarter-finals of the 1917 Pinehurst advertising tournament, Rice opened his story by giving details of three of the four matches. W. O. McGeehan, working the desk when Rice's copy came in, corrected the oversight with an insertion that is a bemused mix of pride and exasperation:

> Of the 126 candidates who started on Monday four still remain to battle for the advertising golf championship. In the last eight today L. A. Hamilton had no trouble winning from R. L. Potts, the entry from National. He took a fast start and picked up a lead early. In the same way Lee Maxwell, of Sleepy Hollow, won handily from R. L. Whitton, of Beverly, by cutting out a steady pace. Don Parker, of Garden City, defeated Fred Ross of Dunwoodie, 4 and 3.
>
> [The fourth man left in the tournament seems to be Grantland Rice, according to the summaries appended. Mr. Rice, you will observe, really does hate to write about himself, and has tried to hide himself and the

story of his game in the agate type. Consequently, it was necessary to pry right into the middle of the story in order to bring out this interesting news feature.—W.O.M.]

The next day, after Rice had won his semifinal match and advanced to the finals, the *Tribune* story was headlined: "GRANTLAND RICE HATES TO ADMIT HE'S A GOLF PLAYER: Results From Pinehurst Show, However, That He Reaches Finals." Rice began his story with an account of the second semifinal and then finally, reluctantly, mentioned his own game. "In the lower bracket your humble correspondent doesn't mind admitting he was extremely fortunate, not to say lucky, to win from Don Parker of Garden City on the last green. Only the raw and horny hand of fate that steered a twenty-foot putt into the cup on the sixteenth kept us in the tournament, Parker at the time being one up and only two inches from the cup for his 4." You then have to read between the lines to learn that Rice won the match on eighteen with a birdie. (He was, perhaps somewhat to his relief, saved from any sort of tap dancing modestly around the facts in his story on the final, where he got whacked 8 and 7 in the thirty-six hole match.)

Rice's most publicized golfing success, and maybe his most satisfying as well, came in the annual Artists and Writers Golf Association tournament in Palm Beach. The Artist and Writers group thrived from the mid-twenties into the Second World War. "Golf and stag conviviality were the bywords" of the group, said Rice, "which, come to think of it, might not be a bad credo for all world organizations."

The group was born during a springtime meeting in 1925 at New York's Algonquin Hotel, site of the famed Algonquin Round Table lunch crowd of Dorothy Parker, Robert Benchley et al.—a pleasingly appropriate venue for the birth of a literary organization. The idea for the group was Ray McCarthy's. He had been a golf writer for the *Tribune* in the early twenties and was then handling the publicity for Henry Flagler's Florida East Coast Hotel and Railway Company. Perhaps he saw in the idea of a winter golf tournament on a Flagler course an opportunity to renew old friendships and serve his new master at the same time; if so, he would seem to have been wildly successful on both counts.

Rice saw the group as a motherlode of "sociable friendship, the richest reward known to man." The number of people from the original group who kept coming back attests to the fact that they shared Rice's warm view of the Palm Beach camaraderie. And, as the original group grew in size and stature and as their considerable renown grew in proportion, McCarthy got quite a lot of ink for Palm Beach. Particularly when people like Rice would come to the tournament with guests like Gene Tunney and Tex Rickard. (Tunney played a decent game of golf, but as for Rickard, Rice said: "As a golfer, Tex remained a top promoter.")

The original Artists and Writers group included Rice and Lardner; artists Rube Goldberg, Clare Briggs and James Montgomery Flagg; novelists George Ade, Clarence Buddington Kelland and Rex Beach; *Vanity Fair* editor Frank Crowninshield, and a dozen or so others. Rice was president and Goldberg, vice president; though how they came to their titles neither man could seem to remember, for there was apparently never an election. Nor was their ever any challenge to their leadership, apparently, for both men were still in office in 1941. "Like other dictators, we expect to die in office (naturally or unnaturally)," said Goldberg.

The stag aspect to the affair was short-lived and by the late 1930s the affair was almost strictly social. "The golfers have grown fewer but the attendance has remained stationary," said Goldberg in 1941. "For every golfer who has dropped out for reasons of old age, arthritis or bankruptcy, a gin-rummy player has taken his place." But in the beginning the golf tournament was the centerpiece, variously played at eighteen, thirty-six or seventy-two holes, and in both medal- and match-play formats. In any given year there were ten or twelve golfers within a half-dozen strokes of par, but Rice and Rex Beach were always the class of the field. Beach was a robust physical speciman—six feet, two inches tall and 220 pounds. He swam in the Olympic Games of 1900, played college and professional football and was impressive enough in bar-room brawls and amateur fights to have friends thinking he coulda' been a contender. His golf game was equally impressive. Long off the tee, his qualifying scores at Palm Beach were invariably 70 or 71, six or seven shots better than Rice. But in the tournament proper he tightened up a bit. "I think what favored me most," said Rice, "was that I seldom worried in any match, particularly against Beach, because I knew he was a better golfer. He worried—I didn't, and we broke about even." Rice's temperament and intelligence—drawn from that 80 percent of the game of his—were always his strengths in tournament play (or even five-dollar Nassau play). He knew he was going to make his bad shot. Walter Hagen once told him he was going to make an average of seven mistakes a round. The secret to winning was to forget them. Beach could outclass Rice shot for shot. But he did not have Rice's long experience as a student of the game.

Rice won the first Artists and Writers tournament; Beach took the second, and the tournament belonged pretty much to the two of them exclusively until the late thirties, when syndicated bridge columnist Hal Sims "at last reported with a better game."

The end of the 1920s pretty much also marked the end of Rice's career as a serious competitive golfer, though hardly the end of his serious playing. His clubs remained a part of his luggage. He continued to play almost daily throughout the thirties and forties, particularly when he was out in East Hampton or when he was on his annual January-February sojourn in Los Angeles. He was an active member of four different clubs—Englewood; the Maidstone Links in East Hampton; Augusta National, his friend Bobby Jones's club in Georgia; and Lakeside in Los Angeles. Lakeside was the "club of stars" as Rice referred to it in a *Golf* magazine feature. His daughter Floncy—who during these years was at the peak of her Hollywood success—was a member. So were Bing Crosby, Humphrey Bogart, Guy Kibbee, Adolph Menjou, Frank Craven, Andy Devine, William Frawley, director Howard Hawkes and studio execs Adolf Zucker and Jesse Lasky. At one time or another over the twenty-plus years he played at Lakeside, it is safe to say that Rice played in at least one foursome with all of them. There were some expensive Nassaus played on the Lakeside course, and actor Guy Kibbee was among the more celebrated of the Lakeside hustlers. He was a frequent Rice partner, and Rice's guests often found themselves paying Kibbee handsomely for the privilege of their round at Lakeside. "Mr. Rice works for Mr. Kibbee on a strictly commission basis, getting, it is said, 10 percent for each victim produced," said Henry McLemore, a United Press sportswriter, Rice protégé, and Kibbee pigeon.

Even though his own game had slipped from its 1920s level, Rice still had to give a lot of strokes away to most of his opponents, and he lost as often as he won. But the money won and the money lost at golf was forgotten as soon as it was exchanged. Winnings were spent in the lounge, losses dismissed as nothing more than a part of the very reasonable price of playing the game—no less onerous or unnerving than club dues or caddy fees. For Rice was not a man who ever measured his wealth in coin. The wealth he cherished in his life—and he was acutely aware of this wealth; he hoarded and protected it—was the abundance of genuine, abiding friends he had. If friends and friendships could be inventoried and appraised like possessions, Rice's wealth would have approached Royal proportions. And to Rice, the golf course was an enormous expansive vault, where he stored these most precious of his riches, and through which he wandered daily to savor their beauty.

But whatever the sport of golf gave to Rice, he more than repaid the debt. Its emergence as a popular sport—both as a participant and as a spectator sport—was given shape and definition in Rice's writing in the years between Francis Ouimet's win in Brookline in 1913 and Bobby Jones's Grand Slam in 1930. To a certain degree Rice did this with all sports. But with golf, at the beginning, he was writing about a sport that was not only absent from the sports pages in general, it was a sport that some readers had never even heard of.

Golf's beginnings go back to Scotland and the fifteenth century, but

its American roots went back only to 1888 and the formation of the St.
Andrew's Golf Club in Yonkers, New York.★ As a social and recreational
activity amongst the wealthy, golf spread quickly. By the time Rice began
playing the game in Nashville in 1909 there were thousands of golf clubs
in America. But it was not a sport that interested the country's sports
editors. The sports pages were a way of selling newspapers to people who
might not otherwise be enticed to buy—the working class, the immigrants, the inner-city poor. Baseball made sense. Boxing made sense.
Football made sense. But golf? What would a twenty-five-dollar-a-week
factory worker just short months removed from Ireland or Italy or Eastern
Europe care about how the swells in the country spend their weekends?

Part of the problem stemmed from the sportswriters' and editors' remarkable ignorance of the game. One of the earliest attempts at coverage
of the sport, an article in a Philadelphia newspaper in the 1890s, got it all
wrong.

> [Golf] is . . . a most aristocratic exercise, for no man can play . . . who
> does not have a servant at hand to assist him. . . . When the word has
> been given to start, the player bats the ball as accurately as possible to
> the next hole, which may be either one hundred or five hundred yards
> distant. As soon as it is started in the air he runs forward in the direction
> which the ball had taken, and his servant, who is called a "caddy," runs
> after him with all the other nine tools in his arms. . . . The one who
> get his ball in the hole at which they began first, wins the game.

Golf news appeared frequently on the society pages during these early
years. It appeared occasionally on the business pages. And when it appeared on the sports pages it was frequently against the editor's will. Rice
had a particularly difficult time in convincing Francis Albertanti, the assistant sports editor of the *New York Mail,* of the game's validity. "[I]n Albertanti's book, golf was something played by unemployed sheep herders
and 'coupon-clipping stiffs.' It didn't belong on the sports pages." The
Mail refused to let Rice cover the 1913 U.S. Open at The Country Club
in Brookline, Massachusetts, where local amateur and erstwhile caddy
Francis Ouimet beat heavily favored British professionals Harry Vardon
and Ted Ray. Nor did they let him cover the 1914 Open in Chicago, here
Walter Hagen won his first national championship. Rice grew so exasperated that he approached managing editor Theophilus England Niles about
Albertanti's intransigence.

"Golf? What's golf?" asked Albertanti when Niles called him in to
ask about its absence from the *Mail's* sports pages.

★The American St. Andrew's is spelled with an apostrophe unlike the Scottish St. Andrews.

"Why it's a game—an important game," said Niles. "A lot of big businessmen are playing it."

"Then put it on the financial page," retorted the unyielding Albertanti.

Francis Ouimet's victory in the 1913 Open is generally heralded as the breakthrough event in American golf. And in Boston, site of the championship and hometown to the champion, Ouimet's victory in an eighteen-hole play-off was indeed front-page news. But only in Boston. Elsewhere, it was relegated to the sports pages, and while it received full-column treatment in papers like the *Tribune* and the *Times* in New York (and in the *Mail* too, where Rice took note of it in his column, even though he had been forced to write the column from New York), in other papers and in other markets it was mentioned in the agate or not at all. Francis Albertanti was not the only sports editor in America with a contempt for the game.

Fortunately for Rice the editors and publisher of *Tribune* did not share the *Mail*'s apathy for the game. Indeed, battling the *Times* and the *Herald* for the better-heeled readers as it was, the *Tribune* sports department made golf a priority. Rice was not only freed to cover the major championships (discounting the British championships, there were really only two in those days: the U.S. Open and the U.S. Amateur), he was given his own Sunday golf feature, "Tales of a Wayside Tee." While the space he devoted to the game in his column was reflective of the growing appeal of golf as a spectator sport, "Tales of a Wayside Tee" defined the growing appeal of golf as a participant sport. There were an estimated 350,000 people playing the game after the war, and "Tales of a Wayside Tee" spoke to them on matters of course membership, course management and construction; golf rules and etiquette; caddies; tips on how to play the game; and tips on how to take tips-on-how-to-play-the-game with a grain of salt.

The advice he offered was sometimes passed on as a result of an interview with a player of some repute; it might be Chick Evans on a putting stance or Bobby Jones on playing out of the sand. At other times, however, the voice proffering the advice was his own. By the end of the 1920s, Rice had acquired an eminence to go along with his prominence. When Clare Briggs, the *Tribune* sports cartoonist, drew a cartoon in 1929 of a duffer spending his off-season immersed in self-help reading, littered about his feet were papers revealing from whom he was taking this advice. The names were a who's-who of the game: Bobby Jones, Gene Sarazen, Francis Ouimet, Jerome Travers, Jess Sweetser, Walter Hagen, Chick Evans, Bob Gardner—and Grantland Rice, the only man in the group who was not a national champion.

He playfully promoted the game at every opportunity. He bet John McGraw and Christy Mathewson that he could hit a golf ball from home plate over the centerfield clubhouse in the Polo Grounds. The two baseball men, who had seen thousands of baseballs die in the vast plateau that was the Polo Grounds centerfield eagerly took him up on his wager. After one Giants game in 1916, Rice came out onto the field with a dozen golf balls. He nervously topped his first two as McGraw and Mathewson kidded him. He then hit the third into the centerfield bleachers and the fourth out over the clubhouse towards the Harlem River.

Rice's syndication took his words on the game across the country. Soon other newspapers began picking up their own coverage of the sport. Tournaments like the U.S. Open and Amateur, the Western and Southern Opens began attracting galleries that numbered in the thousands. This was equal to the size of the crowd at an above-average mid-season baseball game, and short only of the crowds at championship fights and college football games. The game began to witness enormous changes. The U.S. Open began charging an admission fee in 1922. A professional tour gained its foothold in the early twenties and became the dominant force in the game by the end of the decade. In a recognition that golf had an egalitarian as well as an elitist face, the United States Golf Association added a Public Links Championship to its roster of tournaments. Resort hotels, like the Breakers in Palm Beach, the Greenbrier in West Virginia and the Broadmoor in Colorado Springs added championship golf courses to their property in an effort to attract and keep guests. By the close the 1920s more than four million Americans are playing and watching the game. And Rice was at the head of this impressive column in its march down the fairway, sounding a clarion call that was as melodic as it was resonant.

Rice published two books on golf during his lifetime. Together with Jerome Travers, the 1915 U.S. Open champion and four times U.S. Amateur champion, he wrote a how-to book called *The Winning Shot* in 1915. Eleven years later he provided the captions and text for cartoonist Clare Briggs's drawings in *The Duffer's Handbook*. But his single most enduring contribution to golf literature is probably his ten-year editorship of *The American Golfer*, the most widely read golf periodical of its day and an aesthetically pleasing little magazine in its own right, taking its cue more from the likes of classy contemporaries like the *New Yorker, Vanity Fair* and the *American Mercury* than from any sports fan magazine. *American Golfer* had been around since 1907, edited by Walter Travis, three times the U.S. Amateur champion at the turn of the century. Under the former champion's editorship, it was a magazine for golf junkies, heavy with tips and tournament results. When Travis retired in 1921 the owners, Centurion Publishers of New York, approached Rice about taking over. Rice had been a contributor to the magazine; he had even been a featured subject in the magazine back in 1917. But he had little initial interest in running it. "[U]ntil I was told that fresh capital, $200,000 of it, was behind the book," he said. He was also offered a piece of the action. His contract called for a salary of $7500 a year, to be paid as an advance against dividends earned as a percentage of the magazine's profits. It's not known by how much Rice's share of the dividends may have exceeded $7500 a year, but Rice was sufficiently satisfied with the arrangement to remain as editor under Centurion for eight years.

Rice was given a free hand to make over the magazine, and the resulting product was a testament not only to his imagination and vision of the sport but to his wealth of contacts and friends. James Montgomery Flagg drawings frequently graced the cover of Rice's *American Golfer*. Inside, the game enjoyed a rare celebration and intimacy with prose from writers like O. B. Keeler, Innis Brown, John Kieran, George Ade, Rex Beach, Bud Kelland, Henry McLemore, Frank Condon, and Britisher Bernard Darwin, the man who is generally extolled—by Rice among others—as the most eloquent and insightful of the first generation of golf writers.

But the literacy in *American Golfer* went beyond golf. Tennis got regular play in the magazine, most often from John R. Tunis, who was then writing tennis for the *New Yorker* and the *New York Post* and who would in the years ahead gain wealth and fame as the author of such acclaimed juvenile novels as *The Kid From Tomkinsville* and *The Iron Duke*—books with a sports setting and a moral undertone that remain in print and continue to stir young readers to this day. Ring Lardner was also a frequent contributor to his friend's magazine—only seldom, and only peripherally, on the subject of golf. Lardner would write of duck hunting, and of travels in Egypt that never happened, with his peculiar sense of exaggeration and nonsense.

American Golfer reached a circulation peak of 70,000 in 1929, when it was bought out by Conde Nast, publisher of *Vanity Fair* and *Vogue*. Rice agreed to remain on as editor, and no sooner had he lined up some professionals like Walter Hagen and Tommy Armour to help with a big circulation drive than the depression hit, "and twenty pages of automobile advertising immediately fell out—kerplunk," said Rice. "That was the death rattle for *American Golfer*," he remembered. Though it wasn't quite. *American Golfer* lived another seven years, actually outlasting the Conde Nast flagship *Vanity Fair,* which went dark in 1935.

Golf by the mid-1920s, at least in Grantland Rice's space in the newspaper, was a sport with value and prominence equal to baseball, football and prize fighting. The PGA tour had established itself as a serious athletic competition, not a novelty. The U.S. Open and the U.S. Amateur no longer had sports editors questioning their right to prominent coverage in the papers. The importance that both Rice and the *Tribune* placed on golf was evident in the contract he signed in 1925. It was spelled out that, while he was free to cover whatever he pleased and write of it however he pleased, Rice was contractually bound to covering three events every year—the World Series, and the Open and Amateur golf championships.

But the biggest factor in golf's passage to big-time status in American sport was the emergence of a hero—a hero the equivalent of Ruth or Grange or Dempsey. An athlete of ground-breaking talent, with a personality sufficient to compel a vacillating public to care. Golf's hero came from down around Rice's way—Atlanta—and his name, as the world would soon know, was Bobby Jones.

Bobby Jones was fourteen years old when he made his first national headlines at the U.S. Amateur championships at Merion in 1916. Rice had known the young man for eleven years at that point, having first met him when he was three years old and still two years away from swinging his first golf club. Jones's father, Bob Jones, Sr., a prominent Atlanta attorney, had been a friend of Rice's during his *Atlanta Journal* days. The senior Jones had played college baseball at Mercer during the same era Rice played at Vanderbilt, though they had never played against one another (Jones's last year was 1897, Rice's first was 1898).

Rice had maintained his friendship with the elder Jones after leaving Atlanta, and was thus privy to the developing talent of the lad who would become golf's first superstar. Bobby Jones grew up in a home right off the thirteenth green at Atlanta's East Lake Golf Course. At the age of nine, Jones won the junior club championship at East Lake. He broke ninety when he was ten, and shot an eighty a year later. By the time he was thirteen he had shot a seventy at East Lake and was the men's club champion. He was also frequently disgusted with his play and convinced he should have been better. He was throwing clubs across courses all over the South. "Here I was thirteen years old (I reflected), and darn my time, I hadn't won anything yet," he said about he feelings after losing that year in the Southern championship. Earlier that same year, 1915, when Rice was playing East Lake with Alex Smith and Jim Barnes, two former Open champions, he watched Jones hit a mashie (five iron) approach to within thirty feet of the flag and then throw his club in disgust for not having hit it closer. Rice and his two companions were torn as to what they had just witnessed.

"It's a shame," said Smith, "but he'll never make a golfer."

"I disagree," said Barnes, "this kid will be one of the world's greats in a few more years."

Rice, more familiar with Jones's talents and temper than the others, agreed with Barnes. "He isn't satisfied with just a good shot. He wants it to be perfect—stone dead. He has a great ambition to play every shot in the bag right.

"But you're correct about that temper Alex. He's a fighting cock, a hothead. That one fault could prove his biggest hazard. If he can't learn to control it he'll never play the kind of golf he'll be capable of shooting."

On the morning of Jones's first match in the 1916 Amateur, Rice had breakfast with the young man and he cautioned him about his raging temper. Rice had written frequently on the mental side of the game and must surely have feared the long-term damage Jones could do to himself should he let his temper control his emotions and his deportment in his first tournament of this magnitude. If Jones heard Rice he did not acknowledge it. Nor did he take the advice to heart—at least that first day. His first-round match was against 1906 Amateur champion Eben Byers, a man with a similar proclivity for throwing his clubs around. Both men played poorly, both spent the round muttering oaths at themselves and flinging their clubs around the Merion landscape. Byers threw one club out of bounds on the twelfth hole and forbade his caddy to go after it. Jones won, 3 and 1. "I think the main reason I beat him was because he ran out of clubs first," cracked Jones some years later. Rice, even as he was writing up a flowery celebration of the teenager's triumph for *Tribune* readers, had doubts about Jones going very far if his temper continued to rage as it had on that first day. In a way he *had* won because Byers ran out of clubs first.

But on the second day, it was a different Jones on the Merion course, a suggestion of the Jones to come a decade hence, when he dominated tournaments as no one before or since, and dominated the galleries with the panache and the certainty of a Babe Ruth. After a poor start, Jones played much better golf that second day, channeling his temper inward. He played with passion, not anger, in defeating F. W. Dyer, 3 and 2. He also proved that his gallery was a product of more than just simple curiosity about his tender age. There was a rakishness about this kid, a blend of Dead End Kid and royalty. He was a stocky five-feet, four-inches tall and 165 pounds. His head was a might too large for his frame, and he wore his long pants with the cocky defiance of the teenager who has only recently been allowed to put away his knickers. But there was also a twinkle in his deep blue eyes and a luster to his smile that bespoke gentleness and sincerity, which softened and at the same time complemented the harder edges of his desire to succeed. He exuded energy, confidence and style; it was easy to root for this kid. Rice acknowledged all of this in his story, as a way of getting his readers to concentrate on Jones's golf, not his age. He was five down after the first six holes of his second-round match, then played the final twenty-eight holes of the thirty-six hole match in two under par to win. This is what moved Rice to write that "his name is already written on the sporting scroll where only the select, those who combine surpassing skill with lion-hearted courage have any reason to belong."

Jones fell to defending champion Bob Gardner in the quarterfinals but the legend had already been writ. So in demand was Jones that one year later he was one of the featured golfers in a series of Red Cross and war relief exhibitions across the country. He was a fifteen-year-old kid and had never been beyond the second round in a major tournament, had never won anything beyond a state championship. And yet here he was, *the* star amongst a galaxy of national champions and professionals. There was a particularly heavy schedule of these matches in the New York area—matches at Baltusrol, Englewood, Siwanoy and Garden City—and at the request of Bob Jones, Sr., Bobby and traveling companion Perry Adair became wards of Granny and Kate Rice during their New York area visit. Rice said that the Riverside Drive apartment looked like a scene from *You Can't Take It with You* during the boys' stay. "While Floncy scampered about and Kit burst her buttons to entertain Bob and Perry, I'd bang away at the typewriter getting out my column." When the golf and column schedules permitted, Kate and Granny took the boys to visit Coney Island and other New York sights. Jones was fifteen. Rice was thirty-six. The relationship began as the naturally awkward child-guardian mix of deference and paternalism, but within fifteen years it would be a very deep and rich friendship, and Rice credited its genesis to that 1917 visit to New York. "It was during their stay that Bob and I became acquainted in a way few persons with a gap of twenty-odd years between them ever do."

A great deal has been made over the years about the "wilderness years" of Bobby Jones, the seven-year stretch between his auspicious debut at Merion in 1916 and his first win in a major championship, the U.S. Open at Inwood in 1923. So ingrained into the Jones legend is this barren stretch, the seemingly endless, seemingly fruitless Diogenian quest for his own excellence that his 1923 triumph is regarded today—as it was indeed regarded by writers and galleries at the time—as Jones *finally* winning something. This makes for nice romance—the triumph all the sweeter for the years of tortured struggle. But it ignores the plain fact that when Jones won at Inwood he was only twenty-one—a boy still, just finished with college. His wilderness years were nothing more than the simple physical maturing and psychological coming of age of the young athlete. Rice was in Jones's gallery for most of his major tournament appearances throughout his career, walking together with his old Atlanta friend, the golfer's father. His written words on Jones's passage from precocious kid to popular icon alternated between wondering when this kid was ever going to reach his potential and reminding his readers that he was just a kid—that just because he'd been around for four or five years was no reason to lose sight of the fact that he was still but eighteen or nineteen years old.

Sometimes in his column, but mostly in avuncular conversations with Jones, Rice told him that his temper was getting the better of him, that he was not in control of himself on the course, and in a game that relied so much on an athlete's ability to keep a cap on his nerves and blot out all distractions, a Vesuvian temper was as deadly as a chronic hook. Jones listened politely and continued throwing clubs as he finished second in the U.S. Amateur at Oakmont in 1919, second in the Open at Skokie in 1922, and had assorted other top ten finishes in the Open, and quarterfinal and semifinal losses in the Amateur. His father began staying home in Georgia, thinking that his presence put added pressure on his son. Jones himself took to subscribing to the theory of predestination in golf—that the order of finish has been preordained and is out of the golfer's hands. But he still kept throwing clubs. Nothing that his father, or Rice, or *Atlanta Journal* writer O. B. Keeler, his traveling companion and eventual biographer, could say seemed to change him. Nor could their admonitions persuade him to abandon his pie a la mode lunches on days when he played both morning and afternoon matches. Maturity comes from experience, not advice.

For Jones, maturity arrived during the British Open in 1921, when he picked up his ball mid-way through a miserable third round. He got blasted by the British press and spent the rest of his life lamenting the fact that he had quit in a major championship. "I know it's not regarded as reprehensible, in a medal play competition against the field," he said. "I know some great players and some fine sportsmen have done it. . . . but I was a youngster, still making my reputation. And I have often wished I could in some way offer a general apology for picking up my ball. . . . It means nothing to the world of golf. But it means something to me."

It could be argued that if this be the epiphany that matured the emotional Jones, then it hardly "means nothing to the world of golf." Two years later Jones was the United States Open Champion, winning that first big title in dramatic fashion at Inwood on the eighteenth hole of a play-off with Bobby Cruickshank.

The hot-headed boy from Atlanta was suddenly the young gentleman from Atlanta. He had grown four inches to a compact five-feet, eight-inches, and spread his 165 pounds over a leaner frame than the one he had brought to Merion seven years before. The club-throwing kid had metamorphosed into a model of on-course decorum and discipline, with always a kind word for an opponent. Yet he still won with aggressiveness, with a crowd-pleasing confidence. And did he ever start to win.

He took the U.S. Amateur title in '24 and '25; the United States and British Opens in '26; the British Open and the U.S. Amateur in '27; the Amateur again in 28; the U.S. Open in '29. Run down the list—not once between 1923 and 1929 did he fail to win at least one major title. In addition to nine championships during this stretch, he finished second in the Amateur once and in the Open three times. In fact, in the twenty major championships he entered between 1923 and 1929 he finished worse than second only *four* times. On top of all this there is the even more remarkable fact that here was a man who was essentially an after-work and weekend golfer. Throughout his competitive career Jones entered only four or five tournaments a year. The balance of the time he was studying law, and later practicing law full time with his father's firm.

Still, for all of this, the world was not anticipating what he would do in 1930. When he won the British Amateur at St. Andrews in late May the triumph was seen as an end in itself. He had now won all four of golf's major titles at one time or another during his career, the only man ever to have accomplished that. "It's great that the boy now has won all the major titles there are," said Stewart Maiden, the pro at the East Lake Country Club in Atlanta, "but it's sad in a way too, for there's nothing left for him to win. All the rest of his victories will be just repetitions." Jones himself, in a column he was writing for Jack Wheeler's Bell Syndicate, hinted at retiring from tournament golf at year's end.

When he won the British Open at Hoylake three weeks later, however, the golf world was no longer looking back at Jones's career; it was looking ahead to Interlachen in Minneapolis, and the United States Open three weeks hence. And while no one had yet coined the phrase "Grand Slam" (credit would go to O. B. Keeler of the *Atlanta Journal*) for what Jones was about to do, people were beginning to think along these lines: Can Jones duplicate his remarkable British double in the United States? "He has a good chance now," was Rice's assessment immediately after the British Open.

Rice did not see Jones play in Britain; the only American writers on hand for the first two legs of Jones's Grand Slam were the loyal Keeler and Al Laney of the *New York Herald Tribune*. But the paucity of American reporters in Britain did not dull the Jones achievement for the American public. He returned home to New York on July 2 to a ticker-tape parade. Thousands of people, the vast majority of whom had no doubt never set foot upon a golf course, crowded the concrete canyon that was lower Broadway and cheered lustily for this modest young man who had transcended his game to become more folk hero than sports hero. That night, Rice was master of ceremonies for a dinner in Jones's honor at the Vanderbilt Hotel.

The next day, Jones, his parents, Rice, Keeler and a host of other journalists entrained for Minneapolis on the *Twentieth Century*. They arrived early on the sixth, five days before the start of the U.S. Open on July 11.

Handicapping golf tournaments is a flat-out impossible task. It is the most fickle of games, frequently turning on matters as trivial and as uncontrollable as a bad lie on a good shot or a fortunate bounce on a bad one, a spectator's inadvertent shout at the top of a backswing or the inexplicable surge of confidence that can come when a 190-yard approach rolls to within a foot of the pin or a 50-foot putt suddenly takes on a destiny of its own and snakes and rolls and falls into the cup for a bird.

Throughout tournament golf's long rich history, how quickly and how frequently have we seen hot hands cool or steely concentration broken—often only for a single shot—and suddenly, swiftly, cruelly, the champion, the favorite, is simply another supporting player in somebody else's triumphant drama. It has been ever thus, which is exactly why golfers and savvy writers avoid trying to pick winners before the fact, recognizing that golf is different from the World Series or a college football game. Yet for all the circumspection amongst golfers on the matter of tournament favorites, there was not a soul at Interlachen who did not stand in awe of twenty-eight-year-old Bobby Jones, defending Open champion, reigning British Open and Amateur champion. "The remarkable thing about this championship is just this," said Walter Hagen to Rice. "Here is the greatest field ever assembled on any golf course. Here you have the survivors of 1200 entries and yet it is one field against one man—Bobby Jones. Nothing like this has ever happened in golf, from the days of Vardon and Taylor and Braid to the present moment. It is almost unbelievable, but it is true."

Playing with pressure that was as oppressive as the Minnesota heat, Jones shot a one-under 71 in the first round, to trail MacDonald Smith and Tommy Armour by one shot. He was two behind Horton Smith after a second-round 73, and would have been further back were it not for the famous "lily-pad shot." On the 485-yard, par-five ninth, Jones topped his spoon (three-wood) approach shot and sent it on a flat trajectory straight into the pond that guarded the green. Hitting the water just like a flat stone thrown by a child, the ball skipped twice and came to rest in the bank in front of the green, from where Jones pitched to within four feet and made the putt for birdie. Somehow, the writers and the witnesses got it in their heads that the shot had been saved from a dunking by a fortuitously placed lily pad. Saturday newspaper stories sang a song of the lily pad. It was a quarter-century before Jones set the record straight and insisted there was no lily pad, only great good luck.

Jones awoke to face the thirty-six holes on Saturday feeling good about his chances. "[I was not] oppressed by that feeling of having something to protect," he said later of his position two strokes off the lead. For all intents and purposes he put it away in the third round, shooting a four-under 68, the lowest round of the tournament, giving himself a five-shot lead on the field heading into the final eighteen holes after lunch.

The afternoon round might have been anticlimactic, like the play-off in the '29 Open when Jones defeated Al Espinosa by twenty-three shots. But the hoary, resilient British professional MacDonald Smith, who was second to Jones at Hoylake, took another run at Jones, putting up a two-under 70. Jones had given two shots back to par over the first sixteen holes and as he stood on the tee at the unpleasant 262-yard, par-three seventeenth, his lead over Smith was three shots. On seventeen, he then skulled his brassie (two-wood) shot off the tee far off into the right, where it ricocheted off a tree so wildly that the entire gallery of 8000 lost sight of it. A five-minute search proved fruitless, the only logical conclusion was that the ball had caromed into the water off to the right of the trees—some 60 yards off line—though no one had seen it splash. He took a double-bogey five there, and faced the eighteenth needing a par to preserve a one-stroke win over Smith; and when his unsteady approach shot left him 45 feet short of the cup, a par was suddenly problematic. The 45 feet between Jones and his fourth U.S. Open title contained a slight uphill, then a steeper uphill, then finally a small plateau to the cup. A three-putt—and a subsequent thirty-six-hole playoff—was a very real possibility as Jones stood over his putt and studied it for an extraordinarily long time. "And under these conditions Bobby stroked the ball as smoothly and as perfectly as if there were nothing at stake," wrote Rice. According to Jones biographer Dick Miller, the ball "rolled quickly up the gentle slope, slowly up the steeper one, and then six feet from the cup the ball broke just slightly left, falling into the cup for a birdie 3 and the championship."

"I was lucky," said Jones humbly as he accepted the trophy. Nobody else saw it that way.

For maybe the first time in his professional life, Rice was struck darnnear dumb by the Jones achievement. He was left, in opening his morning story, to uncharacteristic understatement:

> Bobby Jones broke all records in the history of golf this afternoon by winning his third major championship in succession and this is a matter of statistical data that goes back 500 years.
> This is a plain statement of fact that no adjective can adorn.

But Rice warmed up as he got into his story. He quoted a spectator who noted that Jones "has caught the fancy of this country as no man, barring Lindbergh, has caught it." And as he closed the piece, Rice, who in his twenty-nine years behind a typewriter had seen Ruth and Cobb and Mathewson and McGraw; Grange and Thorp and Rockne and his Four Horsemen; Dempsey and Tunney and Willard and Firpo, spoke from the heart when he said of the still-young golfer from his own South: "There has been no champion like him that sport has yet given to the game."

Jones's win in the Open unloosed a torrent of breathless, maniacal, consuming fascination with the man as he approached the final leg of the Grand Slam. It was assumed that he would win the Amateur—the tougher challenge, after all, had been the Open, with its tournament-hardened professionals. And it was true in a way—they're were certainly fewer challenges to Jones's dominance in the Amateur than there had been in the

Open. But Jones was never a golfer who concerned himself much with the other golfers in the field; his foe whenever he teed it up in a tournament was "old man par." But for the Amateur in 1930, lying in wait together with old man par was pressure and the weight of expectation the likes of which no single athlete had ever faced. That Jones was able to concentrate on his game amidst the swirl and clamor of press and fans was simply remarkable. Writers around the country had picked up on O. B. Keeler's Grand Slam phrase, and those who didn't tried to go it one better—George Trevor of the *New York Sun* referred to Jones's quest as "storming the Impregnable Quadrilateral of Golf." Al Laney of the *Herald Tribune,* who had seen Jones win the British Open, said later that the "cumulative excitement . . . [was] participated in by millions on both sides of the Atlantic." Rice simply called what Jones was about to achieve "a record beyond the imagination of anyone who has ever tackled this elusive and baffling game." In the ten weeks between the Open and the Amateur, Jones was beset by requests to play assorted rounds for charity, nearly struck by a car as he walked an Atlanta street, crippled at times by nagging spasms in his shoulders and neck, and beset with a summer-long stomach virus that had him drinking paregoric throughout the Amateur championship. He arrived in Philadelphia with his nerves frazzled, struggling against "becoming frantic on the one hand and overconfident on the other," fearing defeat not for its own sting, but because, intelligent and introspective man that he was, he knew that he carried the "special burden . . . of a nation's worship," that, as Dick Miller put it, "the U. S. Amateur [was] more than a golf tournament. It was a battleground of hope for a people experiencing the dread of the Depression. Jones held the promise of a man fulfilling his greatest potential against staggering odds."

Adding a nice poetic touch to the story was the fact that the 1930 Amateur was scheduled for the Merion Cricket Club, the course where Jones's career began in the Amateur of 1916. Reporters swarmed over Merion in a crush that the sport had never seen. The *Philadelphia Evening Bulletin* alone had sixteen writers and photographers covering Jones's every move. The small, stuffy press room at Merion—a long, narrow windowless room on the second floor of a nineteenth-century barn that had been converted into Merion's clubhouse in 1910—became downright claustrophobic as the regiment of reporters churned out more than two million words on Jones during the six days of the competition. Ironically, Rice's syndicated writings on the tournament were carried in fewer papers than usual; so popular was Jones that many of Rice's papers had sent their own staff writers to Merion.

The fans followed the writers' scribblings. More than 5000 people turned out to watch Jones play his *practice* rounds. The would reach 7000 by the start of the tournament on Monday, September twenty-second, some 10,000 by the thirty-six-hole quarterfinals on Thursday, 18,000 by time of the championship match on Saturday. During the competition, Merion would need to call in fifty United States Marines to provide a buffer between Jones and his adoring galleries. Most of the reporters jostled with the gallery for a glimpse of Jones. Rice, by virtue of his stature and friendship with Jones, walked along with him on the fairways as he practiced. On Sunday, after three frustrating practice rounds in which he did not record a single birdie, it all started to fall into place. On his way to a one-under-par 69, Jones could feel it coming together, and waxed philosophical as he and Rice walked down the seventh fairway. "Granny, I've suffered at this game a lot of years. Among other things, I've discovered a man must play golf by 'feel' . . . the hardest thing in the world to describe—but the easiest thing in the world to sense—when you have it completely. I don't have to think of anything . . . just meet the ball. . . .

"Today everything's falling into one piece—perfect. But it seldom happens—at least with me—and very seldom in title play."

It would continue to happen for all of that week, however, and very nearly render the competition anticlimactic. Jones cruised. He shot 69 and 72 to earn medalist honors in the qualifying (bogeying the last two holes to cost himself a U.S. Amateur qualifying record); won his first- and second-round, eighteen-hole matches, 5 and 4 and 5 and 4. He then took time out from his post-match relaxation to spend a half-hour on Rice's weekly radio show on Wednesday night. The next morning he breezed through the quarterfinals 6 and 5; and then on Friday he routed Jess Sweetser—the former champion who had defeated him in the finals at Brookline in 1922—10 and 8 in the semis.*

In a championship match that was absent any drama at all, Jones romped to a seven-hole lead at the end of the morning round, then closed out Gene Homans 8 and 7 as a rowdy and restless Merion gallery whistled and whooped and strained against the cordon of Marines and state troopers who shielded Jones from the affection of his public.

The applause that followed Jones across Ardmore Avenue and back to the Merion clubhouse would echo for the next forty years. He had achieved the perfection we all seek, and his perfection was vicariously ours. And when he announced his retirement from the game seven weeks after

---

*By the time the semifinals came around the writers, eager for stories that would provide a moment's surcease from the Jones hyperbole, seized upon a sturdy eighteen-year-old Californian named Charlie Seaver. Seaver was headed to Stanford on a football scholarship, but in the meantime, his golf continued to improve throughout the week and was being called "the next Bobby Jones." Rice called him "the best all-around athlete in this championship. . . . He can hit a golf ball father than a politician can throw an alibi." Seaver lost to Gene Homans on the last hole, and though he continued to qualify for the Amateur, he never again made it past the third round, never attained the very pinnacle of athletic stardom that many predicted for him at Merion. Thirty-five years later, however, his son, a right-handed pitcher named Tom, did.

he completed his Grand Slam at Merion, he ensured that the aura, the dignity, and yes, the perfection, were forever.

Were he a professional he would have had reason to continue playing the game—it would have been his livelihood—and he would have no doubt gone on winning championships and dominating the game for another dozen years. But golf was not his livelihood; the law was. Golf was his passion. And in the fall of 1930, at the still tender age of twenty-eight, Jones had the intelligence and the insight to recognize that to continue in the maw of championship golf would have inevitably eroded the passion.

"He dislikes the aftermath of glory," noted Rice after the Slam was complete, "which is much harder on such a type as he is than the strain of contest." The aftermath of glory was denying Jones time with his wife and their young children. And while he was always considerate to his public, their affection was never the aphrodisiac it becomes for many. The strain of the contest was very real too; it wrapped his nerves into a tight, raw bundle, and frequently caused him to lose ten to fifteen pounds during a major tournament. Tournament golf had never been fun; by 1930 it was excruciating. Jones's decision to leave it was inevitable, Grand Slam or not.

His quiet announcement—in a letter to the United States Golf Association—prompted bittersweet melancholy. Fans and writers alike recognized that there was a correctness in his exiting now—a competitive epilogue could not possibly enhance and might well diminish what Jones had done in 1930. Yet to be denied forevermore the electricity that Jones brought to the arena was to suffer a great loss, too.

Bobby Jones's vision of golf is not to be found in the gallery-cluttered fairways and clubhouses of Merion or Interlachen or even St. Andrews. It is rather in the pines and magnolias, the dogwood, the azaleas and silent lushness of the fairways and greens of the Augusta National Golf Course in springtime. Augusta National is Bobby Jones's course—he conceived it, he designed it, he brought it into being with the force of his will, his personality and his unique stature in the game. "While born out of Jones's desire to build his dream course," wrote Dick Miller, "Augusta was also seeded by Jones's desire in 1930 to rediscover the joys of golf. In giving up championship golf, he wished again to experience the broader essence of the game—the fellowship that golf offers—under the more remote conditions of being just another member of a foursome, not a celebrity."

In keeping with this "broader essence," Jones gave much careful thought not only to the course at Augusta, but to the people who would make up the club membership who would play on the course as well. Jones wanted a national membership—hence the name, Augusta National—since he had acquired friends from around the country. He sent out letters, inviting these friends to be a part of his new club. Grantland Rice was among the first invited; he swiftly and eagerly accepted.

The friendship between Rice and Jones became largely a friendship of the road. They would frequently travel together throughout the thirties and forties—to prizefights and World Series games no less to the Open and Amateur. The two men shared a lot through those years. They lived through an earthquake together in Los Angeles in 1935. In 1936, together with Kate and Jones's wife Mary, they traveled together to Hitler's Olympics in Berlin. On the way they made a side trip to Scotland, where Jones decided to play a round of golf at St. Andrews, his first since winning the British Amateur there in 1930. He told no one of his plans yet as he walked to the first tee a crowd of 5000 awaited him. Rice stood with Mary Jones, next to a tearful Scot who did not suspect who they were, and proclaimed: "Isn't is ground . . . isn't it ground . . . Bobby's back."

When he had Jones in tow, Rice's presence was an extra treat for his thousands of acquaintances. People were without fail anxious to meet Jones; nobody more so than Sam Breadon, the owner of the St. Louis Cardinals. Jones and Breadon met in the mid-thirties, about the same time Breadon had taken up the game of golf and just after Jones had purchased a share of the Southern League Atlanta Crackers. Breadon wanted tips on his golf game; Jones wanted to pick Breadon's brain about running a ball club. Breadon asked about the grip on a golf club and Jones inquired about a Cardinal farmhand who might help the Crackers' infield. And on and on it went, each man asking questions intensely and answering them distractedly "Honestly, if I had the tongue of men and angels I could not begin to tell you of that incredible and lopsided dialogue between Breaden, the amateur golfer, and Bobby Jones, the newly fledged baseball owner," said writer Cullen Cain, a witness to the Breadon-Jones exchange. Rice watched the bizarre exchange with amusement and allowed afterward that he believed it was the "masterpiece of all time in misunderstanding."

But for all the travels that Rice and Jones enjoyed, the anchor to their friendship became Augusta National. It was a regular stop on Rice's winter odyssey; from the mid-thirties on it was California-Florida-*Augusta*-New York.

Immediately after Augusta National opened in 1932, Jones and his partner Clifford Roberts began beseeching the U.S.G.A. for a chance to host the U.S. Open. The U.S.G.A. balked; they didn't want to change the dates of the Open from their traditional June-July schedule, and Augusta National was at its best in the early spring; the fairways and greens tended to dry out in the summer. So Jones and Roberts instead decided to showcase their course with a tournament of their own—to be called, in the beginning, simply the First Annual Invitation Tournament. They sought Rice's counsel in the matter—he suggested the dates of March 22–25, 1934.

The dates were crucial, said Rice. The baseball writers would be done with spring training and on their way north. It would be easy to persuade them to detour through Augusta. Getting them to out-of-the-way Augusta at any other time was chancy.

Rice's strategy worked. It took little persuading to get a flock of writers to Augusta. There was first of all the promise of a good story—Jones had promised a return to competition. More important, there was the chance to once again see and visit with Jones, a man they held in fond regard, the man Herbert Warren Wind of the *New Yorker* called "that rare sort of hero—in sports or any other field—a man whose actual stature exceeds that of the mythological figure he has been made into." And so was born The Masters. Roberts had wanted to call it The Masters from the outset but Jones dissuaded him. Rice and the other writers immediately took to calling it the Masters, though, and by 1938 Jones had acquiesced and "The Masters" became the tournament's formal name. The dates were changed to the last four days of the first full week in April, when Augusta National is at its most beautiful and many of the nation's sports writers are still making their way back from spring training.

Rice wrote of the Masters with tenderness and energy from its inception in 1934 until his death twenty years later. But amongst the other writers he always seemed more than just a writer at Augusta. By the mid-thirties he was enjoying a certain aura at nearly every event he covered, as new, younger writers came upon the scene replacing the writers Rice had matured with. At Augusta, however, this aura was particularly acute. He was unlike the other writers, and not because of his fame and celebrity. He was, after all, a member at Augusta, and his friendship with Jones—who *was* the Masters—was longstanding and well-known. It was a situation ripe for resentment and jealousy in the often catty world of sportswriting, where younger writers have historically shown a disgust and contempt for writers they perceive as coasting on a reputation earned in another age. But if Rice's industry as a reporter was not in always in evidence in Augusta, his disarming personality was, and pre-empted any such notions. As always, he went out of his way to greet a young sportswriter and introduce him around the press room. He was as much of a natural host as Jones. Far from begrudging Rice his stature, writers basked in it. Fred Russell, sports editor of the *Nashville Banner,* whose affection for Rice was as profound and as enduring as that of a son for his father, remembered Rice's presence at Augusta being a part of what was special about the place in April—ever so much a part of its charm and lure as the stunning golf, Bobby Jones, the magnolias and the eternal azaleas.

12

Dean of American Sportswriters

O n *January 1, 1930,* Rice left the Tribune Syndicate after fifteen years and shifted his syndication to Jack Wheeler's North American Newspaper Alliance. For the most part only the readers in New York had to make an adjustment; around the country virtually all of the hundred or so subscribing papers continued to subscribe to Rice's column after the switch. But in New York, leaving the Tribune Syndicate also meant leaving the *Herald Tribune,* which had been his flagship since 1915. He moved to the venerable old *Sun,* never a paper of great substance but always a paper of great appeal in New York. Its sports section—like the sports section of the old *Evening Mail*—was one of its strengths. It featured solid baseball and fight coverage, as well as columnists Joe Vila and Rice's good friend Frank Graham. In terms of visibility within the city, the *Sun* was probably a step up from the *Tribune.* But what neither Rice nor anyone else could have known in January of 1930 was that the *Herald Tribune,* especially its sports section, was still in its ascendancy, while the *Sun* had already lived its greatest days.

The move to the North American Newspaper Alliance marked the close of Rice's thirty-year career as a newspaperman in a way, for while the *Sun* took his column he was not a part of the staff as he had been at the *Tribune.* He would never again have an assigned desk in a city room.

233

When he was in New York now he worked out of his home, or out of the Sportlight Films office on West 48th Street. But more times than not Rice's column would arrive at Jack Wheeler's office via Western Union, for by the start of the thirties he was living a peripatetic life. As it had been since his Nashville days, it was a life of rhythmic cycles. His year started in December, when he and Kate left New York for Los Angeles to spend Christmas with Floncy. He would hole up at the Beverly Wilshire Hotel there, writing in the mornings and then, on most days, heading off to Lakeside for some golf in the afternoon. He'd remain in Los Angeles throughout most of January and February, then travel to Florida and spring training in March. He preferred the West Coast of Florida, the St. Petersburg-Sarasota area. This was where the Yankees trained but was also convenient to an out-of-the-way little thoroughbred track called Sunshine Park to which he developed a real devotion over the years.

The close of spring training would send him through the South—Augusta National, visits to Kate's family in Americus and his own family in Nashville; and finally a stop in Louisville for the Kentucky Derby before arriving back in New York in early May. Throughout the summer and fall, he would shuttle between Fifth Avenue and East Hampton, though his grip was always packed, for he'd take in the U.S. Open and Amateur every year, most heavyweight championship fights, the World Series, a college football game each week in the fall, as well as assorted polo matches, turf events, and miscellaneous travels associated with Sportlight Films.

Incredibly, he would also find time in this schedule for hunting and fishing vacations. Over the years he went on various excursions to Alberta, New Brunswick, Maryland, Texas and Montana, but his most frequent destination for his hunting adventures was Robert Woodruff's 50,000-acre preserve at Ichauway, Georgia, a lush wilderness that frequently played host to heads of state, captains of industry and the the pedigree of American celebrity. Woodruff was the president of the Coca-Cola Company, the sponsors of Rice's weekly radio show, and over the years he became a devoted friend and heartfelt admirer of Granny's. He kept a picture of Rice behind his desk in his Atlanta office, and in 1951 he endowed—anonymously—a scholarship in Rice's name at Columbia University's Graduate School of Journalism.

Ichauway was stocked with quail and dove and wild turkey. There was an expansive, wood-beamed and leather-chaired lodge, and a staff of savvy guides to escort the tony guests about the property. For Rice, the years of high-rise flats, posh hotels and clubhouse verandas had not so citified him that he had lost his taste for the Southern woods where he had romped as a boy. He once said that the red clay roads of Ichauway seemed more like home to him than any city street. He was no gentleman hunter at Ichauway, spending a couple of mid-day hours on the paths near the lodge and the balance of his time brandishing a glass before the fire.

He would rise early and hunt hard, and one demonstration of his tenacity and perseverance became an Ichauway legend. For years, a battered souvenir from the experience held a place of honor above the fireplace in the Ichauway lodge.

Rice felt that the quail provided "the most dependable diet" for a hunter, "but when it comes to the big thrill, the wild turkey takes over. . . . They are as different from the tame variety [of turkey] as a tiger is from an Angora cat." In pursuit of the big thrill, Rice and his guide rose one morning well before dawn and settled down in a blind near a turkey feeding ground to await their quarry. They waited wordlessly for more than two hours in a steady downpour; the only sound interrupting their vigil was "the far off, dismal call of the mourning dove." Shortly before 6:00 a flock of the wild birds swooped in; Rice fixed one in his sights and fired when the bird was some twenty yards distant.

"He went up into the air like a rodeo bronc," reported Rice. "Then with a great thrashing he was gone." Not about to lose such a prize—the bird weighed over twenty pounds—Rice waded into the swamp after him, following the thrashing of feathers behind some reeds and grass. "It was like I jumped into a threshing machine," said Rice. "He all but beat me to death with his wings. At this point it was either the turkey or me." Swinging his shotgun like a baseball bat, Rice finally subdued the bird and walked unsteadily out of the water, bleeding from gashes from the bird's spurs and bruised by the flailing of the bird's wings, but proudly dragging his kill nonetheless. The taxidermist employed by Ichauway either could not or chose not to remove the vestiges of the struggle from the bird, for the one thing that struck visitors to Ichauway through the years was its somewhat battered and bruised appearance as it hovered nobly above the fireplace, a monument to Rice's undiminished grit, fight and zest for the contest.

And it was the contest of the hunt that attracted Rice, not the kill. The other great tale to survive from his hunting days is the story of the time he couldn't pull the trigger. In late 1950, Rice was hunting wild turkey from a car at the King Ranch in Texas with Red Smith when they came upon a herd of deer that just paused and stared back at them. Rice raised his gun and bracketed what he called a "great white tail with a tremendous rack of horns." But as he sighted the deer he found he couldn't shoot the animal. It just wasn't sport when it was this easy. Smith and the others in the car called him a sissy.

"It's not that exactly," said Rice. "It's just that I never shoot a deer until he pulls a knife on me first."

Writer Henry McLemore once called Rice a "one-man parade." "Everybody stops to talk to him," said McLemore. "Porters ask about Joe Louis. Cabbies, bootblacks and newsboys want dope on the pennant races.

Bankers and brokers inquire about football prospects at their alma maters. One man wants to know whether Waddell or Mathewson pitched three shutouts in the World Series. Another wants a cure for a slice that is ruining his golf. Granny talks to all of them. He is absolutely without the power to end a conversation and walk away from man, woman or child." Rice's wanderlust was matched by his restlessness and need for companionship on his travels. Kate always traveled to California, of course, and generally stayed on to complete the swing through Florida. But she stayed at home on most of Granny's shorter, working excursions.

Yet seldom would Rice travel alone. Ring Lardner and Bill Mc-Geehan and Westbrook Pegler were aboard for most of the early, postwar travels in the 1920s. It was a camaraderie born largely of convenience; the men found that they were frequently covering the same events in the early twenties. But as the years passed they discovered that they were just as frequently assigned to different events. Rice also mildly exasperated the punctual Pegler with his habit of showing up for trains as they were leaving the platform. But Pegler soon found his destiny in writing politics with a particularly venomous bent; McGeehan and Rice—working for the same paper as they were—inevitably found themselves at different events; and even before his illnesses in the thirties, Lardner had curtailed his travel greatly.

So Rice found others, whose spirits burned as bright and whose schedules were more resilient—writers Henry McLemore and Clarence Buddington Kelland; writer and advertising executive Bruce Barton, and actor Frank Craven. Each was remarkable in his capacity to arrange his own life around Rice's needs and travels. Craven, Barton and Kelland—all figures of renown in their own right—would humbly serve as runners or spotters for Rice at a golf tournaments; Barton was once returning from some trip when he bumped into Rice at Grand Central Station, had the porter load his bags onto Rice's train and joined him on his trip. Kelland was almost as spontaneous, though Rice said he needed at least two hours notice. "Bud never seemed to care where we went or what sport we were covering," said Rice.

The four men were forged of different ores and possessed of different personalities. McLemore, younger than the others and in the same business as Rice, regarded him with a measure of awe and deference for much of their relationship. The courtly Barton and the ebullient Craven were likewise calm presences on the road. Barton would serve as a buffer between Rice and those who clamored for a piece of his time. "He was a valuable blocking back," said Rice of Barton, speaking a little bit figuratively and a little bit literally. "He swung a wicked portable through crowds, clearing the way like an icebreaker."

The fiery Kelland, however, was another matter. A popular and prolific novelist and short story writer and a marvelous raconteur—had he been born some years later he would have been a natural guest of Jack Paar's or Johnny Carson's—Kelland was not a man who suffered fools gladly. To Henry McLemore, he complained that he spent "a good third of his life . . . standing first on one foot, then on the other, while Rice talk[ed] long and earnestly to complete strangers." To no one at all, but within earshot of an eavesdropping Rice he once asked himself: "Why do I do it? Why am I here? I didn't have to come. Nothing but drunks and bums. I don't drink, but I rarely see a glass of water. No sleep—no nothing that makes any sense. Never again." But Kelland could be every bit as charming as he was dyspeptic. He was a man of keen wit and intellect, who loved the games as thoroughly as Rice did, and he was Rice's most loyal and regular traveling companion for almost twenty-five years.

Traveling the sports circuit with Rice could be exhausting. Rice's room generally doubled as a hospitality suite for the traveling press and other assorted hangers-on, both worthy and otherwise. "Practically everybody you ever heard of will bounce in during the course of the evening," said one young witness to the scene in Rice's room in St. Louis in the early 1940s, "and those that don't will call on the telephone." Rice would now and again disclaim any responsibility for the chaos in his room, and most of his friends wrote it off as a byproduct of his being unable to say no. But Rice was the instigator of most of these gatherings, not the victim.

At the U.S. Amateur in Cincinnati in 1933, Rice and Bobby Jones checked in to their room at the Hotel Sinton, weary after a long trip from Chicago and determined to have a quiet dinner and retire early. But O. B. Keeler was also a guest at the Sinton and they didn't want to seem rude in not seeking him out to say hello, so they invited Keeler and his wife up for a drink and then dinner. Running into Paul Gallico of the New York *Daily News* and John Kieran of the *Times,* they invited them as well. They were sharing the finest suite in the hotel and the hospitality must have seemed only fitting. Soon word got around that Granny and Bobby had checked in and were having a party in their room. At 10:30, when there were fifteen people there and Rice and Jones still hadn't had dinner, Rice thought it rude not to have invited the hotel manager, an old friend named Dan Horgan. So he did. The noise attracted others; guests who called to complain about the noise were invited; and sometime past midnight Horgan decided that such a crowd deserved to be entertained by the house band, then no doubt concluding their evening's work downstairs. "It was dawn when we swept out the last remains," remembered Rice.

If old friends were not seeking out Rice, he was seeking out new friends. Especially young sportswriters. David Camerer, an ex-lineman for Dartmouth who would later play a large role in Rice's autobiography, was a rookie sportswriter for the *New York World-Telegram* in the late thirties, standing alone at his first football writers luncheon when Rice came up and introduced himself. He then took Camerer around the room and introduced him to the others—"You know Dave Camerer, don't you? The old Dartmouth star?"—and then insisted that Camerer sit next to him at lunch.

Henry McLemore told a similar story. He was waiting for a train after his first big out-of-town assignment for the United Press, and standing alone on the platform when Rice excused himself from the group he was with and came over to introduce himself. "[He] asked if I weren't a newspaperman and if I were alone. He insisted I travel with him. On the trip, he acted as if my opinion of the game we had seen was of the utmost importance in helping him make up his mind what to write about it."

Such kindnesses prompted George M. Cohan to call Rice "a virile saint." Most public mentions of Rice, particularly by his friends, were similarly kind to the point of being cloying.

It was also around this same time, the mid-thirties, that Gene Fowler began calling him the "Dean." So apt was the tag that it stuck for the rest of his life; it became almost an appendage to his name—Grantland Rice: Dean of American Sportswriters. But with all of his friends the relationships themselves remained very human and mischievous. McLemore once told a newspaper interviewer that Rice was the intercollegiate wrestling champion as an undergraduate, once rode the winning horse in the Tennessee Oaks Steeplechase, and was a passionate roulette player. When McLemore topped a drive and sent it squirting some fifteen yards off the tee one day, Rice told him: "I have a hen that can lay an egg farther than that."

As much as his friends loved his company, however, most ultimately wearied of the frenetic and ceaseless travel, joking about the pace and being unable to keep up, and doing it only half in jest. Barton left the road in the thirties, Craven died in 1945, Kelland cut back gradually through the forties and was off the road by 1951.

In New York, the pace was less frenetic but Rice's life was just as social. In town, he would work in the morning, writing either at home or in the Sportlight office, often knocking off a week's worth of columns in one sitting. His favored spot for lunch in the thirties was the Dutch Treat Club, where he would invariably sit at a table in the back row with John Kieran, Theodore Roosevelt, Jr., Rex Beach, and occasional others. They became something of a club within the club, as one member made clear when he walked past the table and asked: "When are you fellows going to join the club?"

If it were baseball season he might take in a Yankees game in the afternoon. The Yankees had long ago replaced the Giants as the darlings of New York, and Rice was developing good friendships with Bill Dickey and Lou Gehrig; Dickey was an off-season hunting partner, Gehrig an in-season bridge partner. Or he might have Charles Goering, his chauffeur, drive him out to Yonkers or Belmont. There were precious few quiet evenings at home while he was in New York—East Hampton was the place for that sort of relaxing. In New York there were banquets and shows and friends, and seldom enough time to accommodate all the demands. Kate would frequently be at his side for his evening's socializing. Very bad with names, he informed his wife that if he became involved in conversation with someone and failed to introduce him immediately, it meant he had forgotten yet another name and Kate was to just walk away, so Granny's friend would not have to be embarrassed that Granny didn't remember who he was.

Rice found comfort in the rhythms of his life. In comparing his love for both sport and verse he said that the rhythm inherent in both was what gave them their enormous appeal. "Rhythm, the main factor in both, is one of the main factors in life itself. For without rhythm, there is a sudden snarl or tangle." So he lived his life according to the rhythms and cycles that had evolved over three decades, and he was a most comfortable creature of his habits.

Habits, however, even when they are infused with the freshness that comes from travel and the energy that comes from bearing frequent witness to compelling events, will eventually exact a toll on the creativity and energy necessary to put out a daily newspaper column. No one who ever wrote a column for as long as Rice did was immune to a diminution of the drive and the spark that give the column its brightest luster. It can take different forms with different columnists, this diminution. Sometimes, as with the energies of life itself, it is a gradual, imperceptible and unchecked erosion. More commonly there are times when events or circumstances— a switch in papers or assignments, perhaps; or the arrival of an athlete, a team, or games of fresh and unique proportions—can excite the passions and for a few weeks, a few months, a few years even, the column can roil and swell and command attention like a crashing surf conjured by an impressive summer storm. Then suddenly, just as the storm-driven surf subsides, the events that gave the column its resurgent majesty pass, and like the sea, it resumes a tranquil, regular, reliable ebb and flow.

This diminution in a column's vitality is often only apparent to someone who goes back to search and scrutinize a columnist's lifetime canon. Here it is unmistakable. Be it Mencken, Pegler, Lippmann, Winchell or Grantland Rice, the later work lacks the exuberance and inhibition of the early work to a degree to which polish and maturity cannot compensate. Yet what is equally apparent in such a scrutiny is that the level at which these writers performed—each in his own decidedly different way—was such that even their ordinary stuff could be pretty arresting journalism. Vitality is also more in the eyes and heart of the reader than it is the writer. In the 1930s, Grantland Rice remained among America's most popular writers. He remained informed, involved and entertaining. Old readers remained loyal and enthusiastic; new readers—essential in the youth-driven world of sports—continued to discover him every year. Yet the column that he brought with him to the North American Newspaper Alliance was notably more predictable, more formulaic, than the column he had written in Nashville a quarter-century before, or the column he first began syndicating before the war.

Part of this can be attributed to having to work with the impossible deadline he faced with his column. His columns were normally mailed to subscribing newspapers, in batches covering a week at a time. Rice would often produce this work in one sitting. A hazard of such a habit is to take a single subject and parcel it out over two or three days, and the inevitable tendency is to masticate each piece of it just a little more thoroughly than is necessary. The other, far greater hazard, is to try to anticipate what the news angle from some forthcoming event will be. This was a folly indulged at great peril in a world as changeable and as immediate as sport. Writing seven to ten days before publication it was virtually impossible to guess what hitters would be hot or which golfers would suddenly begin having difficulty controlling their tee shots.

Sometimes Rice would arrive on the scene of a World Series or a major golf tournament and write daily accounts that were at odds with what he had said in the column that was scheduled to run that day. This would force editors to either kill or prune back Rice's column to a couple hundred words. Occasionally an editor hadn't read the column before ordering it set into type, or the daily story would come in too late to make changes in the column, and Rice would have side-by-side stories saying different things. Readers ignorant of the logistics of newspaper production could only scratch their heads.

For some big events Rice would avoid the problem by telegraphing his column to the subscribing newspapers, as he did with his daily stories. But that was expensive for the syndicator and added an unwanted element of uncertainty for the subscribing newspapers. The easiest way around the problem, and the way Rice came to choose most frequently, was to deliver columns on an entirely different subject for days when he knew he would also be filing a day-of-the-event story. During the World Series, for example, his advance packet of columns would have college football notes; if he was scheduled to be off at a prizefight he might file columns previewing the baseball pennant races.

Nevertheless, though he spread himself thin and a lot of his work has a once-over-lightly feel to it—especially when measured against modern standards and expectations—his output remained enormous and varied; and it is not difficult to understand why he remained as popular as he did. His range was truly impressive. Most sportswriters would be expected to know and write about baseball, football and boxing, and maybe one or two more from a list including golf, tennis, track and field, horse racing, basketball or hockey. The esoteric stuff, like yachting, polo, swimming, wrestling (real and imagined) and all winter sports, was covered strictly on a novelty basis—little was expected from the writers, and little was delivered. Yet Rice would tackle it all.

In 1932 he didn't know a bobsled from a ski jump yet he wrote an extensive series of preview articles on the Winter Olympic Games scheduled for Lake Placid, even did a radio feature on the Games, though he had no intention of going to Lake Placid. He would write winter feature articles on the past heroes of sport, and driver-aviator Eddie Rickenbacker or polo players Tommy Hitchcock or Devereaux Milburn were as likely to be subjects as Ty Cobb or Bobby Jones. These columns, magazine articles and radio broadcasts (and his Sportlight Films as well) had the effect of expanding the definition of sport. That his rendering of these stories was merely adequate and not inspired is not surprising.

As ever, his verse continued to be the soul of his column, and only infrequently did it show the once-over-lightly feeling of the prose sections. There was less of it than there had been, and perhaps there was more of the glib doggerel than of the sad-hearted introspection. But he still regularly brought a tear as well as a smile, as when Knute Rockne was killed in a plane crash in 1931.

> Yes, other teams
> Upon remembered fields will hold their sway;
> But will they bring the same far-lasting dreams
> To span the sunset of an older day?

The other area in which Rice's work showed no signs of boredom or fatigue was his writing from the scene of an event. A compelling contest retained its capacity to stir him throughout his life. Few events charged his batteries as thoroughly as did the 1932 Olympic Games in Los Angeles. An entire community riveted by sport for two weeks, the athletes of the world gathered, the brethren of sportswriting fraternity assembled to witness and record, a suspension of the worry and the pall of the Depression and the growing political powder keg in Europe, and all of it in the name of sportsmanship, brotherhood, competition—Rice would not have changed a thing were he scripting a juvenile novel. And for a happy couple of weeks he must have felt he was living inside a juvenile novel.

He was consumed by the Games in Los Angeles. He even took in the equestrian competition and was moved to call "Empire," the mount of Swedish gold medalist Clarence Von Rosen, "one of the great athletes of all the games—runner and jumper and hurdler." Another of Rice's busy and improbable days during the Games began at the women's diving competition at 8:30 in the morning and concluded at midnight at the boxing

arena. He had moved from diving to swimming and then down the coast
to Long Beach to watch some rowing. "Marco Polo was a piker by comparison," he said of that particular day. "Yet it was all more than worthwhile."

The Games of Los Angeles were an awakening for the American people. Discounting the low-key 1904 Games in St. Louis, an appendage to
the St. Louis World's Fair, this was the first time the Summer Games had
been held on North American soil, and to an America wracked by Depression the Games provided a temporary lifting of the heavy cloud that hung
over daily life. These were also the Games at which the *spectacle* of the
Olympics began to become as much a part of the event as the competition,
as Rice illustrated in his story on the opening ceremonies:

> The gods who sat upon Olympus looked down upon the field of
> Olympia in Greece just 2708 years ago should have been in the Los
> Angeles Stadium this afternoon.
>
> They would have looked upon a sight unbelievable even in the vi-
> sion of ancient gods. They would have seen 105,000 spectators packed
> in the big stadium under the warm and brilliant sunshine of Southern
> California as the picked athletes of the modern world paraded by in all
> the brilliant flare of reds and blues and purples, orange, crimson and
> green shining in a sun that flamed from a blue and cloudless sky. . . .
>
> Only those who looked upon the spectacle today can know what it
> means. One might as well attempt to describe the glory of the Grand
> Canyon or the peak of Mount Everest at dawn. It was something be-
> yond mere words.
>
> It was not the size of the marching platoons that counted. China,
> Colombia, Uruguay had only one lone marcher each under the flag of
> home. Then right after these, the marching legions of Italy, 128 strong
> stirred the crowd with their bright blue coats and white trousers, each
> giving the Fascist salute. . . .
>
> The blended coloring was unbelievable in its rainbow tinting. The
> blues, greens and crimsons continued to flash and flame under the golden
> sunlight until the eye was completely dazzled by the spectacle.
>
> Tomorrow they will remove all this brilliant raiment and go into
> the greatest athletic competition the world has even known in the test
> of speed and power, stamina and brawn, agility and skill. But today was
> something different, the proof that thirty-nine nations could march to-
> gether in good will and friendliness as fifty-seven national flags caught
> the same winds under the same sky in the sportsmanship of a world just
> emerging from the shadows and hatreds. . . .
>
> The Olympic games of 1932 . . . have risen far above all depres-
> sions and economic slumps. They have carved their way through barri-
> cades that halted our financiers and politicians. They are ready to prove
> that the heart of the true sportsman knows no defeat across his path.

If the Games ultimately proved less than what Rice had wished for them in his peroration, what they did prove was Rice's continued enthusiasms and keen eye for a story. In addition to trying to take in nearly every competition of the Games once they were under way, he wandered through the Olympic Village before the Games, in search of the individual who might be the "most interesting" athlete. Among the two thousand "sets of tense human systems now waiting for the starting gun" he found Patrick O'Callaghan of Ireland, a chiseled 240-pounder favored to win the hammer throw but eligible to enter the decathlon and the boxing competition and contemplating doing so. Rice also told his readers of British hurdler Lord David Burghley, "the only athlete in the world who can look at thirty-five Rembrandts hanging in his own home and then take a warm-up jog through 40,000 acres without ever leaving his own domain." In the same piece he introduced his readers to a Canadian sprinter, an Argentine boxer, two Ivy League middle-distance men, and a teenage girl from Texas, a "one-girl track team" named Babe Didrikson. He didn't come right out and say in that first piece who he found to be the most interesting. He didn't have to. His effusive and extensive writing on Didrikson over the next few days left little doubt. He found the Texas teenager to be not only the most interesting athlete at the Games, he placed her among the most interesting and extraordinary athletes he had ever encountered.

Babe Didrikson was an eighteen-year-old clerk-typist at the Employers Casualty of Dallas, though her value to the company stemmed from her contributions to their touring women's basketball and track and field teams. She received a ripple of national attention when she entered eight events in the 1932 Women's National A.A.U. championships in Chicago. She won the javelin, hurdles, shot put, baseball throw, and broad jump; tied for first in the high jump; and took a fourth in the discus to walk away with the team championship for Employers Casualty—all by herself. Still, she came to Los Angeles amidst little fanfare. Rice knew her name, and had heard of her exploits. Still, he was surprised by her impressive athlete's body—the taut, lithe calves and thighs, the broad, muscular shoulders and arms—and by the firm grip of her large hands. But all of it was so natural and in proportion—important characteristics in a time when the thought of a woman athlete still threatened the male order. "She wears no bulging muscles to her wars," Rice felt obliged to note for his readers, as if to reassure them that yes, she is a woman as well as an athlete.

Natural too was Didrikson's poise and insouciance. Meeting Frank Craven along with Rice, she challenged him to a game of golf. Craven hesitated a moment. "If that doesn't suit," said Didrikson, "we'll try boxing or wrestling."

"No, we won't," said Craven, who didn't suspect that the young athlete was setting him up. "You've already broken my right hand just being introduced."

"Well," said Didrikson with a sly grin, "then I'll play dolls with you."

Rice roared. It was the first time he could ever remember the clever Craven being topped in such a repartee. "His mouth flew open but no words came out," said Rice. From that moment on Didrikson was a favorite. He wrote a full article on her before the Games began, and he spoke of her so ebulliently in the press box that his cronies there began referring to Didrikson as "Granny's girl." When she shattered the world record in the javelin by more than eleven feet on her first throw, Granny was exultant: "The Olympic flame in crest of the high tower was outlined even more sharply against a darkening sunset sky as the announcement was made that the famous Babe Didrikson won her first start with a new javelin record."

But Rice's colleagues were unmoved; there was still the tacit prejudice, the innate belief that a women's champion was somehow less of a champion than a men's champion. The ink accorded Didrikson for her gold medal and world record was but a fraction of that received by Eddie Tolan when he tied a world record in the one hundred-meter dash. That particular inequity is understandable, perhaps. Different events have always had different appeal and the women's Olympic javelin throw has never had, nor is it ever likely to have, the cachet of the men's hundred meter. But when Didrikson won her second gold medal, setting another world record in the eighty-meter hurdles four days later, she still wasn't the story. This time even Rice subordinated her achievement to Eddie Tolan's gold medal in the two hundred-meter, *his* second of the Games. She even remained lower-paragraph news on August 7, when she bettered the world record in the high jump with a jump of five feet, five inches, but had to settle for the silver medal when her successful jump at five feet, five and a quarter inches was disallowed—wrongly as it turned out—when judges ruled her head preceded her feet over the bar.

As incredible as it seems from a modern perspective, the sportswriters saw no story in this infusion of tragedy and injustice into Didrikson's already colorful and begging-to-be-written story. They were instead blinded to her story by the surfeit of world records broken that day, and Pavlovian reaction demanded that they all be addressed at the top of the story. Rice too, opened his story with details of the world relay records and the marathon triumph of Argentine Juan Carlos Zabala, but he gave the largest portion of his piece to Didrikson. He talked to Didrikson, analyzed the disqualification, and while he stopped short of saying that the officials were wrong in calling her final jump a foul, he did say they were most certainly wrong in not warning Didrikson earlier, and his article the next day included a most unusual public, in-print questioning of an official.

When Rice spoke with Didrickson that day (this was the day he summoned her from the stadium floor to the press box), it was not the high jump he wanted to talk about. Since meeting Didrikson a week earlier he had imagined her on the golf course. He had coaxed from her a confession that she had hit the ball a few times, could hit it rather long, and was anxious to try the game. The morning after her silver medal in the high jump, Rice had her out at the Brentwood Country Club for a round of golf. He invited Braven Dyer, Westbrook Pegler and Paul Gallico along. "I was almost more excited about [Granny's invitation to play golf] than I had been about the Olympics themselves," said Didrikson.

Both Rice and Didrikson, in their memoirs, insisted that the round of golf at Brentwood was Babe's first. But in reporting the game in the next day's paper, Rice said that she had played about ten times before. Whatever the truth, and the ten-times-before story seems more plausible if less romantic, the golf game at Brentwood demonstrated a couple of things, the most important of which was the Babe's extraordinary athletic gift. Rice didn't count her strokes on the front side, but reported that on the back side she shot a 43, with three, three-putt greens. She hit a half-dozen drives between 240 and 260 yards, he said, "from ten to fifteen yards longer than I ever saw any girl hit a golf ball on any unbaked course where there was no run on the ball over the watered fairways." But this golf match also demonstrated Didrikson's gameness, her will to win, as well as the attitude of some of the nation's male sportswriters, who couldn't accept a female athlete as anything but a freak. The protagonists in this little sub-drama were Didrikson and Paul Gallico.

Rice and Didrikson were standing against Dyer, Pegler and Gallico in a friendly but spirited match. On the sixteenth hole, a par three with a steep down-and-up dip between the tee and green, Rice and Didrikson were two holes up. But both put their tee shots into the trap while Gallico drove the green. Knowing that Gallico was the quintessence of what in later decades would come to be known as male chauvinism, Rice suggested to Babe that she challenge Gallico to a foot race to the green. Gallico, thirty-five years old but still trim and vain about his own athletic wherewithal, accepted. "Paul takes no challenge from any woman and few from men," remembered Rice. "Babe kept two feet in front of him all the way—like Rusty the electric rabbit at a dog track." Gallico collapsed on the green, exhausted. Rice blasted out of the trap and saved par; Gallico, still wheezing from his exertion, four-putted, and Rice and Didrikson took the match. Gallico quietly seethed about his defeat at the hands of a woman for more than five years. Finally, in his 1938 book, *Farewell to Sport,* he struck back, giving voice to a prejudice that Didrikson had to endure throughout her life from sportswriters and fans—the ingrained notion that any woman who excels in athletics is somehow, in some way, less than a whole woman. Gallico charged that beating other women at sports was something that Didrikson had to turn to "simply because she would not or could not compete with women at their own best game—man snatching. It was an escape, a compensation. She would beat them at everything else they tried to do."

There were times in his career when Rice's blind love of sport prevented him from seeing perversions in the game and prompted him to reinforce the stereotype. There were other times, as in the case of Babe Didrikson—whom he continued to see and write about throughout her long career as a golf champion—when his deep-rooted and genuine love of the game caused him instead to be blind to the stereotype, and to celebrate talent that is pure and unique. Babe Didrikson remained a sports page afterthought for most of her career. Though her gifts and achievements as an athlete were the equal of Jim Thorpe's, her celebrity never was. Yet throughout her career, through national championships, an honor acknowledging her as the greatest woman athlete of the half-century, and a courageous and public comeback from cancer in the early fifties, she remained a favorite of Rice. He wrote of her frequently and never had reason to alter the opinion he had voiced when he first met her back in 1932. "It can be said without any extension of the raw truth that the world of sport has never known an athlete even close to her class."

In the abstract, it would have probably humored Rice, and certainly surprised him, to discover that his ingenuous writing on Babe Didrikson constituted something of a real social breakthrough. For in most ways Rice was the consummate "old boy," comfortable in the status quo—certainly he never saw himself as some Quixote, searching out and tilting at the windmills of injustice. But just as he never felt constricted by any accepted consensus of what constituted sport—allowing him to look with as much interest on a polo match as a prizefight—he did not allow himself to close his mind to the possibility that Babe Didrikson could be as appealing a personality, as gifted an athlete, as pleasing a story, and as good a friend as Babe Ruth. His writing on the other prominent women he covered during his career—tennis rivals Suzanne Lenglen and Helen Wills Moody, Olympic swimmer Eleanor Holm Jarrett—is likewise free of any patronizing or stereotyping.

It is, however, a far, far more difficult matter to try to get a fix and put a label on Rice's writing on blacks during this same period. It is certainly not enlightened; it may well be racist. It is, at bottom, complicated and uncomfortable—a veritable mirror of the ugly contradiction that was America in the years of segregation. Rice had never signed on to be a sociologist, but the world of sport was a full two decades ahead of the rest of American society in being forced to confront the reality of blacks and whites drinking from the same cup as equals. Rice's thoughts and writing on the emerging black presence in sport—on Jesse Owens and Joe Louis in particular—give evidence of glib insensitivity and ignorance. But so too does this halting, tentative canon show evidence of confusion and uncertainty, of a man whose senses, and experiences, were shaking and challenging the ingrained prejudices he shared with most Americans of his generation, and had never thought to question before.

When read in the light of contemporary sensibilities, Rice's writing on blacks is patently and sickeningly racist. He routinely referred to Jack Johnson as "The Smoke" and "The Chocolate Champ" and made reference to the hardness of his head. He claimed that Joe Louis's great skills were "a matter of instinct with him, as with most of the great Negro fighters. . . . The great Negro boxer is rarely a matter of manufacture, like many white boxers. He is born that way." At the Berlin Olympics he said "[Jesse] Owens has been like a wild Zulu running amuck." And there was more. In conversations he would frequently refer to Owens or Louis or Jackie Robinson as "the nigger." His All America teams are striking in their whiteness.

Does this make Rice a bigot? Again, as we understand the term today, he was most assuredly a bigot. But in the 1930s so were the rest of the sportswriters. And the newswriters. And the editors and the publishers and the novelists and the poets and the playwrights. So were the athletes and the actors and the fans who came to cheer them. So were the policemen and the shopkeepers and the farmers and the factory owners. So were the teachers who taught the small children, the doctors who tended the sick, the congressmen who made the laws and the jurists who administered justice. We were all bigots. In the 1930s, progress in race relations was measured by the decrease in the number of blacks who were lynched. In 1935 only fifteen blacks were lynched; in 1932 it had been thirty-five. From Maine to Mississippi, from Seattle to Savannah, Americans behaved as if the black man still counted as "three-fifths" of a person, as the Constitution had unfortunately counted a slave. Here and there throughout the period there were people who might have noted the hypocrisy in Boston College leaving its only black player at home when they went to play in the 1941 Sugar Bowl in New Orleans; or the irony of Josh Gibson and Satchel Paige not being in the big leagues. But nobody thought that inappropriate was inconsistent with inevitable.

To measure the behavior of any figure from a past time by contemporary sensibilities is to engage in what historian William Manchester calls "generational chauvinism." We can decry our blindness as a society, but we cannot judge individuals to be ignorant or malevolent simply because they behaved in a way that was wholly appropriate and consistent with the society in which they lived. Grantland Rice's writing on blacks is shameful when viewed from a perspective of more than fifty years later. Less shameful than many; more shameful than some. But so too does it represent mainstream America, mainstream American journalism. As such, it can be served by no apologies or defenses, and deserves damnation less than it demands understanding.

With the possible exception of the nascent art of jazz, sports was the only walk of American life where blacks were crashing the white party in the 1930s. Joe Louis, as the heavyweight champion, and Jesse Owens, as the central figure in the politically and racially charged Olympics in Berlin, were suddenly the two most important athletes in America—black or white. Rice's writing on Louis and Owens was at turns patronizing, gentle and ambiguous. For a modern-day reader it is of little help in understanding either man, or in understanding the struggles they faced or the dreams they dreamed. It is instead a better barometer for seeing the discomfort and confusion that Americans felt as they struggled to be tolerant and accepting, while at the same time crippled by their fears and prejudices.

By the time he won the heavyweight championship in 1937, Joe Louis was getting more ink than any American save perhaps Franklin Roosevelt. He had been in the public eye for more than two years. He was, in the phrase that resonated throughout the time and has echoed down through the years, "a credit to his race." Rice did not originate that phrase (a man named Jack O'Brien was probably the first to use it) though he might have, for his writing carries that tone.

Rice never questioned Louis's right to be in the ring, as a few still did, or veil any such naked bigotry in suggestions that Louis fights were ill-advised on the grounds that they might provoke race riots, as Westbrook Pegler and Hearst columnist Arthur Brisbane did. Rice genuinely liked Louis. He liked the champion's talent and flair; he was comfortable with the man's humility and deference. "Sportsmanship should be the very mortar of an athlete but never an entity in itself for conscious display," said Rice of Louis. "Nobody better exemplified this quality than Joe." Rice's personal relationship with Louis, however, was more complicated than his friendships with most of the athletes he covered. He sat with Louis often at championship fights. Louis called him "Mr. Rice." He called Louis "Joe." They would talk of their respective golf games, and Rice told him to keep his two iron in his bag and use his four wood after Louis admitted to having trouble controlling his long iron. But Rice never invited him out for a round of golf.

Still, for all of this, there is little indication that Rice thought very deeply—or at all—about American race relations or Joe Louis's humanity when he sat down at the typewriter. Like everyone else, he was given to metaphors of the jungle in his writing on blacks. Louis had "the speed of the jungle, the instinctive speed of the wild." In defeating Primo Carnera, he "was stalking Carnera, the mammoth, as the black panther of the jungle stalks its prey." The contemporary eye identifies those immediately as racist stereotypes. But even these lines must be read in context. It is likely they were intended and accepted by his readers as compliments. Rice had, after all, said many of the same things about Dempsey and Tunney.

Rice's writing was also free of the nasty virulence so many of his colleagues indulged in. What follows is a small but representative sampling of the way American journalists saw Joe Louis and the Negro race:

An editorial in the New York *Sun:*

> The American Negro is a natural athlete. The generations of toil in the cotton fields have not obliterated the strength and grace of the African native.

Paul Gallico in the New York *Daily News:*

> Louis, the magnificent animal. He lives like an animal, untouched by externals. He eats. He sleeps. He fights. He is as tawny as an animal and he has an animal's concentration on his prey. Eyes, nostrils, mouth, all jut forward to the prey. One has the impression that even the ears strain forward to catch the sound of danger. He enters the arena with his keepers, and they soothe and fondle him and stroke him and whisper to him and then unleash him. When the leash slips, he fights. He prowls from his corner cruelly and stealthily, the way the lion prowled into the Roman arena when the bars were raised and stood blinking in the light for a moment and then headed for the kill.
>
> He lives like an animal, fights like an animal, has all the cruelty and ferocity of a wild thing. What else dwells with that marvelous tawny, destructive body? The cowardice of an animal? The whipped lion flees. The animal law is self-preservation. Is he all instinct, all animal? Or have a hundred million years left a fold upon his brain? I see in this colored man something so cold, so hard, so cruel that I wonder as to his bravery. Courage in the animal is desperation. Courage in the human is something incalculable and divine. It acquits itself over pain and panic.

When writers tired of the jungle-animal analogies, they reverted to picturing Louis as an indolent and addled, the second of the common racist stereotypes. Bill Corum of the *New York Journal American:*

> He's a big, superbly built Negro youth who was born to listen to jazz music, eat a lot of fried chicken, play ball with the gang on the corner, and never do a lick of heavy work he could escape. The chances are he came by all those inclinations quite naturally.

No one was immune. Even John Kieran of the *Times,* a paradigm of sophistication and enlightenment, whose integrity and sensitivity have never been challenged in any quarter, constantly referred to Louis as "Shuffling Joe" and wrote long columns about his allegedly prodigious sleep habits.

Shirley Povich of the *Washington Post,* who wrote his own share of patronizing copy on Louis, framed the reality of 1930s journalism nicely when looking back on it from the perspective of the 1980s. "To me it was apparent in terms of the times that people insisted that the good thing was a white champion and a bad thing was a black champion," said Povich to Louis biographer Chris Mead. "Simple as that. All people have natural prejudices, this is a prejudice of the American scene."

The irony and difficulty with such prejudices began to show themselves to introspective Americans when Jesse Owens and several of his black teammates stole the Aryan thunder at the Berlin Olympics in 1936. Americans found themselves confronted with an uncomfortable choice between two prejudices—their longstanding racial bias and their continued dislike and distrust of the Germans, their enemy in the Great War and a people who had swiftly and illogically risen from the humiliation of the twenties and were once again goose-stepping across the world stage with an unnerving strut and swagger. For the most part nationalism prevailed over racism. The full extent of the thoroughness and ingenuity of Hitler's evil was still just a suspicion; nonetheless his repugnance was already self-evident, and even those Americans who suffered from a less-virulent strain of his poisonous notion of Aryan superiority found themselves in the unlikely position of cheering for a black man.

Rice was looking forward to the Berlin Games. They were, first of all, to be a vacation. Kit accompanied him to Europe. So did Bob and Mary Jones. Then, the Games themselves held great allure. The festival-like memories of Los Angeles were still warm and acute. Since 1924 he had come to recognize the Olympics for the extraordinary spectacle and story they had become. Because the Olympics were unique in bringing the athletes of the world together, and because they came four years apart, "you remember a little more distinctly what happened in these games than any others. They are stories that write themselves in larger type," he wrote prophetically before the Berlin Games began. Finally, the old soldier in Rice was immensely curious about what had happened in Germany in the eighteen years since the war. He had watched the German people gaze in curious wonder on the American doughboys playing games in 1919. Sport, he had learned then, had been outside the German experience. Now he was going to visit a Germany that was consumed with sport to a degree that they were promising a sporting show that would dwarf and obscure all that had preceded it.

The contrast between Los Angeles and Berlin was readily apparent to Rice—and keenly disappointing too. The distractions from sport began on the *S.S. Manhattan,* the ship carrying to American team to Europe, when Avery Brundage, then the head of the American Olympic Committee and later to head the International Olympic Committee, suspended swimmer

Eleanor Holm Jarrett for breaking training. Jarrett was a beautiful, popular and talented backstroker who had won a gold medal at Los Angeles. She had also won a Hollywood screen test and enormous popularity and renown. Her talent, however, was exceeded by her love of a good time and her contempt for the prissy Brundage and his shipboard rules. "I train on champagne and cigarettes," she happily told reporters. But when she was discovered staggering back to her second-class cabin from the bar in first-class, obviously and delightfully inebriated, the American Olympic Committee called a hasty post-midnight meeting and suspended her.

Rice was not aboard the *Manhattan*—he had sailed on the *Europa* and came to Berlin by way of London—and was thus not privy to the story as it was breaking. His sympathies were clearly with Ms. Jarrett, though they may have been influenced at least in part by his friends in the press; most of Jarrett's revelry aboard the *Manhattan* came in the company of reporters, and coverage of the story had been overwhelmingly on her side. She watched the opening ceremonies from the press box, and she cried when the American team entered the stadium. Rice called her "the most pathetic figure at the Olympic opening," though he was quick to add that her banishment was a result of "the stupidity of American Olympic officials."

"I did nothing really wrong," she told him. "I may have been foolish, but nothing else."

But Jarrett and the commotion over her suspension were but a passing squall in the roil and storm that was pre-Olympics Berlin. It was clear that unlike Los Angeles, these Olympics were not about sport. "I have never seen such a demonstration anywhere at any time before," Rice reported. "The outbreak of national feeling is beyond belief."

The opening ceremonies were held on the twenty-second anniversary of the start of the Great War, and a writer didn't have to be a veteran of the war like Rice to notice that the Games of Berlin were going to be more of an opportunity to showcase German might than German sportsmanship. "Contrary to Kipling, the tumult and the shouting never die, the captains never depart," he reported in his story on the opening ceremonies.

> I passed through more than 700,000 uniforms on my way to the Olympic Stadium—brown shirts, black shirts, black guards, gray-green waves of regular army men and marines—seven massed military miles rivalling the mobilization of August 1, 1914.
> The opening ceremonies of the Eleventh Olympiad, with mile upon mile, wave upon wave of a uniformed pageant looked more like two world wars than the Olympic game.

A tensing and heightening of national feeling were inevitable in such an environment. Especially when the countries that tendered the Nazi sa-

lute—Austria, Italy, France—were greeted with thunderous ovations from
the 110,000 in Olympic Stadium, and the nations whose teams demurred—Great Britain, Australia, the United States—"were . . . marching largely through solid silence." Rice was chilled by the scene in Olympic Stadium when a German athlete triumphed. "It is something to hear
100,000 voices singing 'Deutschland über Alles' as the Swastika flag catches
the wind and 100,000 hands extend in the Nazi salute." He did, however,
find Hitler—conspicuous in his box at the track and field events—more
comical than menacing. "Hitler was rooting for Germany like a Yale
sophomore at a Harvard game."

The Nazi newspapers referred to the blacks on the American Olympic
team (there were sixteen of them) as the "African Auxiliaries." It is, sadly,
rather how Rice and much of America saw them too. "We can beat your
American white men," one German official told Rice, "but not these astonishing Ethiopians." Typical newspaper headlines in the United States
read "NEGRO BREAKS WORLD RECORD" and "NEGROES GIVE
U.S. BIG LEAD IN TRACK AND FIELD." The race of Jesse Owens
and his black teammates was a factor in every story American journalists
wrote; and when the black runners emerged as the dominant athletes of
the Games, writers felt compelled to mention race whenever a white athlete American athlete won a medal, as in "Glenn Hardin, the Mississippi
hurdler, startled the German multitude by proving the United States had
a white man who could win."

Rice reinforced many of the stereotypes and played to the commonly
held prejudices. He continually referred to the success of the American
blacks as "Darktown on parade." He wrote that "America will be okay
until it runs out of African entries. . . . We may have to comb Africa
again for some winners." Another day he concluded his story with a similar thought. "Apparently the race here is to the swift and the black and
sepia are too strong. The white man's burden has broken the white man's
back as far as America is concerned. The United States would be outclassed except for our black-skinned frontal and flanking fire."

This glibness was offset in part by other comments. When Owens set
a world record in the 100-meter trials, running 10.2, Rice's writing was as
neutral as it was graceful. "No European crowd had ever seen such a
combination of blazing speed and effortless smoothness, like something
blown by [a] gale," he said. "You could hear the chorus of gasps as he
left all rivals far behind." After Owens won his long jump gold medal,
his third of the Games, Rice acknowledged that "[t]he weather had been
against him, the crowds have been none too sympathetic, but he has turned
out the finest job ever seen in sport. He will be remembered as the greatest
Olympic artist of all time, including sportsmanship. He stopped to ap-
plaud Don Lash, qualifying in the 5000-meter run, a few seconds before

making his record jump. He apparently was more interested in Lash than in himself." Rice took note of the fact that German newspapers were proffering theories as to why the Aryan race could not defeat the blacks; he dismissed the explanations as "bunk." But then he served up some bunk of his own. He gave voice to a German official in his column who offered this opinion of black athletic success:

> It is the Negro's ability to relax and keep relaxed, plus his inborn sense of rhythm. Games to them are play, while to most others games are largely work. The Negro athlete rarely tightens up or gets too tense. For this reason he is able to produce a greater proportion or percentage of his ability. I watched them at rest around the Olympic village and it seemed to me that most of the time they were dozing or yawning. They are not worn down by too many jumpy nerves.
> Then there is their sense of rhythm, so well known in music.

That was too much even for Rice. "The ease and smoothness of championship is not the property of any one race," he responded. "It is the property of a champion."

Still, for the rest of Rice's life, race remained a matter in which he (and the other sportswriters as well) remained decidedly uncomfortable. His natural awkwardness around the situation was exacerbated by the sad and simple fact that there was not much opportunity to write about black athletes in the thirties and forties. Jesse Owens faded from the spotlight after Berlin, and World War II prevented a successor from emerging in subsequent Olympic Games. Baseball, of course, was completely white, and college football, golf, horse racing and tennis—the other areas where Rice concentrated his energies—were nearly so. Joe Louis remained the single important African-American athlete, and his identity within the newspapers changed little over the years.

For Rice, however, seeing Jesse Owens embraced as he was in Berlin provided testament to the unifying power of sport. "You can't doubt the fact that at Berlin, track and field was a big boom for international sport," he said. "This section of the games was not only loaded up with new world's records, but also with an unusual amount of fellowship and sportsmanship. Losers lost no time congratulating winners and there wasn't even a flare of bad feeling." Jesse Owens's place in this could not be denied. His accommodating nature—decried by more militant blacks during the 1960s—allowed him to be accepted, by his teammates, by the press, by white America. In a competition staged to showcase a society built upon racism, Jesse Owens's inspiring gifts as an athlete, and his humble yet noble bearing as a man, prompted white America to confront their shame. The calcified racist attitudes softened ever so slightly; the threshold of hatred and mistrust was lowered just a little. In temperament and training, Rice was a man much like Owens. His quiet message was the sort that would have had an impact on Rice. "I can't change [the world] with wild words," said Owens in the 1950s, speaking of his attitude and style on civil rights. "But I can bring two people, the other fellow and me, a little bit closer if I am a gentleman. . . . I will listen to him; he will listen to me. Maybe—no guarantee, remember that my friend—but maybe we will part not so far removed in outlook as when we came together."

Just as the Berlin Olympics opened Rice's eyes to some things he might not have had occasion to ponder before they also set him to thinking afresh about matters he had taken up before—the role of sports in the world order. Berlin was a most disquieting competition in this regard. Twelve years before, in Paris, Rice had been rather surprised to discover the crumbling of national identity. Now, in Berlin, national identity was again in the fore. It was all-consuming—to the point where the writers— and undoubtedly some athletes too—could not separate the nationalistic festival from the athletic competition. Winning became the only validation of self and country. Rice cherished winning, and celebrated winning. But he could be unsettled by the cost of winning. Germany had spent $25 million on the Olympics. A *Gleichschaltung* had brought all of German sport under the control of the Nazi party; portions of the German Olympic team had been selected more than a year in advance of the Games and the athletes had spent their time training under state sponsorship. And they had succeeded; the Games were a critical and commercial success and German athletes won more medals than the athletes of any other nation. Now their way threatened to become the way of international sport. Rice reported that Japan was studying German methods in the hopes of emulating German success at their Games in Tokyo in 1940. The Italians were also nationalizing their amateur sport; Olympic officials in the United States and Great Britain were calling for a restructuring of the Games in the light of those developments.

To Rice, this reduced the world of sport to a modern-day Sparta; and made him aware of how badly the world needed its Athens, too. "[T]he human race happens to be developing along competitive lines," he said. "It isn't developing along the line of philosophy painting, sculpture, poetry, statesmanship or other things ranked well up in a world of culture. But it is turning out better and better athletes. . . . "

13

Gathering Twilight

By the late 1930s American sportswriting was a mess. Most of the original generation of eloquence and influence—men like Lardner, Runyon, Broun, McGeehan and Pegler—had either died or moved on to other journalistic and literary endeavors. Most of the coming generation of eloquence and influence—men like Red Smith, Jimmy Cannon, John Lardner, and Dick Young—had either not yet cracked the game or had not yet made their influence felt. The Depression also hampered the credibility of the profession. The ball clubs picked up the tab for the traveling reporters who accompanied them on road trips. Reporters were acutely aware that their newspapers did not have the money or the inclination to assume these expenses should the club's largesse suddenly cease. Such a development would put a very comfortable and hard-to-come-by job in jeopardy. Sportswriters in the thirties had a damn good life and knew it.

"Look at 'em," said Toots Shor to Rice one night as he took in the regiment of sportswriters frolicking in his saloon. "There's not a millionaire in the bunch. They just live like 'em." Reporters thus tread very lightly upon any stories that might compromise their exalted existence—any stories, in other words, that were not in a ball club's best interest.

Exerting pressure at the other extreme, however, was a newspaper's

255

eternal need for stories that might give circulation a jolt. This, of course, was not a product of the Depression; it was merely made more acute by the crunch for cash. Sports, with its immediacy and popularity amongst readers, put a premium on getting it first and getting it exclusive. Getting it right was too often a premium to which editors and undiscriminating readers were happily indifferent.

Sports-page readers in the thirties were thus left with liars, syco-phants, pretenders and imitators. Mostly imitators, since mediocrity is generally born in the vacuum of imagination. Most imitators, as West-brook Pegler pointed out, made a "horrible mess of it."

"The originals have their virtues, but the imitators always are stuff to make men curse and weep," said Pegler. "And the plague is never local-ized but spreads out over the country, so that very soon there are a thou-sand flabby counterfeits appearing in the dailies under the names of young aspirants who would do much better if they would just remember to be themselves."

Imitators generally fashioned themselves after one of two men—Da-mon Runyon or Grantland Rice. Runyon was still very much a newspa-perman, he remained in the employ of William Randolph Hearst, despite the fact that his short stories were among the most widely read in America and the movies that were made from them—films as disparate as Edward G. Robinson's *Little Caesar* and Shirley Temple's *Little Miss Marker*—were making him thousands of dollars per week. Runyon's biographer Jimmy Breslin said that however wealthy his fiction may have made him, Run-yon never felt secure enough in his writing to go off the Hearst payroll. "The payroll is a more important document than the Bible," he said.

Runyon's stories were of the sports underworld—crooked fight man-agers and horse trainers, bootleggers and bookmakers out to corrupt a baseball game, or buy the team if they couldn't. Runyon was a star, a figure of the Broadway he created, not just its chronicler. Young writers just in from the provinces, who envisioned themselves wearing custom-made suits, cashing checks with three and four zeros before the decimal point and sharing a table at Lindy's with Winchell and a bevy of showgirls, began emulating Runyon's street slang and rat-a-tat-tat narrative style. Moreover they tried to cast the players, owners, and hangers-on of the sports world into one of the Runyon wise-guy molds. At best the product was labored, at worst an outright deceit.

The Rice imitators were likewise guilty of committing mayhem on American newspaper readers. Like Runyon, Rice was a celebrity to be envied and emulated. The acolytes in the Rice chapel saw themselves golf-ing with Babe Ruth and wintering in Los Angeles and Florida. So they wrote as they perceived Rice to write, the more ornate the better. If one metaphor was a song, then a dozen metaphors made a symphony. If Rice became larger than life by making the athletes so, then what was wrong with comparing the winners in a high school track meet to the figures in Greek mythology, or giving a college halfback who enjoyed a good day the qualities of a medieval warrior? "Most [of Rice's imitators] get tigers, lions, wildcats, horsemen, crimson tides and yellow perils all gumbled up with purple sunsets," said *New York Herald Tribune* sports editor Stanley Woodward. The result was writing so dense as to be opaque, and it missed reality by as wide a mark as the Runyon-inspired fiction.

Sportswriters also stole from one another—and did it with an efficiency and an absence of conscience that could have earned the envy of a gangster. This gave rise to another problem with Depression era sportswriting, one that persists to this day. Sportswriters—all journalists, really—are swift to attach labels to what they have witnessed or perceived. A fresh or particularly compelling label—never mind if it's accurate or not—can spread through the industry like a cold through an elementary school. Rice was as guilty as anyone. But he was also fond of pointing out that writers were less divine in their inspiration that they might like their readers to believe, and that perception was not always reality:

> He took his turn half-heartedly, outlining an excuse.
> He figured he was beaten—so he couldn't see the use.
> But when he made his little play, it took a lucky swerve,
> A sudden, unexpected hop—a title-winning curve—
> And straightaway they exclaimed about his "courage" and his "nerve."
>
> He started in with bulldog jaw to make a winning fight.
> He started in to see it through, as any stalwart might.
> But when he cut in with his play, it took a hard-luck bound,
> And caromed as it shouldn't have on any sort of ground,
> An so they rose and branded him a "quitter" and a "hound."
>
> Yes, courage is a fancy word that gives us all a brace,
> And yellow is another term we splash about the place.
> But there are things behind the scenes that none of us can see,
> An edict from the gods of chance, whoever they may be,
> Who set the score—and laugh aloud at our philosophy.

All of the imitation tended to make Rice stand even taller above the crowd, and made readers treasure him all the more. A 1938 editorial in the *Dayton* [Ohio] *Daily News* called him the "King of Sports."

"He writes neither 'down' nor 'up' to his readers," it said. "He realizes that among sports fans are the best minds in the country and his style and phrase clearly indicate that in his concept of things the man in the shop doesn't want to be talked to in the language of the barroom. No sportswriter had the literary firmament that Rice possesses. He never makes use of it unless it fits perfectly into his word picture. He is nothing of the pedant; he never attempts to display the real extent of his literary wares."

As Rice's friend and erstwhile colleague Pegler explained it, it came down to a matter of Rice continuing to write for his readers, while nearly everybody else was writing for themselves and their careers.

"Many of our sport journalists have lost interest in sport and taken to night life," said Pegler in a treatise on the subject in 1935,

> and . . . as a result of that, they don't know enough about sport and sportsmen any more to permit even a plausible pretense. . . .
>
> Mr. Rice seemed old fashioned for a while when the vogue for underworld sport copy was coming along, because he continued to write about forward passes, the new tackles at Illinois, Tommy Armour's irons, the swimmers, archers, boxers and tennis athletes strictly as athletes, and let the night life of sport alone. But there is no man in the country now who knows one-tenth as many actual players, past and present, and their records, styles and prospects.

There were literally dozens of such tributes to Rice in newspapers and magazines of the late 1930s. Taken cumulatively, they testify to more than simply Rice's continuing popularity. For they carry a retrospective cast, an unconscious, unknowing acknowledgment that while Rice's career might remain in full flower, his era was beginning to pass. He was nearly sixty, an old man in a young man's world.

There were times when Rice himself must certainly have felt the crush of the encroaching years. He watched with puzzlement as Lou Gerhig stumbled sadly through the opening weeks of the 1939 season before removing himself from the lineup in May. Along with everyone else, throughout that summer Rice refused to believe that Gerhig's affliction was as serious as the prognosis seemed to suggest. But returning from the World Series in Cincinnati aboard the Yankee train that October, playing bridge, as usual, with Gerhig as his partner, Rice was stripped of any illusions and hopes he may have been harboring. Gerhig couldn't even deal the cards. His dexterity had deteriorated to the point where he was forced to clutch the cards with such ferocity that they became bent and destroyed. Rice and the others said nothing; they simply opened a new deck after each of Gerhig's deals, and looked sadly at one another, each man knowing the horror he saw deep in the eyes across the table was a mirror of his own.

Dear friends and colleagues were dying—in a three-year span in the mid-thirties, some eight or ten of Rice's closest and most long-standing friends in the newspaper business died, including, Lardner, McGeehan, and Don Marquis. Rice's best verse during these years returned more frequently, and with more fervor, to the exceedingly fragile and temporary qualities of life:

Knowing how swiftly all the years roll by,
Where dawn and sunset blend in one brief sky.

Rice and some of his friends were asked to be pallbearers so often during these years that it prompted a bit of black humor on their part. Keyed by ex-Hearst sportswriter and Hollywood scenarist Gene Fowler, Rice, together with Gene Buck, Jack Dempsey, W. C. Fields, Rube Goldberg, Herbert Bayard Swope, Bill Corum, Toots Shor and some others formed The Pallbearers Association. "Never Stand a Friend Up" was Fowler's motto. Rice, said Fowler, was an indispensable member of the association—"a pro long known for his military manner and his ability to look straight ahead and not lose his stride when some amateur treads upon his heel."

When Don Marquis died in 1938, Rice wrote one of his best-received and most frequently reprinted pieces of verse. Marquis had shared Rice's coming of age in Atlanta more than thirty-five years before, and his passing hit him harder than perhaps even Lardner's had. The verse was composed for a Writers and Artists function. But he also sent it out to his syndicate papers, where it received wide play. It was called "Via Charon, the Ancient Boatman." It is a simple, tender, clarion-clear testament to friendship, always Rice's dearest treasure. Not a verse likely to move its readers to tears, it nonetheless evokes the subtle and continuing ache that becomes forever a part of us following the loss of a friend.

There are too many gaps in the ranks I knew
When the ranks I knew were young.
When the roll is called, there are still too few
Who answer "Here!" when the call is due—
There are too many songs unsung.
But Charon's boat is a busy barque—
And the dock gets closer as the dusk grows dark.

Pilot—who looks to your river trade
Where the shadowy Styx rolls by
You've taken your pick for the mystic glade,
Lardner—McGeehan—and Hammond's shade
Drifts through a starless sky—
And somewhere deep in the reedy tarn—
Boze Bulger is spinning another yarn.

Charon—answer me this today—
From all the world's corrals
Why do you always look *my* way?
I'm not worried about *your* play—
But why do you pick my pals?

From the Inn we knew where the flagons foam,
One by one you have called them home.

One by one, on a mist-blown eve
Wearing your ghostly hood,
I've seen you plucking them by the sleeve,
Telling them each it was time to leave,
Just as the shot got good.
With a lifted glass, as I looked about,
I've seem them leave as the tide rolled out.

Charon—I'm sorry I failed the test—
You're not the one to blame—
You picked the brightest—you picked the best—
You carried them off to a dreamless rest
That towers above all fame.
Don—Odd—Percy—and Bill and Ring—
No wonder the angels soar and sing!

Listen, Pilot, the last of all
Who knows where the journey ends—
When you have come to the final call
Where the candle flutters against the wall,
Kindly forget my friends.
For friends are all that a little earth
Has yet to give that is any worth.

Fame and Gold? They are less than dust—
Less than April song—
They are less than weeds in the earth's dull crust
When a friendly hand in you own is thrust
And an old mate comes along.
But dock lights flame with a sudden flare—
And Charon beckons—and who is there?

The Flame of the Inn is dim tonight—
Too many vacant chairs—
The sun has lost too much of its light—
Too many songs have taken flight—
Too many ghosts on the stairs—
Charon—here's to you—as man against man—
I wish I could pick 'em the way you can.

"Via Charon, the Ancient Boatman" was included in *Only the Brave,* Rice's fourth volume of verse, published by A. S. Barnes in June of 1941. As with his earlier volumes of verse, *Only the Brave* enjoyed a modestly successful sale and received favorable reviews, though the reviews were fewer in number than they had been for either *Songs of the Stalwart* in 1917 or *Songs of the Open* in 1924. Most of the verse in *Only the Brave* had originally appeared in his column. And some of it, such as "Alumnus Football"—the poem that contained the "how you played the Game" couplet—went back quite a ways.

On December 7, 1941, Rice was on a train to Los Angeles when the Japanese bombed Pearl Harbor, and America entered the war. When he arrived in California on the ninth, he found a population that was not at all sure that the Japanese fleet was not just off the coast and attack imminent. The idea of large numbers of civilians gathering in any one place was immediately seen as illogical by nervous Californians. So the Rose Bowl game was moved to Durham, North Carolina; the Santa Anita racetrack closed.

The national mobilization, and the suddenness of it, immediately put the status of everything that did not directly contribute to the war effort into question. As was the case in 1917, sports were a very visible bit of frivolity. "The future of sport is beyond anyone's guess," said Rice on Christmas Eve. But that didn't stop people from guessing. Rice's own guess was that "sport as spectacle" was doomed, while sport as physical conditioner was essential. "What we need today is not less sport, but more sport," he said.

> By that I don't mean more big sporting spectacles. I mean more people playing games that will bring better health and better physical condition. Those between the ages of twenty and twenty-eight who have drawn the benefits of athletic training, should be giving this advantage to some form of the service. . . .
>
> Outside of those called to action there should be more older men playing golf, more younger people playing football or boxing, or any game that will help develop legs and bodies.

Rice had some support for this position from an Army spokesman he sought out in the days just after Pearl Harbor. "We hope that all those not actively engaged in our service or war will take up any sport suitable for physical fitness," said the unnamed officer. "But there is no physical fitness to be brought about in massing from 50,000 to 90,000 spectators who are out to see others play some game."

Major league baseball was the key to the future of spectator sports during the war. If baseball closed up shop then professional football, college football, horse racing and boxing didn't have a chance. Sportswriters and baseball men—carefully framing their comments in the noble notion that baseball was patriotic and thus vital to American morale—urged Commissioner Kenesaw Landis to lobby the President on behalf of baseball as an "essential industry." Landis refused, probably as much out of his personal and political antipathy for Roosevelt as out of any conviction that the game was indeed expendable. Rice, never very close to Landis, was fully in agreement with him on his decision not to beseech Washington for any special favors. "It might be mentioned that the judge has done a smart job in handling the war baseball situation," wrote Rice. "He has made no claim of any sort that baseball is an essential factor or that it is indeed a morale maker. He has left that part of it to the government. He has asked no favors in any form, or made any suggestion, except along the line of giving all possible aid in any direction."

When Landis refused to approach Roosevelt, Washington Senators owner Clark Griffith, a long-time acquaintance of the President (if not actually a very close friend), did. And Roosevelt settled the matter of spectator sports continuing early and decisively, in a letter to Landis in 1942 that made headlines across the country. In his letter, Roosevelt not only gave Landis the okay to continue the game, he made it seem as though the commissioner would be a traitor if he did suspend play:

> I honestly feel that it would be best for the country to keep baseball going. There will be fewer people unemployed and everybody will work longer hours and harder than ever before. And that means that they ought to have a chance for recreation and for taking their minds off their work even more than before.
>
> Baseball provides a recreation which does not last over two hours or two hours and a half, and which can be got for very little cost. And, incidentally, I hope that night games can be extended because it gives an opportunity to the day shift to see a game occasionally.

But even when it was settled it wasn't really settled. There remained the touchy matter of the ball players draft status. While urging the game to go one, Roosevelt did not go so far as to proclaim it an "essential industry," which would have exempted players from the draft. Landis, to Rice's constant applause, steadfastly refused to request essential industry status. Which meant, as Rice noted approvingly, that draft "classifications will be changed abruptly and on an extended scale." Early volunteers such as Bob Feller, Hank Greenberg and Ted Williams came in for hearty salutes in Rice's column.

Rice followed the war news closely, but his own life changed little in the first weeks of war. On December 26, he left California for New Orleans and the Fordham-Missouri Sugar Bowl. He returned to Los Angeles immediately after the game and spent the month of January in California. He attended the Bing Crosby Pro-Am, in those years played in Del Mar, just north of San Diego, and "quaffed a beaker together" with the host, as the host so playfully put it. He also showed Babe Ruth around town when Ruth came out to take part in the filming of *Pride of the Yankees,* the Samuel Goldwyn film on Lou Gerhig's life. As the weeks passed and the fear of Japanese attack eased, life began to return to normal on the coast. Rice resumed his golf, and though Santa Anita remained closed, he would now and again go out and pass a few hours at the empty track, talking to the trainers and watching the horses work out. In early February he left Los Angeles for Florida, stopping off in Arkansas for a few days of hunting with Yankee catcher Bill Dickey.

When he got back to New York in May, he turned to those matters on his desk that might allow him in some small way to contribute to the war effort. He was much sought-after—for his energies and willingness to roll up his sleeves and do some work no less than for his name and his contacts—by the various organizations and charities sponsoring war-relief fund-raisers. He served on a dozen or more such committees during the war, committees that raised hundreds of thousands of dollars for war-relief projects. Yet none of this pro bono work said as much about the man's sense of right and wrong as did his actions on the one wartime charity he walked away from.

In the summer of 1942, Rice was asked by Secretary of War Henry Stimson to serve as president of the Army Emergency Fund. The proceeds from this fund were destined for Army personnel who needed emergency money faster than the Army or the Red Cross might normally be able to provide it. Rice was involved because the money was to come from a series of athletic contests. Part of it would come from football games between Army personnel—coached by Major Wallace Wade, ex of Duke and Alabama, and Colonel Robert Neyland, ex of Tennessee—and National Football League teams. The rest of the money was to come from a fight between heavyweight champion, Pvt. Joe Louis, and challenger, Pvt. Billy Conn. The football games raised some $230,000, short of the goal of half a million but an impressive collection plate nonetheless. Army officials hoped the Louis-Conn fight would add a million dollars to the coffers. But people got greedy. And the fight never happened.

Louis and Conn had a dandy of a fight in June of 1941, before 55,000 fans at the Polo Grounds. For twelve rounds the agile Conn danced away from Louis's heavy blows and landed punch after deftly placed punch himself. While these blows never really hurt the champion they nevertheless had Conn clearly ahead on points. In the thirteenth round, however, the challenger from Pittsburgh tried to press his advantage, looking for the knockout. Louis's punches finally connected; Conn absorbed a lifetime's punishment in those three short minutes, and fell to the canvas twelve seconds before the bell.

The public immediately began clamoring for a rematch, but the start of the war intervened. Now it seemed as if the war would prove the catalyst for the rematch. Louis had fought two bouts for charity since Pearl Harbor, one for the Navy Relief Fund in January of 1942; another for the Army Relief Fund in March, after his enlistment in the Army.

The stakes on the Conn fight were much higher, however. The gate was projected at more than a million dollars. Mike Jacobs, the fight impresario who handled Louis's fights, was put in charge of arrangements. Louis owed Mike Jacobs almost $60,000. Conn owed him some $35,000. In addition, Louis owed his manager John Roxborough—then serving a jail sentence for his connection with a numbers racket and bribery—$41,000—and he owed the IRS a whopping $117,000. As the fighters made their preparations for the fight, a well-meaning Army officer saw a way to help Joe Louis out. What if Louis and Conn each took some money off the top, this officer asked Mike Jacobs—equal to the amount of their debts to Jacobs, with Louis getting an additional $41,000 to cover his debt to Roxborough? There would still be plenty of money left over for the Army Relief Fund and the Army would be doing something for Louis—who had helped their charities and their image immensely since his enlistment earlier that year. (Neither Louis nor Conn, it must be said, had anything to do with instigating the arrangements that would have allowed them to erase their debts to Jacobs, and both men remained perfectly willing to fight for free, and reiterated their willingness to do so after details of the financial scam became public.)

The deal turned Rice's stomach. Louis and Conn were not private citizens. They were soldiers presumably in the service of their country, soldiers who were being given time off from their service responsibilities to train for this fight. Why should they profit by it? Why should the Army Emergency Fund lose money so that Louis and Conn might settle with Mike Jacobs? He was helpless to do anything about it, though. So he made the only move that his conscience and sensibilities allowed him. He wrote to Henry Stimson and resigned from the committee overseeing the fight, explaining his reasons.

Rice's resignation alone was not enough to scuttle the fight. The War Department asked John Kieran, who had been serving as Rice's vice president, to take over the committee. Kieran, rationalizing that the Army Emergency Fund still stood to profit by three-quarters of a million dollars, agreed. He later admitted that he had been a "sap" to do so. For sportswriters soon caught wind of the adjusted financial arrangements and they became a big story. (Though Rice never wrote a word about either his disgust at the unseemly arrangement or his reasons for walking away from the committee.) Congressman Donald O'Toole of New York, in calling for a Federal inquiry, railed against the hypocrisy of the deal. "What a mockery we are making of our Army and of charity," he said. "Why cannot the doctor or other professional man who is in the service obtain similar leave to pay his obligations? There can be no scolding of the civilian for lack of morale when the Army itself fails to recognize the seriousness of the war." The *Philadelphia Record* opened a scathing editorial with a mocking: "My country 'tis of *me*." In late September, Henry Stimson ordered the fight called off.

There was a bit of talk in 1943 about appointing a three-man commission to advise President Roosevelt on matters involving sport. The talk grew out of a call in the House of Representatives by Wisconsin congressman Lavern Dilweg for just such a committee. The commission never came to pass; it was not at all necessary, and discussion of its formation was probably a delusion on the part of sports people that the status of the industry was something that Franklin Roosevelt actually worried about. Nevertheless, as the concept was being kicked around, Arnold Gingrich of *Esquire* magazine took it upon himself to run a poll to determine who might serve on such a commission. He wired the nation's sports editors, sportscasters, athletic directors and club owners and asked their opinion. There were no nominees or endorsements from Gingrich, his correspondents were left to fashion a committee out of whole cloth.

Rice was the hands-down winner, in a wholly meaningless balloting that is important only insofar as it relates to Rice's status in sport at the time. He was the named on 27 percent of the several hundred telegrams received by Gingrich. Western Athletic Conference director John L. Griffith was second, named on 18 percent. Baseball commissioner Kenesaw Mountain Landis was third, named on 17 percent. Others receiving votes included former Postmaster General and Democratic party honcho James A. Farley, Branch Rickey, Ford Frick, Larry MacPhail, Clark Griffith, Jack Dempsey and Babe Ruth. The only other sportswriters receiving votes were Bill Corum of the Hearst papers, J. G. Taylor Spink of the *Sporting News* and Arch Ward of the *Chicago Tribune*.

While the games continued, the war was never far from Rice's thoughts or his column. Even when he was attending to business there were acknowledgments to the war. In announcing his 1944 All-America team he claimed it as "one of the strongest ever chosen, . . . even if you go for comparison as far back as Walter Camp's first selection in 1889." He could say nothing else, given the fact that two men each from West Point and Annapolis had made the team. He would have seen it as well-nigh treasonous to suggest that the service academy stars had shone brightly only because their constellation was of a dimmer luster.

Though his sense of valor had not changed since the Great War, time and his experiences in France had stripped Rice of any naive notions of romance attached to the fighting of a war. In early 1942, when a newspa-

per article called Rice's attention to the fact that one of the more romantic
figures of his age was set to enlist at the age of forty-three, Rice decided
to pay him a call to dissuade him.

Tommy Hitchcock had been Scott Fitzgerald's model for Tom Bu-
chanan in *The Great Gatsby*. He had flown with the Lafayette Escadrille
during the Great War, was shot down and taken prisoner briefly before
escaping through the window of a moving train. He returned to America
with a *Croix de guerre* and a considerable reputation for pluck. Rice first
met him during the twenties, when he was to the sport of polo what Ruth
was to baseball or Jones to golf—a breakthrough athlete who through the
force of his personality and talent transformed a game. Rice had covered
polo with some zest since arriving in New York in 1911, but he came to
know Hitchcock well in a series of charity softball games arranged by
Teddy Roosevelt, Jr., in Oyster Bay. Rice loved the abandon with which
Hitchcock had played the game of polo. He had once seen him knocked
unconscious during a match, suffer a concussion, and then return against
doctor's orders for the next match. It was the same abandon that had
Hitchcock looking to resume his career as a flier in 1942. Rice saw it as
move more foolish than noble and made an appointment with Hitchcock
to tell him so. "There was no particular reason for my [visit] except my
admiration and affection for a man I thought was going off to die need-
lessly," said Rice.

"Well," said Hitchcock in Rice's account of the visit, "I'm glad to see
you. But what's the story?"

"I understand you are going back in the flying corps."

"That's right. What's wrong with that?

"Nothing," said Rice, "except you've done your part and you're forty-
three years old. I just happen to like you. I don't want to see you commit
suicide."

Rice reported Hitchcock smiling at his friend's concern. "I'm only
forty-three," he replied. "That is still young. I feel sure I am as young as
a man of twenty-four. I am in fine condition. My reflexes are perfect, I
can play polo as well as I could ten or twelve years ago—or twenty years
ago!"

"You broke in as a kid," protested Rice. "This is something else again.
Don't you think you've done enough for your country?"

At this point, Rice's description of the scene takes on the cast of pure
Hollywood melodrama. Hitchcock gazed at Rice for a long time. Finally,
he replied. "Can you ever do enough for your country?" he asked.

Rice had no hesitation in his reply. "Yes," he said, "when you have
passed the peak of usefulness."

The suggestion that he was over the hill, incapable of meeting this or
any challenge, incensed Hitchcock. "I haven't even reached that peak," he
said—"rather violently" as Rice recalled it. "I'm just coming to my peak."

As quickly as it had risen, Hitchcock's displeasure passed, and he and Rice parted with a warm handshake. "What difference does it make—now or a little later," he said to Rice as Rice was leaving. "I'm not worried. Don't you." Rice remembered the smile on the athlete's face as warm and genuine as he took his leave.

Had Hitchcock returned from the war a hero, or even had he merely survived, Rice's memory of his final meeting with him would never have been so poignant or acute. But in April of 1943, Hitchcock's P-51 Mustang crashed during a test flight. For Rice, his death was made all the sadder by Rice's notion that it was inevitable, and by Hitchcock's seeming indifference to it. Yet when he wrote of Hitchcock he celebrated the athlete's fearlessness, quoting Shakespeare's lines from *Julius Caesar:* "Cowards die many times before their deaths; The valiant never taste of death but once," lines which had served as an epigraph to a verse of his own that had appeared in *Only the Brave,* called "To the Last of All":

Whether it's Heaven or whether it's Hell
Or whether it's merely sleep;
Or whether it's Something in Between
Where ghosts of the half-gods creep—

Since it comes but once—and it comes to all—
On the one fixed, certain date—
Why drink of the dregs till the Cup arrives
On the gray day set by Fate?

Is life so dear—are dreams so sure?
Are love and strife so strong,
That one should shrink from the fated step
To a road that is new and long?

The soul—the grave—and the after-trail—
The Mystic River's flow—
How have the living earned their guess
Where only the dead may know?

Who is there left to raise a hand
And send his will to God,
That he should live where others know
The song of spade and clod?

One by one till the line has passed—
The gutter-born—and the crown—
So what is a day—or a year or two—
Since the answer's written down?

What is a day to a million years
When the last winds sound their call?

So here's to the days that rest between—
And here's to the last of all.

Seen from the perspective of nearly a half-century, the end of World War Two is more than simply the close of a chapter in world history. For America, it is a societal demarcation. "Postwar" is an appellation applied to everything from literature to housing stock to an entire generation of Americans, the baby-boomers. The years after the war were the time when forces in American life that were hastened by the war—most notably integration—and forces that were delayed by the war—television, for one—began to impact our lives. Nowhere was the change, the postwar quickening of American society more evident that in the sports world. Games became integrated—and televised. The sixteen teams and ten cities that had comprised baseball since 1901 began to shift and change. The minor leagues withered. Professional football displaced college football as the focus of the fans' world. Basketball began an improbable rise to major-sport status. Television gave boxing such wide spread exposure and popularity that it killed the boxing clubs and damn-near killed the sport.

Nobody embodied the old world, the slower, trolley-car-and-telegram, prewar world more than Grantland Rice did. He lost his dominance; he lost some of his visibility; he lost a large measure of his boundless energy; he lost a lot of his relevance. But he never lost his spirit, his charm, his love of being around press box and a locker room. And to his readers, he never lost his appeal. Entering his sixth decade as a sportswriter he remained surprisingly unjaded. A bit of the child's wonder, a part of the young boy from Nashville, lived with him forever.

What if the drifting years slip by
In grim parade along time's chart,
Or ghosts of winter blur the sky
If April lingers in the heart?

Frank Craven had died in 1945 and Bruce Barton and Bud Kelland curtailed their traveling. Rice's first postwar task was thus to find himself new traveling companions. He found one in his long-time friend and colleague from the *Sun*, Frank Graham, and another in Red Smith, who at the age of forty had finally escaped from the wilderness of Philadelphia and cracked the New York market as a columnist with the *Herald Tribune*.

Rice had known Frank Graham since 1917, when Graham was just a kid that John McGraw thought he could bully. Graham was one of the reporters caught in McGraw's sights when the manager tried to slither out of the lambasting he had given National League president John Tener by saying the reporters who had printed it were liars. Graham, in his first season covering the Giants, had just moved up to the first string of the *Sun*'s sportswriters. Rice, after rebuking McGraw and defending his colleagues strongly in his column, made it a point to seek out the young Graham, reiterate his support and offer encouragement.

Rice first met Red Smith when he and Graham drove up to the Vinoy Park Hotel in St. Petersburg in March of 1946. Rice had been sitting on the front porch, resplendent in a white suit and white hat, awaiting their arrival. His greeting was a robust and warm-hearted: "Where you two little bastards been?"

Frank Graham is a writer who deserves better from history than the footnote to which history has consigned him. He never lived the life of Runyon or Lardner, never had the salary or fame of Gallico, never had the reach or influence of Rice, never had the cell of devoted defenders celebrating his literary style as Smith had. But as a sportswriter, as a newspaper columnist, he just may have been better than any of them.

Graham was thirteen years younger and a head shorter than Rice. When he and Rice began their travels together after the war he possessed the settled body of middle age, close-cropped black hair above a high forehead, and glasses with thick black frames. He was now a columnist for Hearst's *Journal-American,* but for twenty-five years, until 1943, he had been an anchor of the *Sun*'s vital and vibrant sports department. During the 1930s the *Sun*'s sports section ran on facing pages—even and odd. Graham's column, "Setting the Pace," would run down the left-hand side of the even-number, or lead page of the sports section. Rice's column, "The Sportlight," would run down the left-hand side of the facing page.

Graham had spent nearly twenty years as a reporter before taking over the *Sun*'s lead column in 1934. The protracted apprenticeship left him finely tuned powers of observation and a keen ear for the nuance of the language, the economy of expression he found in the dugout and the locker room. His training as a reporter also left him disinclined to proffer windy judgments and homilies to his readers. The result was a decidedly fresh approach to newspaper journalism. It became known as the "conversation piece," and if Graham did not invent it, he most certainly did it—as Red Smith said—"incomparably better than it ever was done before, or had been done since." He did his reporting without ever taking a note, at least not while he was carrying on his conversations. As unassuming in his bearing and mien as he was physically, Graham's greatness was helped by what one friend called his "psychopathic modesty."

"My job," Graham told himself, "is to take the reader behind the scene where the ticket doesn't admit him—into the dugout and clubhouse, the football locker rooms, the jockeys' quarters, the fighter's dressing room—and let him see what goes on there and hear what is said."

Here is an example of Frank Graham at his very best, writing on a

Yankees-Dodgers exhibition game just before the start of the 1939 season, giving dimension and texture to the bafflement that all of baseball felt at Lou Gerhig's vanished proficiency:

> The day was gray and cold and the crowd was small, and in the dugout Joe McCarthy sat with his hands in the pockets of his windbreaker.
> "How's Gerhig?" the reporter asked.
> McCarthy shook his head.
> "I see he hit two home runs in Norfolk."
> "He looked a little better there."
> "Two home runs and two singles," the reporter said.
> "The singles were all right," McCarthy said. "The home runs were fly balls over a short right field fence."
> The Yankees were going out for fielding practice
> "Watch Lou," McCarthy said.
> Gerhig looked very bad. He would go down for a ground ball hit straight at him and the ball would go through him. Or he would come up with the ball and throw it to second base and then start for first base to take a return throw, but he would be woefully slow. Back of first base, some fans jeered at him.
> "Why don't you give yourself up?" one of them yelled. "What do you want McCarthy to do, burn that uniform off you?"
> The reporter turned to McCarthy.
> "He looks worse than I thought he would," he said. "What's the matter with him?"
> "I don't know."
> "Are you going to open the season with him?"
> "Yes."

Anyone who has read an American sports page in the second half of the twentieth century has felt the influence of Red Smith. No American newspaper writer—hell, no American writer period—ever took greater pride in the craft than did Walter Wellesley Smith of Green Bay, Wisconsin. Slight and professorial in appearance, with wire-rimmed glasses and tweed jackets, Smith was the ultimate craftsman, a relentless student of writing. He once quibbled with a passage from *The Great Gatsby*, a reference to "a deft blow." "Didn't sound real to me," said Smith. "If a guy is mad he doesn't hit someone deftly." This quest for precision marked his own work too, and the result was a wonderful tapestry of language, metaphor, humor, observation and insight that made a morning's romp through a Red Smith column a singular delight for the reader. Smith was the man who told us that "rooting for the Yankees was like rooting for U.S. Steel." A particularly graceful outfielder, in Smith's particularly graceful observation, "stayed aloft so long that he looked like an empty uniform hanging in its locker." When he found sports fatuous, Smith's sense of irony had the edge of a stropped straight razor. When a Senate Investigating Committee hearing testimony on baseball's reserve clause indulged in what Smith perceived as some naked grandstanding, he noted: "In an effort to avoid publicity the group started by interviewing that celebrated authority on constitutional law, Mr. Ty Cobb."

In a television age Smith and the disciples he influenced and spawned kept newspaper sports pages relevant beyond the agate. His columns were studied in writing classes at Columbia and Yale. He numbered Ernest Hemingway amongst his most appreciative fans, and Hemingway paid him a most elegant compliment in his novel *Across the River and into the Trees,* when the hero, Colonel Richard Cantwell, refuses to be distracted from his copy of the *International Herald Tribune*: "He was reading Red Smith and he liked him very much."

Any writer, regardless of where or in what form he practices the craft, strives to attain a recognizable *voice,* and no voice in our time has been as distinctive, as resonant, as consistent as Red Smith's. It had taken him twenty years to get to New York, but once there it took no time at all for readers to recognize his greatness. By the end of the 1940s he was the second most widely syndicated sports columnist in America; only Rice appeared in more papers. Here, from 1948, is a striking example of the sort of work that was bringing Smith this audience and its acclaim. It was among the pieces studied in that Yale writing class:

At the Derby, Walter Haight, a well-fed horse author from Washington, told it this way:

There's this horse player and he can't win a bet. He's got patches in his pants from the way even odds-on favorites run up the alley when he's backing them and the slump goes on until he's utterly desperate. He's ready to listen to any advice when a friend tells him: "No wonder you don't have any luck, you don't live right. Nobody could do any good the way you live. Why you don't even go to church. Why don't you get yourself straightened out and try to be a decent citizen and just see then if things don't get a lot better for you?"

Now the guy has never exactly liked to bother heaven with his troubles. Isn't even sure whether they have horse racing up there and would understand his difficulties. But he's reached a state where steps simply have to be taken. So the next day being Sunday, he does go to church and sits attentively through the whole service and joins in the hymn-singing and says "Amen" at the proper times and puts his buck on the collection plate.

All that night he lies awake waiting for a sign that things are going to get better; nothing happens. Next day he gets up and goes to the track, but this time he doesn't buy a racing form or scratch sheet or Jack Green's Card or anything. Just gets his program and sits in the stands studying the field for the first race and waiting for a sign. None comes, so he passes up the race. He waits for the second race and concentrates on the names of the horses for that one, and again there's no inspiration. So again he doesn't bet. Then, when he's looking them over for the third, something seems to tell him to bet on a horse named Number 4.

"Lord, I'll do it," he says, and he goes down and puts the last fifty dollars he'll ever to able to borrow on Number 4 to win. Then he goes back to his seat and waits until the horses come onto the track.

Number 4 is a little fractious in the parade, and the guy says, "Lord, please quiet him down. Don't let him get himself hurt." The horse settles down immediately and walks calmly into the starting gate.

"Thank you, Lord," says the guy. "Now please get him off clean. He don't have to break on top, but get him away safe without getting slammed or anything, please." The gate comes open and Number 4 is off well, close up in fifth place and saving ground going to the first turn. There he begins to move up a trifle on the rail and for an instant it looks as thought he might be in close quarters.

"Let him through, Lord," the guy says. "Please make them horses open up a little for him." The horse ahead moves out just enough to let Number 4 through safely.

"Thanks you, Lord," says the guy, "but let's not have no more trouble like that. Have the boy take him outside." Sure enough, as they go down the backstretch the jockey steers Number 4 outside, where he's lying fourth.

They're going to the far turn when the guy gets agitated. "Don't let that boy use up the horse," he says. "Don't let the kid get panicky, Lord. Tell him to rate the horse awhile." There rider reaches down and takes a couple of raps on the horse and keeps him running kind, just cooling on the outside around the turn.

Wheeling into the stretch, Number 4 is still lying fourth. "Now, Lord," the guy says. "Now we move. Tell that kid to go to the stick." The boy outs with his bat and, as [jockey] Ted Atkinson says, he really "scouges" the horse. Number 4 lays his ears back and gets to running.

He's up to third. He closes the game ahead and now he's lapped on the second horse and now he's at his throat latch and now he's past him. He's moving on the leader and everything behind him is good and cooked. He closes ground stride by stride with the boy working on him for all he's worth and the kid up front putting his horse to a drive.

"Please. Lord," the guy says, "let him get out in front. Give me one call on the top end anyway."

Number 4 keeps coming. At the eighth pole he's got the leader collared. He's past him. He's got the lead by two lengths.

"Thank you, Lord," the guy says, "I'll take him from here. Come on, you son of a bitch!"

Again, as was the case in his relationship with Ring Lardner, Rice frustratingly kept his feelings about the writing of both Smith and Graham to himself. He spoke publicly and affectionately about their companionship. But he uttered not a word about their writing. Yet he most assuredly admired their talent greatly, for both men approached their task as he did—with an emphasis on the people and the contests, not on the results. Smith's piece on the guy, the Lord and Number 4 possessed every quality that Rice admired in fine writing—wit, drama, a sense of lyricism and rhythm, and demonstrable affection for both subject and reader.

From the time that Graham and Smith drove up to the Vinoy Park Hotel in 1946 until Rice's death eight years later, there was scarcely a day in the month of March when the three men were not together in Florida. They would travel together to several events throughout the year, and see one another frequently at places like Toots Shor's in New York, but it was in Florida in March that the bond was formed and the friendships were at their most robust.

The hub of their Florida orbit was not the spring training camps—though they did visit the camps daily, usually in the morning—but a scruffy little thoroughbred track amidst the palmetto and scrub pine in Oldsmar, a little bit northeast of St. Petersburg. It was called Sunshine Park, and in the years that Rice, Graham and Smith went there the snakes and alligators outnumbered the spectators and the horses were closer to a glue pot than they were to the stables at Churchill, Pimlico, or Belmont. Rice loved it. He loved it when the parking lot attendant would tell them "Not too close now. There's plenty of room." He loved it when the boy selling the tout sheets wouldn't sell him one because Graham had already bought one; he knew they were together and didn't need two. He loved the outdoor bar. He loved the restaurant, whose walls were practically falling down around it but which nevertheless served a a first class lunch of Camembert cheese, a dill pickle, tomato juice and a dry martini. "Where else," he would tell Graham and Smith, beaming, "can you enjoy a good lunch, step out of the dining room, walk a few steps to the windows, buy your tickets without having to stand in line, and watch the races from the bar?"

Rice had covered the big stakes races since the twenties, and he had seen and written of nearly all the thoroughbreds of legend—Man o' War, Exterminator, Count Fleet, Seabiscuit, Citation, Whirlaway, War Admiral, Omaha, Native Dancer. Yet for most of his career he would be at the track generally when duty called. Sunshine Park was different. Sunshine Park was recreation. It had replaced golf as his afternoon diversion.

Sunshine Park also afforded him a certain anonymity. He was never fully comfortable with his celebrity. Though he always handled it graciously, he would have rather not had to deal with the admiring throngs in public. "He walked unrecognized [at Sunshine Park]," said Frank Graham. "Or so he thought," for at Sunshine he could stand at the rail and whip-ride his horses home without interference from fans and strangers.

Graham and Smith claimed that Rice made Sunshine Park famous by writing about it so glowingly. In truth, both Graham and Smith wrote of Sunshine Park more often than Rice did, always building their pieces around the fact that this was "Granny's favorite track." Rice did do his part in spreading the Sunshine Park word, though. One winter he was invited to appear on a national radio broadcast from Santa Anita during some major stakes race there. "And what it your favorite track?" asked the trusting young California announcer, expecting to get a plug for Santa Anita.

"Sunshine Park," said Rice without hesitation.

"Sunshine Park?" asked the stunned announcer. "Where on earth is Sunshine Park?"

"Oldsmar, Florida," said Granny happily, no doubt thinking ahead to March and his forthcoming visits there.

Rice was a sucker for tips. He bought every tout sheet ever published and asked everyone from trainers to ticket sellers for their advice. Arnold Rothstein, the gambler and gangster, had told him once during the twenties that he went to the trainers and told them he would bet a thousand dollars for them if they gave him sound information. "But don't double cross me," he told them, as if they had to be reminded. "In this way," said Rothstein, "I get the best information. Information, sound information, is what you need at the race track."

Rice agreed, but unlike Rothstein, he had trouble discriminating which information was sound. And he would have considered it an insult to a friend not to buy a ticket on a tip. He claimed to have a system on the Daily Double. His system was to take the horse he liked in the first and match him up with his top three picks in the second. Then take the second and third horses in the first and wheel them with the field in the second. Then fill in the holes by buying tickets on whatever horses he'd been given tips on. In short, his system was: the more daily double tickets you buy, the better your chances of hitting it. "Granny always had the daily double," said Dan Parker. "How could he lose when he bought practically the whole rack?" Often he would pass out extra daily double tickets to friends, so that they might share in his obsession; he couldn't understand why they didn't have a dozen tickets of their own. The double was serious business to him. Frank Graham once expressed a desire to be buried in the infield at Saratoga. And every day at post time for the first race, he said, he would poke his head out of the dirt and see what was going on. "Post time for the first race!" said Rice with genuine incredulity, as though Graham were serious. "Why, you'd miss the daily double."

Daily doubles aside, Granny's partiality was for long shots. So when he cashed tickets, he cashed big—he once wired Kate about having had fifty dollars on a 35–1 shot that came in. But he didn't cash often. Still, for all of the money he left at Sunshine Park, Santa Anita and other tracks about the country he never lost more than he could afford; he stayed away from the thousand-dollar windows. There are no recorded occasions of Granny going home with his pockets turned inside out, though on one occasion he did need to be pulled away from a roulette table in Miami.

Late in the evening, after not having seen Rice for a couple of hours at a party, John Wheeler found him at a roulette wheel with four or five others—including two men Wheeler recognized as house players. Rice was $2500 down. Wheeler's instinct was to get him out of there, but Wheeler knew Rice and knew the gambler's instinct—pull a guy away from the table when he's down and he's gonna' believe he woulda' won if he'd had more time, and resent the chap that yanked him away. So Wheeler sat and watched. Granny's luck changed and when he had recovered all of his loses save sixty-five dollars' worth, Wheeler told him to cash in and get out. "Hell no. We've just got 'em on the run," said Rice. But Wheeler was insistent. But it was not often that Rice had be pulled away. When the people left, Rice left too, for as much as he enjoyed the action it was the camaraderie that brought him out. If there weren't enough of his friends there on any particular day, he'd like as not make new ones. One of the seminal stories of the Rice legend was the day he saw Yogi Berra, then a Yankee rookie, was standing forlornly off to the side at a St. Petersburg dog track, hands stuck in his empty pockets, looking for all the world like a dead-end kid with nothing to live for, instead of someone who'd just made the Yankees. "Who do you like in the next race," said Rice as he approached Berra. "I don't like nothing," said Berra, "I just lost my last two bucks."

"Well, you and I each have a deuce on Number 8 dog," said Rice, handing Berra a ticket. The Number 8 dog, of course, came in, and paid a neat $108.

Stories about Rice abounded in the late forties. As at the close of the thirties, the close of another decade set people to writing about him—tales of kindnesses like the Berra story, tales from his past like the Ty Cobb "discovery," stories of his befuddlement at things mechanical, heartfelt testaments to his honesty, integrity, humility and warmth, and the requisite acknowledgment of his authorship of the Four Horsemen lead, and the "how you played the Game" couplet. "He is the a model not only for all sports writers, but for all men," said of of these profiles unabashedly, echoing the tenor of them all.

Publicity begets more publicity, of course, and inevitably, these newspaper and magazine notices attracted wider notice. At the end of 1948, Ralph Edwards featured Rice on the network radio version of *This Is Your Life*. Kate was in on the planning. So was Rice's secretary Catherine Mecca, and the people at Vanderbilt. The program was scheduled for December 28, and Rice was completely duped by the invitation to come by the NBC studios in Hollywood to talk a bit with Ralph Edwards about the New Year's Day bowl games. *This Is Your Life* was less than a month old and was not yet Edwards' signature show; he was a sort of jack-of-all-shows at NBC at the time. Rice arrived at the studio at around four o'clock Los

Angeles time, as per his instructions; he had about an hour before air time. He was no stranger, of course, to the chaos in a studio in the moments before air, but he was a little puzzled that everybody seemed to be leaving him alone; no one had seen fit to get him a script or debrief him on his thoughts. "This is the daffiest show I've ever been on," mumbled Rice to a *Los Angeles Times* reporter who was on hand. "Here it's about to start and I haven't seen a script yet."

Just prior to five o'clock he got his first hint that this was not about any bowl games when he was led, not to a small studio with a couple of microphones, but onto a large stage in front of a packed auditorium. He sat calmly and quietly on stage but friends could tell he was nervous, the jaw muscles in his cheek were bulging—his teeth were clenched.

When the show began, Edwards started with the litany of teases and surprises that would become familiar to American radio and television audiences over the next two decades. "Do you remember anything about a cherry tree?" he asked Rice.

Rice scratched his head and spent a couple of quick moments in hard thought. "No, I don't think so," he said.

"You didn't cut one down?" teased Edwards.

"No, that was somebody else."

Then, from off stage came the voice of Rice's brother John. "Mama! Grant's fallen out of a tree and I think he broke his arm." Unfamiliar with the show's format and not knowing what to expect, Rice's mouth dropped open when his brother came on stage. Next came Professor C. B. Wallace from the Wallace School, still puckish at the age of eighty-nine, who told Edwards that Rice had been an obedient student, he'd never had to "excoriate his epidermis." Wallace also reminded Rice of the letter he had sent him some years before saying that if he achieved any success as a writer, he owed it all to Wallace. "And I meant every word of it," grinned Rice.

The show's most arresting moment came when Edwards asked Rice whether he remembered the words "Outlined against the blue-gray October sky. . . . " and from behind the curtain came Stuhldreher, Crowley, Miller and Layden, the Four Horsemen of Notre Dame.

Then came a parade of a dozen men whom Rice had named to his All America team over the years, and then finally, at the close of the thirty-minute show, Jim Thorpe and Amos Alonzo Stagg, whose brushes with Rice went back forty-eight years, to the March weekend during Rice's junior year at Vanderbilt, when Stagg brought his University of Chicago baseball team to Nashville to play the Commodores. Stagg brought with him word that Philip Morris, the sponsors of the program, had donated a trophy to Vanderbilt University, to be given annually to that Vanderbilt student judged "most proficient in both athletics and scholarship."

At the post-show party at the Hollywood Roosevelt Hotel, Rice remained flabbergasted by all the attention, and deeply touched that so many people—and Stuhldreher, Crowley, Miller and Layden in particular—thought enough of him to come to California to show it. Perhaps, as one newspaper editorial put it, in considering his life that evening in Hollywood, "even he must have been surprised at its color and drama." Perhaps *especially* he.

The balance of his life would be filled with such tributes and honors. Scholarships in his name at Vanderbilt and the Graduate School of Journalism at Columbia; a quite-serious call from Henry McLemore to erect a monument in his honor; and a persistent call from Vincent X. Flaherty of the *Los Angeles Examiner* to vote him into the baseball Hall of Fame. Flaherty's admiration was unbounded. He also wrote a column urging President Eisenhower to issue a proclamation for a national testimonial. Fred Russell lobbied publicly in his column and privately with the Vanderbilt administration to have some sort of annual Grantland Rice Day at Vanderbilt. It all amounted to a continuing retirement party for a man who had no intention of retiring. It all must have been a little overwhelming, too; but Rice let it all pass by like a pleasing summer's zephyr. He continued to do his job, carrying his own typewriter up into the press box, generally plopping it down in an empty space next to some wide-eyed kid writer and giving the kid a lifetime's memory by ingenuously sticking out his hand and smiling and saying: "Hi. My name is Grantland Rice."

In the press boxes he was an imposing and as popular a figure as he had even been. Even brash and cocky kids like Jimmy Breslin of the *Journal-American,* who showed no elder any deference as he began to carve out his own considerable niche in the New York newspaper world, admitted to being "in awe" of Grantland Rice. On the playing fields and in the locker rooms, players would come up and shake his hand. Angelo Bertelli of Notre Dame said that when he won his Heisman trophy it was a bigger thrill sitting on the dais with Rice at the awards dinner than it was in receiving the trophy itself.

But in the nation's newspapers he was beginning to lose both his presence and his relevance. The quickened pace of journalism and sportswriting had passed his leisurely style by. His column, unchanged from the 1920s, had become an anachronism as the world moved into the 1950s. Newspapers that once relied on syndicated writers like Rice to provide their stories could now afford writers of their own to cover the big stories. And they couldn't afford not to have writers of their own. Readers in Chicago and Boston and Los Angeles and Kansas City wanted the immediacy and intimacy of a story tailored precisely to a paper's geography and readers.

And so too had the business changed. Led by the example of Dick

Young of the New York *Daily News,* sportswriters were now storming
the locker room after games, incorporating quotes and reactions from players
and coaches in their morning stories. Stacked up against such fare, Rice's
days-old ruminations on general topics and themes seemed rather like the
oatmeal that was hot and fragrant at breakfast and now, at lunchtime, sits
cold, leaden and neglected. The number of papers subscribing to Rice's
column never dwindled; it remained right around one hundred until the
day he died. But the number of subscribing papers that printed his column
in full every day shrank steadily from the end of the war on. The white
space was just too precious; sports editors used it for a column from one
of their staff writers, or coverage of some emerging sport like basketball.
Even in Nashville, the *Banner* would often run Rice's column only a couple of times a week, and sometimes they'd run just one or two paragraphs.
But neither the *Banner* nor any other paper ever thought of surrendering
Rice's column, much to the annoyance of Rice, who would sometimes
have a paper eager to take him and give him prominent play. In Los Angeles, the *Mirror,* a scrappy afternoon also-ran owned by the *Times,* wanted
to acquire the Los Angeles rights to his column but the *Herald-Examiner,*
the afternoon competition—which seldom ran the column—refused to let
it go. For however indifferently the subscribing papers may have treated
his column in the years after World War II, they still recognized an association with Rice as a valuable asset. Papers would trumpet his name among
those providing coverage of some big event in their promotional material.
He had, in these markets and these papers, become something of that peculiar twentieth century American phenomenon—the hollow celebrity, his
name more valuable than his work.

He was also physically not a well man. Susceptible to colds and the
flu, he was felled by pneumonia for several weeks in 1950, and again in
1953. He smoked too much, and too many afternoons in New York he
would linger over too many lunchtime martinis at Toots Shor's. His friends
worried about him and looked after him. The task of scurrying down to
the locker room in search of quotes after games and then back up to the
press box to file his story was physically beyond him. Graham and Smith,
and sometimes people like Fred Russell, would ask him what he needed
and feed him quotes when they returned to the press box. Sometimes their
concern would be groundless. Fred Russell was worried one day that Rice
had had too many martinis at a luncheon before the Keeneland Stakes in
Kentucky, when he was scheduled to give a speech. Russell fretted over
whether to take Rice away so he wouldn't embarrass himself. He seemed
to Russell clearly incapable of delivering any lucid remarks. But when the
bell rang Rice rose and delivered a positively charming talk about carving
out a new state from southern Kentucky and Middle Tennessee—pure
Bluegrass, he told them.

At other times, however, there was genuine cause for concern. At the 1948 World Series, Fred Lieb found him in the lobby of the Hotel Cleveland before the final game. "He was alone and appeared confused and bewildered," said Lieb. He was just a few blocks from Municipal Stadium, and Cleveland was a city he knew—he had spent a year there as a young writer, remember. But he told Lieb that present-day Cleveland was a puzzle to him. As they walked to the ballpark Rice leaned heavily on Lieb's arm, "and when I steered him safely through the press gate and to his seat in the press box he was profuse in his gratitude," Lieb reported.

"Thank you, thank you, Freddy," he told him. "I couldn't have got here without your help." Lieb had tears in his eyes as he found his own seat.

And yet . . .

The column remained curiously relevant. For together with the immediate, the sports fan treasures the remembered. And Rice was the reader's connection with the past. He had written of and partied with the Babe and Dempsey and Jones. And he was still in the press box typing as he watched DiMaggio, Marciano and Hogan. There were roughly seventy-five men in the Baseball Hall of Fame in the early 1950s. It was said that with the possible exception of a couple of nineteenth-century figures like Abner Doubleday and Henry Chadwick, Rice had seen and known them all. He had known every heavyweight champion from John L. Sullivan to Rocky Marciano. (Though he had never seen John L. fight.) He would frequently use his column to talk of the past—not to reminisce so much as to put things in perspective. A reader would submit an all-time baseball lineup with all the usual suspects—Ruth, Cobb, Gerhig, Wagner, Mathewson—and challenge Rice to get up a team that could challenge it. Rice would resurrect the likes of Nap Lajoie, Tris Speaker, George Sisler and Cy Young and explain how his team and the reader's would be more evenly matched than one might suspect. Or he would run through the list of the heavyweight champions from John L. Sullivan on, and limn each man's fighting personality and list his attributes. Jim Corbett was his pick as the greatest. Rice was providing a course in twentieth-century sports history, and for the reader, getting this history from Rice was rather like taking military history from Pershing, or diplomatic history from Cordell Hull. Rice may have lost a step in his coverage of last night's game, but for the young boy coming to the sports page he provided something more unique—stories of his father's heroes. Yesterday has always been a large part of sport's sweet allure.

There were other times when he was as timely and as forceful as he ever been in his prime, as when he blasted Notre Dame coach Frank Leahy for having his players feign injury in the waning moments of close game, stopping the clock and giving Leahy precious seconds, to which Rice's sense of sportsmanship and propriety said Leahy wasn't entitled.

And then there were other times when he was just so damned eloquent as to be irresistible, as in the *Sport* magazine article on Bob Feller and his father that he opened with the haunting and timeless line: "Every father who ever lived has had a dream for his son."

14

Charon's Boat

T_he New York_ **Sun,** after one hundred and nineteen proud years, published its last issue on January 4, 1950. It was a melancholy day for New York. The _Sun_ had introduced America to a brave new world of affordable (one cent) journalism for-the-masses back in 1833. It was the paper that gave the world the timeless "Yes Virginia, There is a Santa Claus" editorial in 1897. Despite the fact that it never approached the substance or impact of the _Times_ or the _Tribune_ or the _World_ or even the Hearst papers, even more than these papers it was a part of the fabric, the fiber, the charm of the city—something uniquely New York, like Pennsylvania Station, Delmonico's, the Metropolitan Opera House or Ebbets Field.

Scripps-Howard, publishers of the afternoon _World-Telegram,_ bought up the name, the assets and the goodwill of the paper. Grantland Rice's column was among those assets, and his column was in the _World-Telegram_ the day after the _Sun_ folded. Indeed his column was among the assets that the paper capitalized on most heavily in their promotional material in the days after the _Sun_ closed. The _World-Telegram_ told its readers to look for all their favorite _Sun_ features in the new _World-Telegram and Sun._★

★ It was the _sentiment_ surrounding the _Sun_ that the _World-Telegram_ most wanted, so they somewhat halfheartedly appended the _Sun_'s name to theirs, in full-size letters at first, but quickly shrinking them to letters one-quarter the size of those which read _World-Telegram._

Grantland Rice's column was among those features prominently mentioned in these promotions. But it was never a particularly happy fit. In
the weeks just prior to the *Sun*'s demise, with rumors of the Scripps-
Howard interest in the paper swirling like flakes in a blizzard, Rice spoke
to Roy Howard at a social function. Howard was mute on the subject.
After the deal was consummated he apologized to Rice for his silence,
saying "I'm sure that you understand enough about how these things go
to realize the vital importance to us of maintaining the secret of the contemplated purchase."

Whether he understood, or whether he was slightly put off by Howard's not thinking him worthy of his trust, Rice nevertheless wrote Howard from California and asked what he might do to help out the cause.
But the column was never a good fit. Rice was somewhat disappointed
that the *World-Telegram* had no place for his old friend Rube Goldberg.
And Roy Howard was somewhat upset at Rice's references to the paper
as the *Telegram and Sun*. "Our emphasis, logically I think is on the *World-
Telegram*," said Howard in a note to Rice. "Despite the *Sun*'s highly ad-
mirable ancient past and the fine traditions associated with that past, and
without in any way wishing to jar your loyalty to the old paper, the fact
of the matter is that news-wise around the world, and advertising-wise
around the agencies, the *World-Telegram* has I think a position of value and
prestige which it is not our intention or desire to dilute or dissipate."

Rice seldom got the opportunity to write anything but the column
for the *World-Telegram*, and when the contract for his column was up in
January of 1951, Scripps Howard did not renew. Rice and John Wheeler
then discovered that the column was not in great demand in this new era;
for a time it appeared that Rice would be without a New York flagship.
But Dan Parker, sports editor of the Hearst *Mirror*, the saucy morning
tabloid—responding to some intensive lobbying from Bruce Barton—persuaded the powers at Hearst to take Rice on. He would remain in the
Mirror for the three and a half years until his death.

In July of 1953, Rice was standing at the bar at Toots Shor's, waiting
for the start of a luncheon honoring Ben Hogan on his British Open
triumph. Rice was to be the toastmaster. "I don't know why they wanted
me," he said. "I haven't seen him play. I've been out of touch. I don't
know what I'm going to say." The protestations were merely the pre-
show butterflies of a seasoned performer. He was, witnesses remember,
superb in his role—witty, tender, polished.

Rice was talking to Lowell Pratt, the president of A. S. Barnes publishers, and Dave Camerer, a former Dartmouth football star who had
recently joined Barnes as an associate editor after ten years in the newspaper business. In 1949, after a decade's cajoling, Pratt had finally persuaded Rice to sign a contract to write his autobiography. Now, four

years later, Rice had yet to write a word, and Pratt gently prodded him on when he might get around to beginning.

"I'm afraid the ship has sailed on that one, Lowell," said Rice, meaning that it was time to admit that at the age of seventy-two he had neither the energy nor the passion for a project that had really always been Pratt's idea.

"Well how about if this guy here were to help you," said Pratt, gesturing to Camerer. "Well . . . ," answered Rice, thinking it over. "I don't know." He had met Camerer several times during the younger man's newspaper days and had always liked him. When Camerer called a few days later, Rice agreed to give it a try. So began a year long collaboration between Camerer and Rice—the last major undertaking of Rice's long career.

Camerer was more than a research assistant, less than ghost writer; not precisely an amanuensis, much more than than just an editor in the conventional book-world sense of the job. He went through Rice's files—Kate Rice's files really—old clippings and profiles on Granny dating back to the Atlanta days. He went through Granny's poetry scrapbooks and visited with two dozen of his friends and associates—everyone from Rube Goldberg and Gene Fowler to Ty Cobb and Babe Didrikson—pulling from their memories images of Rice. He presented the result of this research to Rice, and it proved a boon in "routing out old and sometimes fading memories," as Rice put it. Through the fall of 1953 and the winter and spring of 1954, Rice pecked away at the typewriter. He wrote with no particular organization—aside from the fact that he wrote thematically, not chronologically—and sent the chapters off to Camerer when they were finished. He made no attempt to make any grand sense out of his life and career, or give the book any central theme. He merely told stories. "This isn't, praise be, a formal book," said Red Smith. "This is Grant Rice talking, rambling happily along, telling again in his wonderful way the wonderful stories he loved to tell."

Camerer and Rice spoke a couple of times a week. They would speak by phone, and when they met it was generally in Rice's office at Sportlight Films on West 48th Street. Camerer wrote or called with specific suggestions on chapters he knew Rice was currently working on—"you should pass along The Grantland Rice Method of Beating the Daily Double"—and once sending along a Barnes book on horse racing, suggesting "it might jog your memory re some wonderful anecdotes of your own," and also doing a bit of marketing on behalf of A. S. Barnes: "a mention of it in your column [would] help a great deal." When the chapters came in Camerer would make sense of them, sometimes getting back to Rice with specific editorial questions only Rice could handle—"the ranking of all-time horses . . . could stand expansion." But just as often Camerer would fill in holes himself, on the basis of his own research, carefully—and expertly—preserving Rice's rambling, easy-going, pull-a-chair-up-by-the-fireside-and-give-a-listen voice. He showed the chapters on Dempsey, Tunney and Bobby Jones to the three athletes so that they might reconcile their memories and Rice's. He then ran large sections—some eighteen chapters—of the book past both Gene Fowler and Frank Graham.

Rice received the galleys in late June, and must have been pleased, for he had been well served by his editor-assistant author. Camerer also provided Rice with a list of those he had contacted, so that Rice might have a record of their names should he wish to include them in the acknowledgments. Camerer also modestly made one request. "Granny, if I were to be included with this list, by you, in whatever assistance I've been able to bring forward," he wrote, "it would mean more to my never-to-be-forgotten association with you—and my pride in that association—than anything that cash could ever buy." Recognizing the debt he owed to Camerer, when Rice typed out his brief acknowledgments in East Hampton on July 1, he began:

> I wish to extend deepest thanks and appreciation to Dave Camerer, former Dartmouth football tackle, and sportswriter, for the fine work he contributed in routing out old and sometimes fading memories from over fifty years ago. His assistance was invaluable.

As to everyone else, he merely thanked "the many." The exclusion of any other names was partially an insurance against forgetting anybody amidst the jumble of people both he and Camerer had spoken to. It was also, in typical Rice fashion, a heartfelt and public kindness to a man who had bestowed a great kindness upon him. He returned the acknowledgments and the galleys to Camerer and the book went into final design and production. Barnes set a publication date for November 1, 1954—Granny's seventy-fourth birthday.

Among the last projects for the book was getting a contemporary photo of Rice to go with the pastiche of photos from throughout his career. A photographer came to Rice's office at Sportlight Films and shot a series of pictures with Rice sitting at his typewriter. When the proofs came in, Rice chose an over-the-shoulder profile shot. For the cut line he wrote: "I tire easily these days. Sometimes I think, perhaps, I've lugged too many typewriters to the top of too many stadiums."

In the fall of 1953 he didn't feel up to traveling to Nashville for a celebration marking Fred Russell's twenty-fifth anniversary as sports editor of the *Nashville Banner*. He sent along a verse instead. While he made his annual pilgrimage to Los Angeles and Florida, he didn't feel up to traveling to the Kentucky Derby in May. In June of '54, despite the fact that it was being held just across the river at Baltusrol, he missed the U.S.

Open—the first time since the end of World War I that he had been absent. He by-passed the baseball All Star game in Cleveland. But he seldom groused or complained about the maladies of growing old. "Kate and I are just sitting in the sun now," he said in a letter to Fred Russell on July 10. "Listening to your arteries harden isn't such bad sport after all."

Time-Life was planning a new magazine called *Sports Illustrated,* scheduled for an August debut. Herman Hickman, the former Yale football coach, contacted Rice about providing a story on Bobby Jones. Rice obliged. He also continued to get up each morning and write his column. After having fashioned more than 17,000 newspaper columns over the preceding fifty-three years, it had become as much a part of his morning as breakfast.

On the day of the All Star game, Tuesday, July 13, he rose early, and after reading the papers, worked on the column at home, in his Fifth Avenue apartment. He finished up a piece on Willie Mays, just back from two years in the Army on his way to leading the National League in hitting. "[W]e can say that no ballplayer ever reached such heights at twenty-three years [of age]—," he wrote, "not even Wagner, Cobb, Ruth or Speaker. . . .

> There is no reason why Willie shouldn't go on. He has youth power, speed, and unbounded enthusiasm. But even if he doesn't hold this pace, he has already contributed in just a little over half a season more than any other player has shown in such a short span.
>
> It is too early yet to compare him with Ruth, Cobb or any other twenty-year star. Time is also a big factor in the building of a reputation. Willie, at least, has a golden start.

But the column was about much more than Willie Mays. Indeed, he didn't get around to talking about Willie Mays until the last three hundred words. Looking up where Mays had come from—Westfield, Alabama— set Rice to musing about just where baseball stars did come from. He talked about home towns—Factoryville, Pennsylvania (Christy Mathewson); Humboldt, Kansas (Walter Johnson); Donora, Pennsylvania (Stan Musial); Hubbard, Texas (Tris Speaker); Cario, Georgia (Jackie Robinson); Roxboro, North Carolina (Country Slaughter).

From places on a map, Rice moved to places in our hearts—the nut of this particular and strikingly vintage column. He recalled wondering, when he had been a cub, how anyone might replace the stars of his youth. "We had Wagner, Cobb, Mathewson, Cy Young, Fred Clarke, Rube Waddell, Bill Dineen, Chief Bender, Eddie Plank, Tinker to Evers to Chance—and others—from 1901 to 1906," he wrote. "Who could ever replace such lost stars?"

The happy answer, he pointed out, were men like Johnson, Ruth,

Eddie Collins, Joe Jackson, and Tris Speaker. "These were bound to leave a big gap," he typed. "Who could ever take their places?"

He then recited a litany of the the next generation's hall of famers— Frank Frisch, George Sisler, Charley Gehringer, Lefty Grove, Dizzy Dean, Jimmy Foxx, Hank Greenberg, Mickey Cochrane, Bill Dickey, Joe Gordon, Carl Hubbell, Mel Ott, Ted Williams, Stan Musial. " 'The game has run out of stars,' someone wrote around 1950. Outside of Joe DiMaggio, William and Musial we had no headliners left," noted Rice. But that should not be the case for long, he pointed out, for it was about this time that Duke Snider began to emerge. And he was joined forthwith by Roy Campenella, Yogi Berra, Jackie Robinson, Mickey Mantle, Eddie Mathews, and "the most exciting ballplayer of them all"—Willie Mays. It was a wholly satisfying newspaper column.

Rice moved next to a column on discussing which sports were most hazardous to an athlete's health. He submitted that it was horse racing, and began building a column around a conversation he had had at some point with jockey Eddie Arcaro. But he was having trouble typing. Generally a clean, precise typist, he was having trouble with the shift key and the space bar; sometimes he would run two or three words together, at other times he had three and four spaces between each word. there were letters suspended between lines, and strikeovers and cross-outs well beyond the realm of ordinary composition and revision. He got about four hundred words into the piece before stopping in mid-sentence. He left the column in his typewriter and left his apartment for the Sportlight office sometime after eleven o'clock.

It was an oppressive summer's day in New York; the afternoon temperature was in the mid-nineties. In Cleveland, Whitey Ford of the Yankees was set to oppose Robin Roberts of the Phillies in the All Star game; but it was to be a hitter's day in Cleveland. Al Rosen hit two home runs before the hometown crowd, and set an All Star game record with five runs batted in. Larry Doby and Nellie Fox keyed a three-run American League eighth, giving the American League an 11–9 win in the highest scoring All Star game ever. Rice was planning to watch the game on television after taking care of the some brief business at the office.

Charles Goering drove Rice through the late-morning heat to the Sportlight office on West 48th Street. Inside, Rice gave Catherine Mecca his column to send to the syndicate, and then sat down with her at her desk to begin going over the mail. At some point in the conversation, Rice's speech began to slur. Mecca asked if he was all right. He said yes, and excused himself to go to the men's room.

He was suffering a stroke. Mecca alerted Jack Eaton, the Sportlight Films producer, and when Eaton checked he found Rice dazed and incoherent. Catherine Mecca called Rice's doctor and the doctor called for an ambulance. Rice was taken to Roosevelt Hospital on West 59th Street, where he slipped into a coma. He died shortly after six o'clock, with Kate and Catherine Mecca at his bedside, and Floncy en route from Los Angeles. Catherine Mecca returned to the Sportlight office and called the newspapers.

Newspaper people generally receive pretty handsome obituary space. It is their brethren, after all, who are the custodians of the space. But it generally stops at a column or so in the newspaperman's home paper and a few inches in assorted other papers. Rice's obituaries, however, rivaled those of a head of state. His death was front-page news in New York and in most of the other cities where his column was syndicated. The full Associated Press obituary ran more than two thousand words. *Time* and *Newsweek* each gave him two columns. The *Times* of London made note of his passing. New York congressman, Stuyvesant Wainwright eulogized Rice on the floor of the House and the remarks were entered into the *Congressional Record.*

While the front page notices of his death bespoke the industry's regard for his career and his achievement, elsewhere in the paper—throughout the paper—there were editorials and columns that bespoke the tremendous affection in which he was held by those with whom he worked. "He's greatest who's most often in men's good thoughts," was the quote that haunted Fred Russell when he thought and wrote of Rice. And throughout American newsrooms on the summer's evening that word came of his death, Granny Rice was in men's good thoughts. Editorials in four or five dozen different newspapers spoke of things like his "high sense of propriety and a lofty code of ethics which provided a standard for the newspaper profession"; of his having "never lost the fine bright rapture of the true amateur"; of being a man "who left the sports world a better place than he found it."

"He came to manhood when the spell of Rudyard Kipling had seized the imagination of ambitious young writers," said the *Providence Journal* editorial, "and his outlook and his verse (especially his verse) echoed the *If* standard of conduct as the right guide to playing the game."

The most passionate editorial came in his hometown, in the pages of the *Nashville Banner:*

> He chronicled events in the realm of sports, but gave to this business the majestic touch of a great mind and an understanding heart. He wove words in the pattern of beauty. There was in everything he wrote the personal touch. . . .
>
> He was the creative writer, with the power to inspire respect for those standards of dynamic faith and purpose and kindred virtues of manliness and sportsmanship which his own life exemplified. His were the concepts of character which despised duplicity, unfairness, coward-

ice. He magnified nobility and imbued his readers with regard for an uncompromising moral code.

Though with his death a star has fallen, its light will continue to shine as long as memory lingers. His niche was in America's heart, and the people who loved him will not easily forget.

The Atlanta *Constitution* ran an editorial cartoon that showed Rice's typewriter atop a thick and towering pedestal. At the base, with heads bowed, were boxers, jockeys, golfers, referees, football, baseball, basketball and tennis players, and other assorted just plain folks—the likeness of the fan, no doubt. In New York, Willard Mullin sketched Granny in bow tie and fedora, mural sized, while at the table in front of him stood a dozen men with glasses up-raised; and Scripps-Howard papers around the country ran it. In a signed editorial in the New York *Journal-American,* William Randolph Hearst, Jr. gave a rather pedestrian account of Rice's life and career, and then concluded with a sentence that, in its frankness and directness, was as poignant as anything that was written. "I have never tried to write an obituary before and I know this is woefully inadequate," he said, "but I just felt like doing it because Granny was a friend and I will miss him."

The depth of personal feeling for Rice was most evident, of course, in the columns of his friends. "When the very best man dies," wrote Henry McLemore, "and you are a real good friend of his, [and] you are a newspaper columnist to boot, you must write a story about the best man's death." Rice had more "real good friends" who had newspaper columns than perhaps any man who ever lived. And every single one of them wrote a column. McLemore, Smith and Graham, Fred Russell, Arthur Daley, Bill Corum, Dan Parker, Jimmy Powers, Jimmy Cannon, Bob Considine, Vincent X. Flaherty, Arch Ward, Ralph McGill, Sec Taylor in Iowa, C. E. McBride in Kansas City; Bill Cunningham in Boston; Ed Danforth in Atlanta; Warren Brown in Washington; Paul Zimmerman, Bill Henry and Ned Cronin in Los Angeles, and maybe a hundred others across the country. Many wrote more than one column; the tributes continued for over a week. They were all witty and tender and filled with love and melancholy—pieces clearly written from the heart and with a lump in the throat.

Perhaps the most powerful and moving of these came from Jimmy Cannon. You didn't expect it from Cannon; he was a man who trafficked in sentiment about as often as Babe Ruth laid down a bunt. He was very much the antithesis of Rice, a wise-guy from Greenwich Village who went to work as a copy boy at the age of fourteen and never finished high school. He was the embodiment of what sportswriting had started to become in the 1950s and would continue to become in the years ahead—

confrontational, hard-edged, unconcerned with the blood he might spill in typing out his eight hundred words a day. He would shout out the things that others would whisper—stories on racism and fixed-fights. And he was an intensely vain, competitive writer; in short he could not have been more different from Rice. Among the nation's sportswriting fraternity he would have seemed to have been least inclined to appreciate Rice's style or following, and least reluctant about saying it. It was a shock almost— *is* a shock, reading the words nearly forty years later—to find Jimmy Cannon dewy-eyed over Rice's death and the close of his age:

> The music Grantland Rice made was small but it filled his life. The great sports writer . . . died young although he spent seventy-three years on earth. He didn't seem old because he held onto the illusions of his youth. The dream remained pure and glowed with an obsolete splendor because of his faith in the goodness of men.
>
> He cherished decency above all and searched for it in the characters of those he knew. It was often hard to find but Granny didn't become discouraged. It was his failing that he judged all of us according to his own standards and there were few such as Grantland Rice.
>
> This sad piece is inadequate because I haven't the melodious compassion that was his. I suppose I envied him more than any one who came my way. My life was enriched because of the time I spent in his company.
>
> I apologize for this column even as I write it. There should be in it some of the stately grief Granny expressed in the poem he composed for Ring Lardner's obituary. But I'm not Grantland Rice and he's the only one who might appreciate the worth of such a beautiful guy. All the tributes Granny squandered on others apply to him.
>
> Granny was a glorious man and the human race is running short of his kind. Meaner men are not capable of measuring him. The ornate style that was his would have seemed reserved if he had used it to describe himself as he truly was.
>
> It is not bragging to describe him as my friend because Granny was generous with his affection and also with his praise. He worked in a competitive profession. But Granny recognized no rivalry among sports writers. . . .
>
> Those who flagrantly copied his style failed in their imitations because they couldn't steal from him his unshakable belief in the integrity of the human race. It is true his popularity suffered in his last years. Granny's was a fragile talent that was not meant for an age of turbulence. But he was an original and he influenced every sports writer who came after him.
>
> The world changed but Granny didn't. Of course, this is a flaw in any journalist. Realism became the fad in sports reporting. But most of us who work at this job are grateful he was loyal to the old ideals.
>
> We, who still depend on his philosophy in some way, toiled to

disturb his mood of clean excitement. All his young men were gay and spectacular. But we, who consider ourselves blase, still look for such athletes.

We, who owe him so much, turned crabbed and became proud of our suspicions. House detectives worked at a poet's chore. It may be that Granny made games more important than they should be. But remember he became famous in the years of Jack Dempsey, Babe Ruth, Bobby Jones, Red Grange, Bill Tilden and Man O'War. We were a different country then. We laughed more. We didn't ask our heroes to die.

The stories Granny wrote then were filled with a massive imagery that seemed slightly ludicrous now. But this was during the joyous years when there was a cause for happiness and a time for boisterous foolishness. All of us in this generation of sports writers are improved because we borrowed some of his techniques. Few handle the language with as much grace. Many of us croak because we can't sing.

Granny shall achieve an immortality denied some poets who were fascinated by greater themes. All of us on this beat are in his debt because on every sports page in the land a part of him sings. We, reader, writer, relative or friend, are better because we knew him in some way.

They spilled out onto the sidewalks outside the Campbell Funeral Home on Madison Avenue—standing under umbrellas on sidewalks glistening with a summer's rain, waiting to pay their last respects to Granny. When someone dies in the fullness of years as Rice did there is sadness and emptiness, but seldom is there a real searing grief. But Rice had so touched the lives of those he knew that the pain was as if he had died tragically in the flower of his youth. Bob Considine—whose introduction to Rice twenty years before had come when he was standing on the fringes of a conversation in which Rice was at the center, when Rice noticed and drew him him with his characteristic: "And what do your think, Bob?"—stood silently for a few moments before Rice's bier. Then, unself-consciously, he stepped back, knelt on the floor, and, by himself, began saying a rosary. His gesture so touched Catherine Mecca that she knelt beside him and joined him.

A tower of roses, gladiolus, chrysanthemums and other flowers overflowed the sanctuary of the Brick Presbyterian Church on Park Avenue for the funeral on Friday, July 16. A sports world who's-who was among the four hundred mourners who crowded into the church. Jack Dempsey and Gene Tunney walked side-by-side at the front of the procession as honorary pallbearers. Harry Stuhldreher, Jimmy Crowley, Don Miller and Elmer Layden, the Four Horsemen of Notre Dame, were there. Bobby Jones, already crippled by the degenerative disease that would take twenty years to kill him, walked in leaning heavily on a cane. Others in attendance included Toots Shor, Frankie Frisch, Carl Hubbell, Bill Dickey, Moe Berg, Craig Wood, Ken Strong, Al Schacht, Giants owner Horace

Stoneham, Mrs. John McGraw, baseball commissioner Ford Frick, Army football coach Red Blaik, former Yale football coach Herman Hickman, and former Postmaster General and then New York state boxing commissioner James A. Farley.

Helen Rogers Reid, the publisher of the *New York Herald Tribune,* the paper for which Rice did perhaps his finest writing, was there. So was William Randolph Hearst, Jr., the publisher of the paper for which Rice did his final writing.

Newspapermen comprised the biggest contingent of mourners. They included Frank Graham, Red Smith, Ralph McGill, Bill Corum, Bob Considine, Dan Parker, Tim Cohane, Rube Goldberg, Arthur Daley, John Kieran and Stanley Woodward.

As they stared at the coffin covered in roses and thought back upon the fullness of Rice's years, they listened to a tender and emotional eulogy rendered by Bruce Barton. "To believe that such a life is ended is to say that human life is meaningless and the universe itself a ghastly joke," said Barton in his concluding remarks. "No one of us believes that. Grant is not lost to us.

"Gainsborough, the artist, cried exultantly: 'We are all going to Heaven, and VanDyke is of the company.'

"We are all going to Heaven, and Grant is there already—telling his stories, talking his wisdom, cracking his jokes, and, we may be sure, encouraging play. Already they have learned to love him. And he is waiting for us—still with his joy in living and his eternal courtesy."

Following the thirty-minute service the long procession made its way through the crisp, radiant summer day to the Woodlawn Cemetery in the northern Bronx, where Rice was laid to rest with a final prayer: "Oh grave, where is thy victory?"

Many of the mourners made their way back to Toots Shor's after the burial. There, in Granny's lair, they swapped stories and remembrances of their friend, and summoned from their memories some lines of his verse. So many of them had quoted his verse in their columns—lines from "Charon's Boat," and lines from this verse, "A Message from a Front Trench," written during World War I, brave and hopeful thoughts on what lay beyond:

> When my time has come and all farewells are said
>   To what few friends may still survive the fight,
> I shall not shrink to hear the ghostly tread
>   That signals Death is stalking through the night
> To lead me forth across the Mystic Moor
>   Unto the Tavern of the Silent Land—
> But I shall smile—and through the open door
>   We two shall go, as good friends—hand in hand.

There I shall meet the friends who've gone before,
  And we shall gather in a room apart,
And, cup to cup, shall pledge the days of yore,
  Soul unto soul and silent heart to heart;
And there beneath the crimson rose that nods
  And sways above us, free from toil and strife,
We'll quaff to you—forgotten by the Gods—
  Poor souls who linger at the Inn of Life.

On Halloween night they were all back at Toots Shor's—two hundred and fifty of them. Dempsey, Tunney, the Four Horsemen, and just about every other prominent American athlete of the last fifty years, running the gamut alphabetically from Eddie Arcaro to Johnny Weissmuller. Politicians like James A. Farley and Manhattan District Attorney Frank Hogan; shows business types like Jackie Gleason, Ed Sullivan and Douglas Fairbanks, Jr. And, like a living scrapbook of Rice's past came the writers he had known. From the clatter of the sports department of the old *Evening Mail,* circa 1912, came Rube Goldberg. From the lawns of Long Island in the twenties came Herbert Bayard Swope. From out of a hundred smokey Pullman cars crisscrossing the country in the thirties and forties came Tom Meany and John Kieran and Bruce Barton and Jack Wheeler. From a hundred hotel rooms where the glasses tinkled with ice and the conversation sometimes greeted the sunrise came Bill Cunningham, Bill Corum, Dan Daniel, Bob Considine, Tim Cohane and Fred Russell. From the porch of the Vinoy Park Hotel in St. Petersburg, awash in the March sunshine, came Red Smith and Frank Graham.

A five-foot-high blow-up of Willard Mullin's sketch of Granny looked down on the gathering from high on the dais. It was the eve of what would have been Granny's seventy-fourth birthday, and the eve of the publication of his memoirs, *The Tumult and the Shouting*—soon to become a best seller, breaking all records for sports books, selling more than 150,000 copies.

Lowell Pratt of A. S. Barnes had gathered them together to help launch the book. But Toots Shor wouldn't let Pratt pick up the tab. A few months before he died, Shor had promised Rice that on his seventy-fifth birthday he was going to throw him the biggest party he had ever seen. It was a year early and Rice was notably absent, but this was going to be the party that Shor had promised. The crowd spent the first hour of the party crowded around the television set in the bar, watching Ed Sullivan introducing the Four Horsemen, seated in his audience, then holding up a copy of Rice's book and telling his viewers that *The Tumult and the Shouting* was a *"must."* As they watched Douglas Fairbanks, Jr., read "The Ghosts of the Argonne" from Rice's book, Rube Goldberg muttered: "If Granny were here now, he'd be talking."

It was that sort of evening, tinged with sentiment yet bathed in spirit and fun. The speeches were short and generated grins, not tears. "This is the sort of party Granny would've liked," somebody noted.

"I've been offering eight-to-five all night that Granny will show up before the party's over," said Caswell Adams "and I ain't gettin' no takers. Everybody's afraid to bet against it."

Acknowledgments

Oone of the treasures of a project like this is the kindnesses of others.

At Vanderbilt University in Nashville, Marice Wolfe and her staff in the Special Collections Division made my weeks of sorting through the Grantland Rice papers a delight. (Whether for inspiration or not, they also put me in a room where a bust of Grantland Rice looked down upon me as I worked.) The Rice papers, given to Vanderbilt by Kate Rice in the early sixties, are short on letters and other personal documents, and the collection of professional papers is frustratingly incomplete. Still they were an indispensable resource, and I thank Ms. Wolfe for her help in making sense of them.

A heartfelt and particular thanks to Fred Russell, sports editor of the *Nashville Banner* for some fifty years, who still, deep into his eighties, types out a regular column for the paper. Russell's respect and affection for Rice is undiminished by the passing of so many years; his memories are personal and colorful and touched with sentiment, and my interviews with him gave this story an immediacy and a human edge that I simply could not have found anyplace else. Like Rice, Russell is a Vanderbilt man and an American original; and I shall always be grateful for his courtesy, his enthusiasm and his confidence.

Catherine Mecca was Grantland Rice's personal secretary for more than ten years, and though I have yet to meet her, we have corresponded and spoken on the telephone and her mark is on this book more than she probably suspects.

I have also yet to meet Jane Luthy Champion, Kate Rice's niece; but we have corresponded and she has shared with me her elegantly written memoir of her family. This memoir is rich in details of Granny and Kate Rice's courtship and wedding, and also contains Ms. Champion's own wonderful memories of summers spent with Granny and Kate in East Hampton during the 1920s.

David Camerer had the unique opportunity of not only knowing Grantland Rice, but also of working with him when he was assembling his memoirs in the last year of his life. This brought Camerer a perspective on Rice's life and career that went well beyond his personal experiences with the man, and he was thus able to provide context as well as anecdote; and he provided it all with courtesy and care.

To these four people in particular, and to all the others who knew Grantland Rice and shared their memories of him with me, I would hope that the man you find in these pages is the man you knew and admired.

At Oxford University Press I found the perfect editor in Sheldon Meyer. To this project he bought the skilled hand of the consummate professional and the enthusiasm of a man who read Grantland Rice as a boy. Leona Capeless did the line editing with a keen eye and good cheer, resisting the urge to grow exasperated at the many idiocies she found in the original manuscript. My agent, Robert Lescher, found these good people for me, which increases my considerable debt to him.

Colleagues at Northeastern University have not only indulged my preoccupation with Grantland Rice for nearly four years, they have also provided real help. LaRue Gilleland, Nick Daniloff, and Bill Kirtz have all provided counsel and encouragement. Jim Ross, his toils on his own book notwithstanding, has been a critic, sounding board and, above all, a good friend.

Finally, my thanks to Cathy, whose smile is still as beguiling as it was the day I first saw it more than twenty years ago, and to whom this book is lovingly dedicated.

Duxbury, Massachusetts   Charles Fountain
April 19, 1993

Selected Bibliography

Alexander, Charles. *Ty Cobb*. New York: Oxford University Press, 1984.
————. *John McGraw*. New York; Viking, 1988.
Ashley, Perry J., editor. *Dictionary of Literary Biography, Vol. 25: American Newspaper Journalists 1901–1925*. Detroit: Gale Research, 1984.
————. *Dictionary of Literary Biography, Vol. 29: American Newspaper Journalists 1926–1950*. Detroit: Gale Research, 1984.
Ashley, Sally. *F. P. A. The Life and Times of Franklin Pierce Adams*. New York: Beaufort Books, 1986.
Asinof, Eliot. *Eight Men Out*. New York: Holt, Rinehart and Winston, 1963.
Baker, William J. *Jesse Owens: An American Life*. New York: The Free Press, 1986.
Berkow, Ira. *Red: A Biography of Red Smith*. New York: Times Books, 1986.
Breslin, Jimmy. *Damon Runyon: A Life*. New York: Ticknor and Fields, 1991.
Champion, Jane Luthy. *Echoes of the Past*. Privately published, 1986.
Churchill, Allen. *Park Row*. New York: Rinehart, 1958.
Cobb, Irvin S. *Exit Laughing*. Indianapolis: Bobbs-Merrill, 1941.
Conkin, Paul K. *Gone with the Ivy: A Biography of Vanderbilt University*. Knoxville: The University of Tennessee Press, 1985.
Corum, Bill. *Off and Running: The Autobiography of Bill Corum*. New York; Henry Holt, 1959.
Creamer, Robert W. *Babe: The Legend Comes to Life*. New York: Simon and Schuster, 1974.
Davis, Richard Harding. *Gallegher and Other Stories*. New York: Charles Scribner's Sons, 1917.
Dempsey, Jack, with Barbara Piattelli Dempsey. *Dempsey*. New York: Harper and Row, 1977.
Elder, Donald. *Ring Lardner; A Biography*. Garden City: Doubleday, 1956.

Ellis, Edward Robb. *Echoes of Distant Thunder: Life in the United States, 1914–1918*. New York: Coward-McCann & Geoghegan, 1975.

Fowler, Gene. *Skyline: A Reporter's Reminiscences of the '20s*. New York; Viking, 1961.

Fowler, Will. *The Young Man from Denver*. Garden City: Doubleday, 1961.

Gallico, Paul. *The Golden People*. Garden City: Doubleday, 1965.

Gardner, Martin. *The Annotated Casey at the Bat,* second edition. Chicago: The University of Chicago Press, 1984.

Geismar, Maxwell, editor. *The Ring Lardner Reader*. New York: Scribners, 1963.

Graham, Frank Mr. *A Farewell to Heroes*. New York; Viking, 1981.

Hart-Davis, Duff. *Hitler's Games: The 1936 Olympics*. New York: Harper & Row, 1986.

Hecht, Ben, and Charles MacArthur. *The Front Page,* New York: Charles French, 1928.

Holtzman, Jerome. *No Cheering in the Press Box*. New York: Holt, Rinehart and Winston, 1973.

Hynd, Noel. *The Giants of the Polo Grounds*. New York: Doubleday, 1988.

Johnson, William Oscar, and Nancy P. Williamson. *Watta Gal: The Babe Didrikson Story*. Boston: Little, Brown, 1977.

Jones, Bobby, with O. B. Keeler. *Down the Fairway*. New York: Minton, Balch, 1927.

Kieran, John. *Not Under Oath*. Boston: Houghton Mifflin, 1964.

Lardner, Ring, Jr. *The Lardners: My Family Remembered*. New York: Harper and Row, 1976.

Lieb, Fred. *Baseball As I Have Known It*. New York: Coward-McCann & Geoghegan, 1977.

Lipyte, Robert. *Sportsworld: An American Dreamland*. New York: Quadrangle, 1975.

Littlewood, Thomas B. *Arch, A Promoter, not a Poet: The Story of Arch Ward*. Ames: Iowa State University Press, 1990.

Lubow, Arthur. *The Reporter Who Would Be King: A Biography of Richard Harding Davis*. New York: Charles Scribner's Sons, 1992.

Mandell, Richard D. *Sport: A Cultural History*. New York: Columbia University Press, 1984.

―――――. *The Nazi Olympics*. Urbana: University of Illinois Press, 1971.

Marsh, Irving T., and Ehre, Edward, editors. *Best Sports Stories of 1944*. New York: E. P. Dutton, 1945.

―――――. *Best Sports Stories: 1947 Edition*. New York: E. P. Dutton, 1947.

McGaw, Robert A., and Reba Wilcoxon. *A Brief History of Vanderbilt University*. Nashville: Vanderbilt University, 1983.

McLemore, Henry. *One of Us Is Wrong*. New York: Henry Holt, 1953.

Mead, Chris. *Champion: Joe Louis, Black Hero in White America*. New York: Charles Scribner's Sons, 1985.

Mead, William B. *Even the Browns: The Zany True Story of Baseball in the Early Forties*. Chicago: Contemporary Books, 1978.

Meredith, Scott. *George S. Kaufman and His Friends,* Garden City: Doubleday, 1974.

Meyer, Karl E. *Pundits, Poets & Wits: An Omnibus of American Newspaper Columns*. New York; Oxford University Press, 1990.

Mott, Frank Luther. *A History of American Magazines, Vol. IV.* Cambridge: Belknap Press of Harvard University Press, 1957.

Peper, George, general editor, with Robin McMillan and James A. Frank. *Golf in America: The First One Hundred Years.* New York: Harry N. Abrams, 1988.

Peterson, Theodore. *Magazines in the Twentieth Century* Second Edition. Urbana: University of Illinois Press, 1964.

Rice, Grantland. *Baseball Ballads.* Nashville: The Tennessean Company, 1910.

——————, and Jerome Travers. *The Winning Shot.* Garden City: Doubleday, Page, 1915.

——————. *Songs of the Stalwart.* New York: Appleton, 1917.

——————. *Songs of the Open.* New York: Century, 1924.

——————. *Sportlights of 1923.* New York: Putnam, 1924.

——————, editor, with Hanford Powell. *The Omnibus of Sport.* New York: Harper and Brothers, 1932.

——————. *Only the Brave.* New York: A. S. Barnes and Company, 1941.

——————. *The Tumult and the Shouting.* New York: A. S. Barnes and Company, 1954.

——————. *The Final Answer.* New York: A. S. Barnes and Company, 1955.

Ritter, Lawrence S. *The Glory of Their Times.* New York: Macmillan, 1966.

Robinson, Jackie, as told to Alfred Duckett. *I Never Had It Made.* New York: Putnam, 1972.

Rockne, Knute. *The Autobiography of Knute Rockne.* Indianapolis: Bobbs-Merrill, 1930.

Russell, Fred. *Bury Me in an Old Press Box.* New York: A. S. Barnes and Company, 1957.

Schoor, Gene. *100 Years of Notre Dame Football.* New York: Morrow, 1987.

Seymour, Harold. *Baseball: The Early Years.* New York: Oxford University Press, 1960.

——————. *Baseball: The Golden Age.* New York: Oxford University Press, 1971.

——————. *Baseball: The People's Game.* New York: Oxford University Press, 1990.

Smith, Red. *The Red Smith Reader.* New York: Vintage Books, 1982.

——————. *To Absent Friends.* New York: New American Library, 1982.

Stuhldreher, Harry A. *Knute Rockne: Man Builder.* New York: Grosset & Dunlap, 1931.

Thayer, Ernest Lawrence. *Casey at the Bat: A Centennial Edition,* Afterword by Donald Hall. Boston: David R. Godine, 1988.

Walker, Stanley. *City Editor.* New York: Frederick A. Stokes, 1934.

Wallace, Francis. *Knute Rockne.* Garden City: Doubleday, 1960.

Waller, William, editor. *Nashville in the 1890s.* Nashville: Vanderbilt University Press, 1970.

——————. *Nashville, 1900–1910.* Nashville: Vanderbilt University Press, 1972.

Wheeler, John. *I've Got News for You.* New York: E. P. Dutton, 1961.

Wind, Herbert Warren. *Following Through.* New York: Ticknor and Fields, 1985.

Yardley, Jonathon. *Ring: A Biography of Ring Lardner.* New York: Random House, 1977.

Zaharias, Babe Didrikson. *This Life I've Led: My Autobiography.* New York: A. S. Barnes, 1955.

Notes on Sources

Abbreviations

AJ	Atlanta Journal
BG	Boston Globe
CN	Cleveland News
GR	Grantland Rice
KHR	Katherine Hollis Rice
NDN	Nashville Daily News
NT	Nashville Tennessean
NYEM	New York Evening Mail
NYHT	New York Herald Tribune
NYT	New York Tribune
SAS	Stars and Stripes
Sun	New York Sun
Tumult	*The Tumult and the Shouting,* GR's posthumously published memoirs (New York: A. S. Barnes, 1954).
VUH	Vanderbilt University *Hustler*
VUL	GR papers, Special Collections division, Vanderbilt University Library

1. How He Played the Game

Page #

4. Sixty-seven million words: *Tumult,* p. xv.
5. Didrikson visiting press box: Johnson and Williamson, *Watta Gal,* p. 136.
6. "Every time GR meets": Irvin Cobb "Is He James Whitcomb Riley's Successor?" *American Magazine,* August 1917.
5. "most unoriginal remark": Richards Vidmer, quoted in Bob Cooke, "A Tribute to Granny," *NYHT,* July 15, 1954.
5. "pure courtesy": *Tumult,* pp. xi–xii.
5. Ticket from scalper: Red Smith, "They Call Him Granny," *The Sign,* December 1954.
5. Lipsyte quote: Lipsyte, *Sportsworld,* pp. 172–73.
6. "GR was the greatest man": Red Smith, quoted in Arthur Daley's review of *Tumult, New York Times Book Review,* November 21, 1954.
6. "learned more from defeat": *Tumult,* p. xv.
6. "Couldn't say something nice": Victor O. Jones, "GR," *Boston Globe,* July 16, 1954.
7. "even jealous or disdainful": Lipsyte, p. 172.
8. Reid quote: *Tumult,* p. 147.
8. "evangelist of fun": *Tumult,* p. xi.

2. The Blue-gray October Sky

10–11. New York weather, Broadway shows, etc: *New York Times, NYHT,* Oct. 10–18, 1924.
12. "much too exciting": Wallace, *Knute Rockne,* p. 143.
12. Tickets available: *New York Times, NYHT,* October 18, 1924.
13. "as if a building had fallen on me": Gallico, quoted in Holtzman, *No Cheering in the Press Box,* p. 64.
14. "Wherever GR sits": Red Smith, quoted as caption in Willard Mullin cartoon, *New York World-Telegram,* July 14, 1954.
14–21. Background on Rockne, The Four Horseman, and Notre Dame football comes from Wallace; *Tumult,* pp. 178–80; Schoor *100 Years of Notre Dame Football*; and Rockne, *Autobiography.*
21. "shock troops," "with a light backfield . . .": Rockne, pp. 119–20.
21–24. The details of the game are reconstructed from newspaper accounts published on October 19, 1924.
22. "can't run through a broken field until": GR, *NYHT,* Oct. 19, 1924.
23. "Is that the great Mr. Garbisch?": Rockne, p. 144.
24. Party at GR's apartment: *Tumult,* p. 189.
25. Associated Press account: *Boston Herald,* October 19, 1924.

25. Danzig account: *NYT,* October 19, 1924.
26. Runyon account: *New York American,* October 19, 1924.
26. Broun account: *New York World,* October 19, 1924.
27–28. GR account: *NYHT,* October 19, 1924.
28. "From what angle had he been watching": Lipsyte, *Sportsworld,* p. 171.
28. "as tottering as the Rock of Gibraltar": *NYHT,* Oct. 19, 1924.
29. "Strickler claims to have given": Holtzman, p. 147.
29. "Rice remembered it differently": *Tumult,* p. 178.
30. Four Horseman photo: Holtzman, pp. 148–49.
31. "Granny, . . . the day you wrote us up": *Tumult,* p. 181.
32. "with Rice in the carriage behind": Wallace, p. 159.

3. Tennessee: Where the Softer Dreams Remained

33. Historical Commission marker: Assorted clippings, *NB, NT, New York Journal American,* April 26–May 16, 1956. VUL.
34. "I'm going home some day": GR, "Some Day," *NT,* August 18, 1908. Reprinted frequently throughout GR's career.
34. Rice family genealogy: Letter to GR from Albert Hughes Rice and Frierson Hopkins Rice (first cousins to Rice's father, Bolling), January 6, 1951, VUL.
34–35. "100-proof individualist"; waterlogged cotton: *Tumult,* pp. 4–5.
35. Details on the Grantland Cotton Company, as well as Henry Grantland's and Bolling Rice's careers are taken from various Nashville street directories, 1880–1900, Nashville Public Library.
36. "Hog brains . . . hominy grits": etc, *Tumult,* p. 6.
36. "Yes, I knew hard work": *Ibid.*
37. "Pied Piper in my march through life": *Ibid.,* p. 5.
37. Games Nashville boys played: Edward F. Webb, "Boyhood Recollections," included in Waller, *1890s,* pp. 226–33.
38. Broken arm: Waller, *1890s,* p. 233.
38–39. Details of nineteenth-century baseball are drawn principally from Seymour, *The Early Years,* pp. 13–46.
39. North Brookfield, Mass.; Gilmore, Ohio; Factoryville, Pa.; Narrows, Ga; Humbolt, Kan. and Nashville, Tenn., are the home towns of Connie Mark, Cy Young, Christy Mathewson, Ty Cobb, Walter Johnson, and GR.
39. "Just at this time every season": GR, "The Champs of the Alley League," *NT,* March 7, 1909.
39–40. Boys baseball in Nashville: Waller, *1890s,* pp. 218, 228.
40. GR hitting fly balls: *Ibid.,* p. 218.
40. GR "split his school time between": *Tumult,* pp. 6–7.
40. "mental, moral and physical education": Waller, *1890s,* p. 31.
40. Did not teach English grammar: *Ibid.*

40. "I owe it all to you," "Now right there's": *This Is Your Life*, NBC Radio, December 28, 1948. Quoted in untitled clipping, VUL.

40. "Doctor Wallace": Waller, *1890s*, p. 31.

41–42. Vanderbilt University: Unless otherwise cited, the background and details on Vanderbilt University are taken from Conkin, *Gone with the Ivy*.

42. GR's new home: Nashville Street Directory, 1896, Nashville Public Library.

43. "Upon the new men we would urge": *VUH*, September 24, 1897.

43. GR football injuries: *Tumult*, p. 7.

43. "Because football calls for courage": Quimby Melton, "Famed Sportswriter Hated to See Rules Intentionally Broken. *Griffen* (Georgia) *Daily News Magazine*, August 26–27, 1961. VUL.

43. "earned praise from the *Hustler*": *VUH*, October 28, 1897.

44. Near riot at Vanderbilt–University of Nashville game and aftermath: Conkin, pp. 139–141.

44. "Let the yell go up": *VUH*, September 24, 1897.

46. "participated with a vengeance": *The Comet*, Vanderbilt University yearbook, 1898.

46. "active player": *VUH*, March 24, 1898.

47. "The game was really a rough one": Waller, *1890s*, p. 224.

47. "all students . . . will avoid the game of basketball": *VUH*, February 3, 1899.

48. "sensational play of the game": *VUH*, April 4, 1899.

48. GR's statistics and the Vanderbilt record was pieced together from accounts and box scores in the *VUH*, April-June 1898.

49. "will live in Vanderbilt football history": *VUH*, December 2, 1899.

49. "this unsportsmanlike breach": *Ibid.*

51. GR's stats and the accounts of the '00 baseball season are taken from various issues of *VUH*, March–June 1900.

51. "I had fifteen assists": *Tumult*, p. 7–8.

51. Broken shoulder blade: *Ibid.*, p. 7.

51. GR elected captain: *VUH*, February 7, 1901.

52. *Outing* magazine charges and reaction: *VUH*, February 7, 21, March 7, 1901.

53. "to abstain from . . . any . . . form of dissipation": *VUH*, March 14, 1901.

53–54. The 1901 baseball season is reconstructed from various issues of *VUH*, March–June 1901.

55. "prepared and read the class history": *The Commencement Courier*, June 18, 1901. VUL.

55. "honored . . . as one of thirteen": *Ibid.*, June 19, 1901.

55. The speculation on the offer to pay professional baseball is drawn mainly from an author's interview with Fred Russell, who spoke to many people, including Rice, on the matter and wrote of it on several occasions during his career with the *Nashville Banner*. Rice was evasive on the matter throughout his life but did say in his memoirs that Nashville manager Newt Fisher had offered him a contract for $250 a month, which he had to pass up because of his sore arm.

56. Job in the dry goods store: *Tumult,* p. 8.
57. Job at the Nashville Daily News: *Ibid.*

4. Apprentice Sporting Writer

58–59. Pre-Civil War sports and sportswriting: Mandell, *Cultural History.*
   59. "The regatta of the Palisade": *NYT,* July 5, 1863.
59–60. The evolution and development of the sports page is taken from read-
       ings from the 1860s through the 1890s in the *New York Times, NYT,*
       New York Herald, New York World, New York Journal, New York Sun,
       Brooklyn Eagle, Boston Globe, Boston Herald.
   60. "unlettered chroniclers": Walker, *City Editor,* p. 116.
   61. "Dada school of sportswriting": Walker, p. 120.
   61. "Saloons and pool halls . . . installed": Mandell, *Cultural History,* p.
       184.
   62. The scene at the debut of the *NDN* is taken from the account in the
       NDN, July 20, 1901.
   63. Colonel Jere Baxter; railroad bond issue: Waller, *1900–1910,* p. 34.
   64. "wonder what the score was": *Tumult,* p. 9.
   64. "Did you ever hear of the battles": GR, *NDN,* August 14, 1901.
  64n. Footnote: Dan Jenkins, *Fast Copy,* New York: Simon and Schuster, 1988.
       pp. 176–77.
   65. Darwinian reference: *NDN,* September 11, 1901.
   65. "call his antagonist a jackass": *NDN,* October 6, 1901.
   66. Southern League meetings; pennent in dispute: *NDN,* October 19–21,
       1901.
   66. "EGGS BRING CENT APIECE": *NDN,* August 14, 1901.
   66. Traded assignments with Louis Brownlow: Brownlow quote in Waller,
       1900–1910, p. 201.
   66. Society ball: *Tumult,* pp. 9–10.
   67. "didn't know a Christmas tree from": *Tumult,* p. 10.
   67. *NDN*'s perilous finances: Waller, *1900–1910,* pp. 34, 199, 306.
   67. appendicitis: *Tumult,* p. 10.
   67. Dinner for GR and Brownlow: Waller, *1900–1910,* p. 285.
   68. "Great Umpire of the game of life": *AJ,* April 27, 1904.
   69. "took ballplayers on tour of Hermitage": *AJ,* May 11, 1903.
   69. Southern League cities: *AJ,* June 8, 1904.
   69. "recognized as leading authority": "Rice Leaves Atlanta," undated clip-
       ping, VUL.
   69. "Use just a little more backbone": *AJ,* May 18, 1904.
69–70. Ty Cobb writes GR; *Tumult,* pp. 17–31; Alexander, *Ty Cobb,* pp. 18–
       20.
   70. "taken rooms at the Aragon Hotel": *Tumult,* pp. 11–12.
   70. Don Marquis: *Tumult,* pp. 11–15; Sam G. Riley, "Don Marquis," in-
       cluded in Perry J. Ashley, *1900–1950,* pp. 189–96.

72. Frank Stanton: *Tumult,* pp. 14, 324–25, 352; Bruce M. Swain, "Frank L. Stanton," included in Perry J. Ashley, *1900–1950,* pp. 252–68.
73. "riding sidesaddle on the merry-go-round": *Tumult,* p. 33.
73. "Since this hyper-torrid weather": Verse is included in Champion, *Echoes of the Past,* p. 81.
74. Horse and buggy ride: *Tumult,* p. 33.
74. "Listen to how many times": Champion, p. 81–82.
74. "What joy!!! Will it last?": *Ibid.,* p. 82.
74. "Willis is in a rage": *Ibid.*
75. "went to Cleveland": "Sporting Editor of *Atlanta Journal* is a Cleveland Visitor," undated clipping, VUL.
75. Mr. Rice endeared himself": *Ibid.*
75. "some . . . felt . . . job should not go to Rice": *Tumult,* p. 32.
75. "That was real money": *Ibid.*
75. "that ballplayer": Champion, p. 83.
75. Florence Davenport Hollis: *Ibid.,* pp. 9–26.
76. GR letter to Mother Hollis: *Ibid.,* pp. 83–84.
77. "I scampered like a . . . schoolboy": GR, *CN,* March 18, 1906.
77–78. GR and KHR wedding: Champion, pp. 84–86.
78. "huge barrel of china": *Tumult,* p. 34.
79–80. History of "Casey at the Bat": Gardner, *The Annotated Casey,* pp. 1–16.
80. "None of the triumphant sequels will do": Donald Hall, "A Ballad of the Republic," an Afterword to Thayer, *Casey at the Bat,* pp. 31–32.
82. "The Man who Played with Anson on the Old Chicago Team": GR, *Base Ball Ballads,* pp. 73–78.
82. He Never Heard of Casey": GR, *NYHT,* June 1, 1926.

5. Cumberland's Calling: Hometown Sporting Writer

83. "Broad, liberal and forgiving": Yardley, *Ring,* p. 90.
84. Charley Dryden: *Ibid.,* pp. 81–82; Walker, *City Editor,* pp. 116–19.
84. Nashvillians claimed English spoken with purest accent: Waller, *1900–1910,* p. 3.
84. Tennessee State Fair: *NT,* various dates, May–September, 1907.
85. "Once again among your people": GR, "Home and Tennessee," *NT,* September 3, 1907.
85. "carrying Floncy . . . like a football": *Tumult,* p. 35.
85. Luke Lea: Waller, *1900–1910,* pp. 21, 84, 93–94, 342; *NT,* May 12, 1907.
85. "God's Country": *NT,* May 12, 1907.
86. *Bowen Blade* tribute: Undated clipping, VUL.
86. "The Fan's Revery": *NT,* June 9, 1907.
87. Dan McGugin: Conkin, *Gone with the Ivy,* 214–17. Also, letters from McGugin to GR, VUL.

88. Lack of VU student spirit: *NT*, October 7, 10, 1907.
88. Trip to Annapolis: *Ibid.*, October 10–16, 1907.
90. "the shooting was good": *Ibid.*, December 17, 1907.
90. "GR worked at . . . desk overlooking": Waller, *1900–1910*, pp. 190–91.
90. Rice's working day: *Ibid.; Tumult*, pp. 35–36; Louis Davis, "At the Top, But Dreaming Still," a profile of GR in *The Nashville Tennessean Magazine*, February 6, 1949. Rice used the Davis piece as a prod to his memory when writing his memoirs; the similarity between Ms. Davis's research and what Rice included in *Tumult* five years later is too acute to be dismissed.
91. "How can I think of songs": GR, "The Disturber," *NT*, June 9, 1909.
91. Took up golf: *Tumult*, p. 37.
92. Refered football: *Ibid.*; also, various football box scores, *NT*, 1907–1910.
92. "I will get me a new hat": Champion, *Echoes of the Past*, p. 87.
92. "From the Commodore point of view": *NT*, April 1, 1908.
95. "It isn't so much 'Did you make a hit' ": GR, "Out on the Lines," *NT*, August 1, 1907.
96. Naming the Volunteers: *NT*, February 14, 17–18, 29, 1908.
97. Nor can Prof. Holmes be depended upon": *NT*, Nov. 12, 1907.
97. "wide of the mark": GR quoted letter in his *NT* column, November, 16, 1907.
97. "Dudley censored Georgia"; "Georgia fired": *NT*, November 21, 23, 1907.
97. L.S.U. scandal: *NT*, November 4, 10, December 3, 6, 8, 1908.
97. Ad for haberdasher: *NT*, November 20, 1910.
99. "There was a lightheartedness": *Tumult*, p. 317.
99. Poetry scrapbooks: Now a part of the GR papers, VUL.
99. "Grantland could write a poem . . . almost as quick": Anonymous quote in Davis, "At the top . . ."
99. "How or why I fell": *Tumult*, p. 9.
99. "Sometimes I can write one in twenty minutes": Davis.
100. "rarely rose above . . . rhyming roadside signs": Johnson and Williamson, *Watta Gal*, p. 91–92.
100. "near poet": *NT*, July 20, 1907.
100. "The average life is all I care": GR, "The Average," *NT*, April 29, 1909.
100. "Story of Two Poems": *NT*, September 22, 1907.
101. "When you ponder with acidity": GR, "Taking Comfort," *NT*, January 23, 1908.
102. Irvin S. Cobb quote: Cobb, "Is He James Whitcomb Riley's Successor?"
102. "his art, simple, glowing and precise": *Boston Evening Transcript*, January 2, 1918.
102. "truly American", "A New Poet of Americanism": Anonymous and undated clipping, VUL.
103. "The Return": *NT*, March 10, 1909.
104. "Wouldn't you be willing": *NT*, December 3, 1910.
104. "At the annual track meet": *NT*, June 13, 1907.

104. "Shakespeare and Beethoven": *NT*, June 8, 1907.
105. "If a baserunner": *NT*, June 10, 1907.
105. How to write verse: *NT*, May 29, 1907.
106. "may not be a wide open town": *NT*, October 9, 1909.
106. "Sherlocking": *NT*, August 26, 1909.
106. "The Hills of Fame still beckon": Gr, "The Hills of Fame," *NT*, May 16, 1908.
107. Job offer at the *Mail;* Stoddard quote: *Tumult*, p. 37–38.
108. Marmaduke B. Morton; "can you write rhymes?": Ralph McGill, "Too Many Vacant Chairs," *Atlanta Constitution*, undated clipping (written shortly after GR's death), VUL.

6. Big City Sporting Writer

109. Journalism's three eras: Cobb, *Exit Laughing.*
110–111. Richard Harding Davis sketch is drawn from Lubow and from Churchill.
110. "young boys dreams of becoming": Churchill, *Park Row*, p. 169.
111. "All Gallegher knew": Davis, *Gallegher*, p. 4.
111. "Peeking through keyholes!": Hecht and MacArthur, *The Front Page*, p. 36.
112. "Readers constant and fickle": *NYEM*, January 12, 1911.
112–114. F.P.A.: Sally Ashley, *F.P.A.*
113. "genesis of the New York City myth": *Ibid.*, p. 11.
113. "raised her from a couplet": *Ibid.*, p. 129.
113. "Dozens of things he said": Meredith, *George S. Kaufman and His Friends*, p. 31.
113. "Nothing is so responsible"; "ninety-two percent of the stuff": *Ibid.*
114. "Tinker to Evers to Chance": Sally Ashley, p. 65.
114. "Frank may write a better piece": *Tumult*, p. 316.
114. G.S.K.: *NYEM*, January 12, 1911.
115. Rube Goldberg; "simple bookmark": *Time*, December 7, 1942.
115. "Goldberg greeted him warmly": *Tumult*, p. 38.
115. "Adams arranged for quarters": *Ibid.*
116. Morningside Heights neighborhood: *Ibid.*, pp. 38–39.
116. "we have set our soul ahead": *Ibid.*, p. 40.
116. "box score doesn't absorb much space": *NYEM*, March 29, 1911.
116. F.P.A. poem on language: *NYEM*, March 16, 1911.
117. "we record the demise of 'baseball slang' ": Undated *Cleveland News* clipping, VUL.
117. "baseballically speaking"; "keep weather orb"; "crack slabman": *NT*, various dates, 1907.
118. "Are the scribes slowing up?": *NYEM*, February 27, 1913.
118. "The movement to eliminate slang": *NYT*, undated 1923 clipping, VUL.

118. "foremost freshman class": Holtzman, *No Cheering in the Press Box*, p. 47.

119. Ty Cobb's salary: *NYEM*, April 21, 1913.

120. "had become Giant-conscious": Hynd, *The Giants of the Polo Grounds*, p. 178.

120. "great to be young and a Giant": *Ibid.*, p. 155.

121. "box seats would be filled with": *Ibid.*, pp. 158–159.

121. "New York of the brass cuspidor": *Ibid.*, p. 171.

121–122. Mathewson: *Ibid.*, pp. 115–17; Alexander, *John McGraw*, pp. 100–102.

121. "What a grand guy": Ritter, *The Glory of Their Times*, p. 17.

121. "a little bit better at games": *Tumult*, p. 45.

122. "Always have an alibi": *Ibid.*, p. 46.

122. McGraw and Dreyfuss: Alexander, *McGraw*, pp. 113–15; Hynd, pp. 130–32.

123. "courted dislike": *Tumult*, p. 291.

123. "only fun when I'm winning": Hynd, p. 188.

123. McGraw and Mathewson: Hynd, p. 120.

123. "combines in a rare degree": Alexander, *McGraw*, p. 171.

124. McGraw blasting John Tener: *Ibid.*, p. 199; Hynd, pp. 202–3.

124. "don't give a damn": Alexander, *McGraw*, p. 199.

124. "did not make these statements": *Ibid.*

124. "Everyone connected with the signed statement": GR, *NYT*, June 21, 1917.

125. McGraw gives GR off-day story: *Tumult*, p. 291.

125. New York prices, *c.* 1911; Hesse & Loeb's: Churchill, pp. 220–24.

126. Floncy visits F.P.A.: Sally Ashley, p. 87.

126. GR contributes to F.P.A. play: *NYEM*, March 5, 1913.

126. "FLIER WRECKED / GIANTS LOSE": *NYEM*, July 11, 1911.

126. "was a time when sportive life had intermissions": *NYEM*, December 8, 1913.

127. "Feds are getting more": *NYEM*, January 20, 1914.

127. Jim Thorpe case: *NYEM*, January 24–February 22, 1913.

127. "know of at least four": GR, *NYEM*, February 4, 1913.

128. "one who can get away with it": *Ibid.*

128. GR lobbies on behalf of restoring Thorpe's medals: *Tumult*, pp. 235–36.

128. "What right did the AAU": *Ibid.*, p. 235.

129. "fools rush around and score": *Collier's*, May 3, 1913.

130. "Over the uncharted way that 1915 has": *NYEM*, December 31, 1914.

131. "Puffed Rice": *NYT*, January 5, 1915.

132. "drab outline against dull gray sky": *NYT*, May 11, 1917.

133. Against multi-year contracts: *NYT*, July 1915.

134. "at heart a sentimentalist": Walker, *City Editor*, p. 127.

135. "No amount of build-up could convince": *Ibid.*

135. "Considering the personal equation": *NYT*, September 7, 1916.

135. "At his age the game has never": *NYT*, September 10, 1916.

135. "Adventures of Beatrice Buggs": *NYT*, September 22, 1916.

7. Of Clotted Gore and Dreamless Sleep

139. "at least 250,000 golfers": *NYT*, January 23, 1915.

139. "I'm going to have the sorest arm": *NYT*, March 18, 1916.

139. "no touch of sport in war": *NYT*, April 10, 1917.

140. *"among those who are left": NYT*, March 2, 1915.

140. U.S. regaining sports championships after war: *NYT*, January 18, 1917.

141. "Sergeant Gibson 'greaty surprised' ": *NYT*, March 25, 1917.

141. "curtain will be drawn": *NYT*, May 5, 1917.

141. College baseball cancels season: *NYT*, April 9, 1917.

142. GR suggestion to play football with new men: *NYT*, August 8, 11, 1917.

142. Canadian vacation; *NYT*, May 30–June 1, 1917.

143. "When the big debt is due": *NYT*, April 10, 1917.

143. GR posing as Max Foster: *Tumult*, pp. 343–44.

144. "With all its misery and death": GR, "The Great Adventure," *NYT*, undated 1917 clipping, VUL.

144. Roosevelt, Fitzgerald, Kilmer, Barrymore: Ellis, *Echoes of Distant Thunder*, pp. 347–48.

145. "left for Camp Sevier": *Trench and Camp*, Camp Sevier newspaper, December 18, 1917, VUL. Also, *Tumult*, p. 89. Rice fixes the date in *Tumult* as December 5. The contemporary account seems the more reliable.

145. Snow at Camp Sevier: *Ibid.*

146. "cold lasted until Armistice": *Tumult*, p. 89.

146. Y.M.C.A. Building: *Trench and Camp*, December 18, 1917.

146. "annoyances and discomforts": *NYT*, January 5, 1918.

146. Clearing the field: *Tumult*, pp. 90–91.

147. Birth of *SAS*: *SAS*, February 7, June 13, 1919.

147. Harold Ross; Meredith, *George S. Kaufman and His Friends*, pp. 282–83.

148. "is but one Big League today": *SAS*, July 26, 1918.

148. "With thousands of their countrymen": *SAS*, August 2, 1918.

148. GR reported from the front: Unsigned typescript profile of GR, VUL, p. 14.

149. Poker games: Merdith, p. 162–63

149. "cover the comic side of war": Yardley, *Ring*, p. 199.

149. "orders reassigning him to 115th": Unsigned typescript profile, VUL, p. 14.

149. "enough equipment to quartermaster": *Tumult*, p. 93.

149. "infernal gas mask": "shed stuff like a moulting": *Ibid.*

150. "generals didn't want to go back to being": Unsigned typescript profile, VUL, p. 14.

150. GR's journey to the front: *Ibid.*

150. Baldridge sketch: Picture is in the Rice papers, VUL, inscribed: "Flirey, France. In memory of the day the Yanks got St. Mihiel." Rice's caption to the sketch tells the story of the dugout and the dead German.

150. "flaming spirit of the company"; "marched to war": Edwin Alger, typescript profile of GR for radio program "Who's Behind the Name?", December 29, 1930, VUL, p. 11.

150. "would downplay his role": *Tumult*, p. 93; various profiles of GR, post 1918, VUL.
150. "incoming round landed five yards away": Unsigned typescript profile, VUL, p. 14.
151. "with their ideals and idols exactly the same": *Tumult*, p. 94.
151. "moved to Third Corps Headquarters": *Tumult*, p. 95.
151. "alerted reporters to . . . German prisoners": Unsigned typescript profile, VUL, p. 15.
151. "Teutonic land of gray fogs": GR to KHR, December 14, 1918, VUL.
151. "It was a strange sight": *NYT*, March 4, 1919.
152. "startled Teutonic faces": *Ibid.*
152. Complaints about cantonment life: GR to KHR, December 14, 1918, VUL. GR's thoughts on officers who mistreated enlisted men are detailed in a passionate column after his return, *NYT*, March 27, 1919.
152. Kate and her sister: Champion, *Echoes of the Past*, p. 42–43.
152. Sent to Angers. *Tumult:* pp. 95–96.
153. Shipping home: *Ibid;* Also, John Wheeler, "My Most Unforgettable Character," *Reader's Digest*, December 1965.
153. Lake Placid: *Tumult*, p. 97.
154. "One at a time they come": GR, "Six Years Later," *NYHT*, November 11, 1924.
154. Lost all his money: *Tumult*, p. 97.
154. Bought flowers for funeral: Various profiles of GR, post 1918, VUL.

8. Black Clouds on a Golden Dawn

155. "get together on a peace programme": *NYT*, undated 1919 clipping, VUL.
156. "those . . . patriots who ducked service": *NYT*, March 9, 1919.
156. "A.E.F. takes its time": *NYT*, April 4, 1919.
156. Spring Training; GR, "In the Training Camp," a daily column from Florida, *NYT*, March 23–April 4, 1919.
158. GR's first writing on Babe Ruth: *NYT*, May 17, 1917.
158. "once my swing starts": *Tumult*, p. xvi.
158. "His is the perfectest": *Ibid.*, p. 102.
158. Shipbuilding league; Ruth holdout: Creamer, *Babe*, pp. 163, 187–89.
159. Ruth's home run show in Gainesville: *NYT;* April 9, 1919; *Tumult*, pp. 101–2.
159. The onion: *Tumult*, p. 102.
159. "sharp-breaking curve of lefthander": *NYT*, April 9, 1919.
159. "bum will hit into a hundred double plays": Creamer, *Babe*, p. 190.
159. Intensity of Sox-Giants series: *Ibid.*, pp. 190–192.
160. Jim Thorpe looking to wrestle: *NYT*, April 10, 1919.
160. "I've seen a few I thought could hit": *Tumult*, p. 104.
160. "Ruth, the man-boy": *Ibid.*, p. 114.

160. Ruth on the radio: *Ibid.*, pp. 112–13.
161. Ruth visiting child: *BG*, August 17, 1948.
161. "Get out your golf clubs": *Ibid.*, June 4, 1935.
161. "Game called by darkness": *Ibid.*, August 17, 1948.
162. GR ridicules Jess Willard: *NYT*, May 1917; August 23, 1917.
162. "If boxing . . . is brutal": *NYT*, March 30, 1916.
162. 400 reporters in Toledo: *NYT*, July 1, 1919.
162. "with nothing in either workout": *NYT*, June 28, 1919.
163. "peculiar human menagerie": *Ibid.*
163. "one con trying to out-con": Dempsey, p. 94.
163. "crafty son-of-a-bitch"; "sneaky no-good": *Ibid.*
163–64. Dempsey-Willard background: Dempsey, *Dempsey*, pp. 96–111.
164. "captured the public fancy": Joseph Durso, Introduction to Dempsey, p. xiii.
164. Doc Kearns: *Ibid.*, pp. 61–87.
165. "gotta sell them good, kid": *Ibid.*, p, 73.
165–66. Rice account of fight: *NYT*, July 5, 1919.
167. "blow that hurt more than Willard's punches": Dempsey, p. 122.
167. "not the champion fighter": *NYT*, July 5, 1919.
168. "it hurt like hell": Dempsey, p. 123.
168. "I'll set Rice straight": *Ibid.*
168–72. Background on the 1919 White Sox and the Series fix comes primarily from Asinof, *Eight Men Out.*
170. Hal Chase: Seymour, *The Golden Age*, pp. 288–93.
172. "will be no slaughter": *NYT*, September 27, 1919.
172. "This Series is fixed": *Tumult*, p. 105.
172. Mathewson and Fullerton: Asinof, *Eight Men Out*, pp. 46–47.
173. Mathewson nodded: *Ibid.*, p. 62.
173. "bottom of the order starts hitting you": *Ibid.*, p. 67.
173. "Cicotte will only be beaten": *NYT*, September 27, 1919.
174. "keep the Reds hitting until darkness fell": Asinof, pp. 68–69.
174. "forever blowing ball games": Asinof, p. 94.
174. "GR TELLS HOW": *NYT*, October 2, 1919.
175. "was wilder than Tarzan": *NYT*, October 3, 1919.
175. "ancient wing . . . , ancient bean": *NYT*, October 5, 1919.
176. "only past performance chart that counts": *NYT*, October 16, 1919.
176. "easiest sport to fix": *NYT*, October 28, 1919.
177. Comiskey statement: Asinof, p. 129.
177. Fullerton article: Asinof, pp. 132–36.
178. "I believe the series . . . was absolutely honest": *NYT*, December 22, 1919.
178. "accused the National Commission": *NYT*, September 30, 1919.
178. "prison sentences for every ballplayer": *Ibid.*
178. "anyone who would extend a welcome"; "crooks . . . are gone forever": *NYT*, October 14, 1920.
179. "clean sweep at the top": *Ibid.*
179. "proposed William H. McCarty": *NYT*, October 21, 1920.
179. "called for national figure": *NYT*, October 5, 1920.

180. "had forgotten what terrible things worlds series were": Ring Lardner, Jr. *The Larnders*, p. 145.
180. "kicked in the stomach": *Tumult*, p. 106.
180. "just how much of it is sport":, *NYT*, October 24, 1920.

9. A Golden Age for Friends

181. "I have seen your verses": Warren G. Harding to GR, March 9, 1921, VUL.
181. "not very much of a player": Warren G. Harding to GR, March 17, 1921, VUL.
182. Seated next to Vice President: Ring W. Lardner, "The Presidential Golf Might Be Better," The Bell Syndicate, April 29, 1921, VUL.
182. "wife doesn't like Great Neck": *Tumult*, p. 327.
182. "I and the 1st lady": Lardner, "Presidential Golf."
182. "got a servant problem": *Ibid.*
182. "mine will be very risque": *Ibid.*
183. GR on Harding golf game: *NYT*, April 6, 1921.
184. "rather be Rice than President": Lardner, "Presidential Golf."
184. "Harding was a poor President": *Tumult*, p. 327.
184. Yardley quote: Yardley, *Ring*, p. 200.
184. Ring Lardner, Jr. quote: *The Lardners*, p. 168.
185. "never quite knew that I knew him"; "wrapping up a wraith": *Tumult*, p. 326.
185. "How can you write if": Ring Lardner, Jr., p. 175.
185. "realism"; "viewpoint and philosophy": Untitled, unsigned essay, *Kansas City Times*, August 12, 1938, VUL.
185. "the literary firmament Rice possesses": "King of Sports", unsigned essay in *Dayton Daily News*, October 12, 1935.
185. "one of the first poets": H. G. Salsinger, *Detroit News*, February 28, [year missing], VUL.
186. "doubt whether Mr. Lardner will survive": *Ibid.*
186. "of colored parents": *Tumult*, p. 326.
186. "Kate and Ellis who really forged": Ring Lardner, Jr., p. 168.
187. "Rices have been with us right along": *The Lardners*, p. 170.
187. "genuine pulchritude": *New York World Telegram*, undated clipping (1928–29), VUL.
187–88. "I Can't Breathe": Geismar, *Ring Lardner Reader*, pp. 214–23.
188. "exactly caught the self-indulgence": Yardley, p. 312.
188. Toots Shor looking out for Floncy: Interview with David Camerer.
189. *June Moon*: Yardley, pp. 333–36.
189. "like a duck to golf": *New York Daily News*, October 6, 1929.
189. Comments on Floncy's acting: Florence Rice file, VUL.
190. "won't be known as the daughter": Radio script, WOR, May 8, 1934, VUL.

190. "no indication of property line": Ring Lardner, Jr., p. 171.
190. "on a porch on our dune": *Tumult*, p. 336.
190. Parties in East Hampton: *Ibid.*, pp. 338–39; Champion, pp. 90–93.
191. GR's love of watermelon and bourbon: Champion, *Echoes of the Past*, p. 101.
191. "have been many other friends": *Tumult*, p. 339.
191. Storm damages East Hampton homes: *Ibid.*, pp. 337–38; Lardner, Jr., p. 204.
192. Lardner death: *Ibid.*, pp. 339–340; Lardner, Jr., p. 226; Yardley, p. 381.
192. "Charon—God guide your boat": *Tumult*, pp. 340–41.

10. Fame, and Other Earthly Rewards

193. GR forgets press credentials at Forest Hills: Henry McLemore, "Sports Writer No. 1" *Look*, January 27, 1942.
194. GR forgets credentials at World Series: John Wheeler, "My Most Unforgettable Character," *Reader's Digest*, December 1965.
194. "Who the hell hit me?": Robert H. Davis. "People You Read About", *Chicago Daily News*, December 31, 1924, VUL.
194–96. GR broadcasts World Series game: *NYT*, October 1, 5, 1922.
196. "voice ill-suited to radio success": There are some Rice broadcasts and appearances that survive in the collection at the Museum of Radio and Television in New York.
196–98. Sportlight Films: Interview with Catherine Mecca; *Tumult*, pp. 257–65; "Puts the Spotlight on Every Light-O-Sport," undated magazine article, VUL.
198. McGeehan and moose hunting: *Tumult*, pp. 260–61.
198. "sturdy clarion of fellowship": *New York Post*, October 25, 1924.
198. "poetry of the diamond and gridiron": *Boston Evening Transcript*, November 1, 1924.
199. "somewhere in this blighted land": GR, *Sportlights of 1923*, p. 112.
199–201. Dempsey-Gibbons: *Ibid.*, pp. 33–44; *Tumult*, pp. 123–27; *Dempsey*, pp. 152–55.
201. Dempsey-Firpo: GR, *Sportlights of 1923*, pp. 49–60; *Tumult*, pp. 120–22; Dempsey, pp. 155–62; *NYT*, September 15, 1923.
202. *The Kick Off:* W. O. McGeehan, "GR and Craven Smash Drama Line with 'The Kick Off,' " *NYHT*, November 25, 1925.
202. "No one could take his place": *Tumult*, p. 331.
203. "to name the world's best sprinter": *NYT*, July 5, 1924.
203. "No one would have believed": *NYT*, July 8, 1924.
203. Eric Liddell: *NYT*, July 9, 1924.
204. Parvo Nurmi: *NYT*, July 11, 1924.
204. "the Olympics' greatest star": *NYT*, July 17, 1924.
205. "Sport needs the revival": *NYT*, July 19, 1924.
205. GR on Walter Johnson: *NYHT*, October 11, 1924.

206. Jeweled gold watch: Frederick G. Lieb, "Scribes Pick GR for Fourth Spink Citation," *The Sporting News*, November 12, 1966.

207. New contract: A copy of the contract is in the GR papers, VUL.

207–8. All-America teams: James M. Young, "How the Annual All-America Football Teams Are Selected," *MacGregor Goldsmith Sportsvue*, undated, VUL; *Tumult*, pp. 206–8.

208. "most scholarly sport scribe": Otto Floto, "GR the New Camp, A Fine Selection," *Denver Post*, undated, VUL.

209. Football was his favorite: Quimby Melton, *Griffen* [Ga.] *Daily News Magazine*, August 26–27, 1961.

209. "Due to the ingredients": *Tumult*, p. 219.

209. "soul of the college spirit": *Ibid.*, 217.

210. Exchanged letters and telegrams: VUL

11. A Game for the Age—A Champion for the Ages

211. "twenty percent mechanics and technique": Peper, *Golf in America*, p. 264. This book, published in 1988, also named Rice one of "100 Heroes of American Golf" over the previous one hundred years.

212. "gives you insight into human nature": *Tumult*, pp. 53–54.

213. McGeehan inserts in GR golf story: *NYT*, January 26, 1917.

214. "your humble correspondent doesn't mind": *NYT*, January 27, 1917.

214. "Golf and stag conviviality": *Tumult*, p. 310.

214. Birth of Artists and Writers; *Ibid.*, p. 309.

215. Rice was president: *Ibid.*, p. 310.

215. "Like other dictators": Rube Goldberg, "Artists and Writers Tournament", *Golf*, May 1941.

215. "old-age, arthritis or bankruptcy": *Ibid.*

215. GR and Rex Beach dominate: Various news accounts, VUL; *Tumult*, pp. 310–11.

215. "I seldom worried": *Tumult*, p. 311.

215. Hagen quote: *Ibid.*, p. 60.

216. Lakeside Golf Club: GR, "Club of Stars," *Golf*, April 1938.

217. "a most aristocratic exercise": Peper, *Golf in America*, p. 156.

217–18. Francis Albertanti resists golf: *Tumult*, pp. 60–61.

218. Briggs cartoon: Peper, p. 163.

218. Hitting golf balls out of Polo Grounds: Undated clipping, pre-World War I, VUL.

219. Golf's emergence: Peper, pp. 8–23.

219. "I was told that fresh capital": *Tumult*, p. 308.

220. "$7500 a year": Contract between GR and Centurion Publishers, VUL.

220. "automobile advertising immediately fell out": *Tumult*, p. 309.

221–22. Jones background is taken from Miller and from Jones.

221. "Here I was thirteen years old": Jones, *Down the Fairway*, p. 53.

221. GR watches Jones throw club: *Tumult*, pp. 74–75.

222. Breakfast with Jones: *Ibid.*, p. 77.
222. "he ran out of clubs first": Jones, p. 66.
222. "sporting croll where only the select": *NYT*, September 7, 1916.
223. Jones stays with GR in New York: *Tumult*, p. 79.
224. Jones picking up in British Open: Jones, pp. 106–8.
225. Stewert Maiden quote: *BG*, June 1, 1930.
225. Jones hinted at retirement: *Ibid.*
225. "has a good chance now": GR, *BG*, June 21, 1930.
225. Ticker-tape parade: Miller, pp. 117–18.
226. "remarkable thing about this championship": GR, *BG*, July 10, 1930.
226. "lily-pad shot": Miller, pp. 120–26.
226. "having something to protect": *Ibid.*, p. 127.
227. "Bobby stroked the ball": GR, *BG*.
227. "quickly up the gentle slope": Miller, p. 140.
227. "I was lucky": *BG*, July 13, 1930.
227. "Bobby Jones broke all records": GR, *BG*, July 13, 1930.
228. "old man par": Jones, pp. 39, 49–51.
228. "Impregnable Quadrilateral of Golf": Miller, pp. 10–11.
228. Al Laney quote: *Ibid.*, p. 13.
228. "frantic on the one hand"; "burden of nation's worship"; "Amateur more than golf": Miller, pp. 8–9.
228. Merion press room; crowds at practice rounds: Miller, pp. 12–20.
229. Jones talks to GR during practice round: *Tumult*, p. 84.
229–30. 1930 U.S. Amateur is taken from Miller and from news accounts.
230. "dislikes aftermath of glory": GR, *BG*, September 29, 1930.
230. Jones retirement: Miller, pp. 145–46.
230–32. Augusta National and The Masters: Miller, pp. 195–208.
231. "lived through earthquake together": O. B. Keeler, "How Bobby Jones and Family Went Through Quake," *AJ*, march 12, 1935, VUL.
231. Jones returns to St. Andrews: *Tumult*, pp. 86–87.
231. Jones and Sam Breadon discussion: Cullen Cain, "Bobby Jones and Sam Breadon in a Lively Hobby Mix-Up," *Miami Daily News Sunday Magazine*, June 5, 1949.
231. GR picks Masters dates: Miller, p. 199.
232. GR at the Masters: Fred Russell interview.

12. Dean of American Sportswriters

234. Worked out of home and Sportlight Films: Interviews with David Camerer and Catherine Mecca.
234. Ichauway: *Tumult*, pp. 344–45.
234. Woodruff, GR picture and scholarship: Champion, *Echoes of the Past*, pp. 87–88.
235. Wild turkey: *Tumult*, pp. 344–45; Ralph McGill, *Atlanta Constitution*, undated clipping (published shortly after GR's death), VUL.

235. GR can't shoot the deer: *Tumult*, p. 347.
235. "one-man parade": Henry McLemore, "Sports Writer No. 1," *Look*, November 1942.
236. Travel companions: *Tumult*, pp. 308–47; various profiles of GR, especially McLemore, "Sports Writer No. 1" and McLemore columns, VUL.
237. "Why do I do it?": *Tumult*, p. 329.
237. "everybody you ever heard of": Virginia Tracy, "Reporter Interviewing GR Gets the 'Facts,' " *St Louis Globe-Democrat*, October 2, 1942.
237. Party with Bobby Jones: *Tumult*, pp. 85–86.
237. GR befriends Camerer: Camerer interview.
238. GR befriends McLemore: McLemore, "Sports Writer No. 1."
238. "virile saint": McLemore, undated column, VUL.
238. "McLemore told interviewer": Tracy, "Reporter Interviewing GR Gets the 'Facts.' "
238. Lunch at Dutch Treat Club: Kieran, *Not Under Oath*, p. 251.
239. GR tells KHR to walk away: McLemore, "Sports Writer No. 1."
239. "Rhythm, the main factor in both": *Tumult*, p. 352.
241. "Yes, other teams": GR, "Rockne," *Sun*, undated clipping, VUL.
241. "Empire": *BG*, August 15, 1932.
242. "Marco Polo was a piker": *BG*, August 11, 1932.
242. "gods who sat upon Olympus": *BG*, July 31, 1932.
243. Candidates for "most interesting": *BG*, July 27, 1932.
243. "wears no bulging muscles"; Didrikson and Craven: *BG*, July 29, 1932.
243. "new javelin record": *BG*, August 1, 1932.
244. GR questions official's decision on Didrikson: *BG*, August 8, 1932.
245. Golf at Brentwood: *BG*, August 9, 1932; *Tumult*, pp. 239–40; Zaharias, *This Life*, pp. 59–61.
245. "almost more excited about golf": Zaharias. p. 57.
245. "Paul takes no challenge", *Tumult*, p. 240.
246. "man-snatching": Gallico, quoted in Lipsyte, *Sportsworld*, p. 219.
246. "never known an athlete even close": *BG*, August 9, 1932.
247. "matter of instinct with him": Chris Mead, *Champion*, p. 64.
247. "wil Zulu running amok": *BG*, August 5, 1936.
247. "generational chauvinism": William Manchester letter to *New York Times Book Review*, 1991.
248. "credit to his race": Chris Mead, *Champion*, p. 54.
248. Pegler, Brisbane comments: *Ibid.*, p. 57–58.
248. "Sportsmanship should be the mortar": *Tumult*, p. 248.
248. "Mr. Rice"; "Joe": *Ibid.*, pp. 248–49.
248. "speed of the jungle"; "stalking Carnera": Chris Mead, *Champion*, p. 63.
249. Gallico quote: *Ibid.*, p. 68.
249. Corum quote: *Ibid.*, p. 119.
249. Kieran quotes: *Ibid.*, p. 69.
250. Povich quote on prejudice: *Ibid.*, p. 73.
250. "remember a little more distinctly": *BG*, July 28, 1936.
250. Difference in German reaction to sport: *BG*, July 31, 1936.
251. Jarrett suspension: Mandell, *Nazi Olympics*, pp. 242–49.

251. "most pathetic figure"; "stupidity of American officals": *BG*, August 2, 1936.
251. GR on opening ceremonies: *Ibid.*
252. "Hitler rooting like a Yale sophomore": *BG*, August 4, 1936.
252. "can beat your . . . white men": *BG*, August 5, 1936.
252. "United States had a white man": *Ibid.*
252. "until it runs out of Africans": *BG*, August 4, 1936.
252. "something blown by a gale": *BG*, August 3, 1936.
252. "weather has been against him': *BG*, August 5, 1936.
253. "ability to relax . . . plus inborn sense of rhythm": *BG*, August 27, 1936.
253. "unusual amount of fellowship and sportsmanship": *BG*, August 29, 1936.
254. Owens "quiet message": Baker, *Jesse Owens*, p. 175.
254. "better and better athletes": *BG*, August 26, 1936.

13. Gathering Twilight

255. Toots Shor quote: *Tumult*, p. 318.
256. Pegler quotes: Pegler, "Fair Enough," syndicated column, October 18, 1935, VUL.
257. Stanley Woodward quote: Woodward, "Lovable Old Characters," undated 1943 clipping, *NYHT*, VUL.
257. *Dayton Daily News* editorial; Undated 1938 clipping, VUL.
258. "have lost interest in sport": Pegler, "Fair Enough."
258. Gerhig dealing cards: Robinson, *I Never Had It Made*, p. 256.
259. Pallbearers Association: Will Fowler, *The Young Man from Denver*, p. 257.
259. "Via Charon": GR, *Only the Brave*, pp. 59–60.
261. "sport as spectacle": *Sun*, December 24, 1941.
261. "any sport suitable for physical fitness": *Sun*, December 17, 1941.
262. "the judge has done a smart job": *Sun*, March 24, 1942.
262. Roosevelt letter to Landis: William B. Mead, *Even the Browns*, pp. 35–36.
262. "classifications will be changed abruptly": *Sun*, December 13, 1941.
262. "quaffed a beaker together": Bing Crosby to KHR, March 19, 1947, VUL.
263. Army Emergency Fund: John Kieran told the story to Earl Banner of the *Boston Globe* following Rice's death. *BG*, July 16, 1954.
264. Conn's and Louis's debts: Chris Mead, *Champion*, pp. 227–29.
265. Kieran admitted he was a "sap": *BG*, July 16, 1954.
264. O'Toole quote: Editorial, *Philadelphia Record*, September 25, 1942.
265. Discussion of wartime committee on sports and GR's proposed role: Stanley Woodward, *op cit.*
265. 1944 All-America Team: Marsh and Ehre, *1944*, p. 97.
266. Tommy Hitchcock story: *Tumult*, pp. 172–74.

267. "Whether it's Heaven of whether it's Hell": GR, *Only the Brave,* p. 4.

268–69. Frank Graham, John McGraw and GR: Frank Graham, Jr., *A Farewell to Heroes,* pp. 18–20.

269–71. The backgrounds on Graham and Smith are taken primarily from Frank Graham, Jr. and Berkow, *Red.*

269. "Where've you little bastards been?": Berkow, p. 105.

270. Graham story on Gerhig: Frank Graham, Jr., pp. 146–47.

271. Smith story on the guy, the Lord and Number 4: Berkow, pp. 117–18.

273. Sunshine Park: This sketch is built primarily from undated Smith and Graham columns, 1947–1958, VUL.

274. Arnold Rothstein's advice: *Tumult,* pp. 279–80.

274. Daily double system: *Ibid.,* pp. 277–78.

274. "Granny always had the daily double": Dan Parker, "Sunshine Park Misses Best Friend Race Track Ever Had," *New York Mirror,* undated clipping, VUL.

275. GR at the roulette table: John Wheeler, "The Old Second Guesser," syndicated column, February 9, 1954.

275. GR gives Yogi Berra winning ticket: Dan Parker, "Win or Lose, GR Always Played the Game," *New York Mirror,* July 15, 1954.

275. "model . . . for all men": Dick Hyland, "Hyland Fling," *Los Angeles Times,* December 30, 1948.

275–76. Details on *This Is Your Life* taken from newspaper and magazine clippings, VUL.

277. "erect a monument": McLemore, *One of Us Is Wrong.*

277. "vote him into Hall of Fame": Vincent X. Flaherty, five different, undated columns, *c.* 1950–1954, VUL.

277. Fred Russell lobbied publicly and privately: Various *Nashville Banner* columns, 1947–1955. Russell to Vanderbilt Chancellor Oliver Carmichael, February 3, 1941, VUL.

277. Jimmy Breslin "in awe" of GR: Breslin interview.

277. Angelo Bertelli thrilled at sitting next to GR: Bertelli to GR, January 4, 1949, VUL.

278. Subscribing papers not printing his column: Fred Russell interview.

278. *Los Angeles Mirror* wanted his column: Fred Beck, "On the Town," *Los Angeles Mirror,* undated clipping, *c.* 1948, VUL.

278. Physically not well; friends fed him quotes: Fred Russell interview.

279. Fred Lieb helps him in Cleveland: Lieb, *Baseball As I Have Known It,* p. 210.

279. GR on heavyweight champions: *Sun,* January 7, 1942.

280. "Every father who ever lived": Marsh and Ehre, *1947,* p. 90. Article originally appeared in *Sport,* September 1946.

14. Charon's Boat

282. "sure that you understand": Roy Howard to GR, January 16, 1950, VUL.

282. "Our emphasis . . . is on": *Ibid.*

282. Bruce Barton lobbies *Mirror* on GR's behalf: David Camerer interview.
283. GR conversation with Lowell Pratt and David Camerer: *Ibid.*
283. Camerer's role in *Tumult:* Camerer and Russell interviews.
283. Red Smith quote.
283. "you should pass along": Camerer to GR, undated, VUL.
284. Camerer passes manuscript along to others: Camerer to GR, June 22, 1954; Camerer interview.
284. "if I were to be included": Camerer to GR, June 22, 1954.
285. "Listening to your arteries harden": Fred Russell, "GR, Virile Saint," *Nashville Banner,* July 15, 1954.
286. Rice's final column was published on Sunday, July 18, 1954. The typed copy of the unfinished final column was slugged: "For Release Wednesday, July 21"; VUL.
286. GR suffers stroke: Catherine Mecca interview.
287. Wainwright's eulogy: *Congressional Record, 1954,* pp. 11483–84.
287. "who's most often in men's good thoughts": Russell, *Bury Me in an Old Press Box,* p. 203.
287. "high sense of propiety": *Patterson* [N.J.] *Evening News,* July 14, 1954, VUL.
287. "fine bright rapture of true amateur": *Providence Journal,* July 15, 1954, VUL.
287. *Nashville Banner* editorial: July 14, 1954, VUL.
288. William Randolph Hearst, Jr. tribute: *New York Journal-American,* July 18, 1954.
289. Jimmy Cannon column: *New York Post,* undated clipping, July 1954, VUL.
290. Bob Considine saying rosary: Catherine Mecca interview.
290–91. Details of GR funeral are taken from various newspaper accounts.
291. Bruce Barton's eulogy was reprinted as the foreword to *Tumult.*
291. "Oh grave, where is thy victory?": Ralph McGill, *Atlanta Constitution,* July 17, 1954.
292. Publication party at Toots Shor's: Russell and Camerer interviews, newspaper accounts, VUL. Guest list is taken from the dinner program.
292. Shor wouldn't let Pratt pay: Fred Russell interview.

Index